Mobile Media in the Asia-Pacific

This century has been marked by the rapid and divergent uptake of mobile telephony throughout the world. The mobile phone has become a poignant symbol for post-modernity and the attendant modes of global mobility and immobility. Most notably, the icon of the mobile phone is most palpable in the Asia-Pacific in which a diversity of innovation and consumer practices – reflecting gender and locality – can be found. Through the lens of gendered mobile media, *Mobile Media in the Asia-Pacific* provides insight into this phenomenon by focusing on case studies in Japan, South Korea, China and Australia.

Despite the ubiquity and multi-layered nature of mobile media in the region, the patterns of female consumption have received little attention in the growing literature on mobile communication globally. Utilising ethnographic research conducted in the Asia-Pacific over a six-year period, this book investigates the relationship between gender, technology and various forms of mobility and immobility in the region. This book outlines the emerging modes of gendered performativity that make the Asia-Pacific so distinct to other regions globally.

Mobile Media in the Asia-Pacific is a fascinating read for students and scholars interested in new media and gender in the Asia-Pacific region.

Larissa Hjorth is Lecturer in the Games and Digital Art programs at RMIT University, Australia.

Asia's Transformations
Edited by Mark Selden,
Binghamton and Cornell Universities, USA

The books in this series explore the political, social, economic and cultural consequences of Asia's transformations in the twentieth and twenty-first centuries. The series emphasizes the tumultuous interplay of local, national, regional and global forces as Asia bids to become the hub of the world economy. While focusing on the contemporary, it also looks back to analyse the antecedents of Asia's contested rise.

This series comprises several strands.

Asia's Transformations

Asia's Transformations aims to address the needs of students and teachers. Titles include:

Debating Human Rights
Critical essays from the United States and Asia
Edited by Peter Van Ness

Hong Kong's History
State and society under colonial rule
Edited by Tak-Wing Ngo

Japan's Comfort Women
Sexual slavery and prostitution during World War II and the US occupation
Yuki Tanaka

Opium, Empire and the Global Political Economy
Carl A. Trocki

Chinese Society
Change, conflict and resistance
Edited by Elizabeth J. Perry and Mark Selden

Mao's Children in the New China
Voices from the Red Guard generation
Yarong Jiang and David Ashley

Remaking the Chinese State
Strategies, society and security
Edited by Chien-min Chao and Bruce J. Dickson

Korean Society
Civil society, democracy and the State
Edited by Charles K. Armstrong

The Making of Modern Korea
Adrian Buzo

The Resurgence of East Asia
500, 150 and 50 year perspectives
Edited by Giovanni Arrighi, Takeshi Hamashita and Mark Selden

Chinese Society, second edition
Change, conflict and resistance
*Edited by Elizabeth J. Perry and
Mark Selden*

Ethnicity in Asia
Edited by Colin Mackerras

The Battle for Asia
From decolonization to globalization
Mark T. Berger

State and Society in 21st Century China
*Edited by Peter Hays Gries and
Stanley Rosen*

Japan's Quiet Transformation
Social change and civil society in
the 21st century
Jeff Kingston

Confronting the Bush Doctrine
Critical views from the Asia-Pacific
*Edited by Mel Gurtov and
Peter Van Ness*

China in War and Revolution, 1895–1949
Peter Zarrow

The Future of US-Korean Relations
The imbalance of power
Edited by John Feffer

Working in China
Ethnographies of labor and workplace
transformations
Edited by Ching Kwan Lee

Korean Society, second edition
Civil society, democracy and the State
Edited by Charles K. Armstrong

Singapore
The State and the culture of excess
Souchou Yao

**Pan-Asianism in Modern Japanese
History**
Colonialism, regionalism and borders
*Edited by Sven Saaler and
J. Victor Koschmann*

The Making of Modern Korea, second
edition
Adrian Buzo

Asia's Great Cities

Each volume aims to capture the heartbeat of the contemporary city from multiple perspectives emblematic of the authors' own deep familiarity with the distinctive faces of the city, its history, society, culture, politics and economics, and its evolving position in national, regional and global frameworks. While most volumes emphasize urban developments since the Second World War, some pay close attention to the legacy of the longue durée in shaping the contemporary. Thematic and comparative volumes address such themes as urbanization, economic and financial linkages, architecture and space, wealth and power, gendered relationships, planning and anarchy, and ethnographies in national and regional perspective. Titles include:

Bangkok
Place, practice and representation
Marc Askew

Shanghai
Global city
Jeff Wasserstrom

Hong Kong
Global city
Stephen Chiu and Tai-Lok Lui

Singapore
Wealth, power and the culture of control
Carl A. Trocki

Representing Calcutta
Modernity, nationalism and the
colonial uncanny
Swati Chattopadhyay

The City in South Asia
James Heitzman

Asia.com

This series focuses on the ways in which new information and communication technologies are influencing politics, society and culture in Asia. Titles include:

Japanese Cybercultures
Edited by Mark McLelland

Asia.com
Asia encounters the Internet
*Edited by K.C. Ho,
Randolph Kluver and
Kenneth C.C. Yang*

**The Internet in Indonesia's
New Democracy**
David T. Hill & Krishna Sen

Chinese Cyberspaces
Technological changes and political
effects
*Edited by Jens Damm and
Simona Thomas*

Mobile Media in the Asia-Pacific
Gender and the art of being mobile
Larissa Hjorth

Literature and Society

Literature and Society is a series that seeks to demonstrate the ways in which Asian Literature is influenced by the politics, society and culture in which it is produced. Titles include:

The Body in Postwar Japanese Fiction
Edited by Douglas N. Slaymaker

**Chinese Women Writers and the Feminist
Imagination, 1905–1948**
Haiping Yan

Routledge Studies in Asia's Transformations

Routledge Studies in Asia's Transformations is a forum for innovative new research intended for a high-level specialist readership, and the titles will be available in hardback only. Titles include:

**The American Occupation of Japan and
Okinawa***
Literature and memory
Michael Molasky

Koreans in Japan*
Critical voices from the margin
Edited by Sonia Ryang

Internationalizing the Pacific
The United States, Japan and the
Institute of Pacific Relations in war and
peace, 1919–1945
Tomoko Akami

Imperialism in South East Asia
'A fleeting, passing phase'
Nicholas Tarling

Chinese Media, Global Contexts
Edited by Chin-Chuan Lee

Remaking Citizenship in Hong Kong*
Community, nation and the
global city
Edited by Agnes S. Ku and Ngai Pun

Japanese Industrial Governance
Protectionism and the licensing state
Yul Sohn

Developmental Dilemmas
Land reform and institutional change
in China
Edited by Peter Ho

**Genders, Transgenders and Sexualities
in Japan***
*Edited by Mark McLelland and
Romit Dasgupta*

**Fertility, Family Planning and Population
Policy in China**
*Edited by Dudley L. Poston,
Che-Fu Lee, Chiung-Fang Chang,
Sherry L. McKibben and
Carol S. Walther*

Japanese Diasporas
Unsung pasts, conflicting presents and
uncertain futures
Edited by Nobuko Adachi

How China Works
Perspectives on the twentieth-century
industrial workplace
Edited by Jacob Eyferth

**Remolding and Resistance among
Writers of the Chinese Prison Camp**
Disciplined and published
*Edited by Philip F. Williams and
Yenna Wu*

**Popular Culture, Globalization and
Japan***
*Edited by Matthew Allen and
Rumi Sakamoto*

medi@sia
Global media/tion in and out of context
*Edited by Todd Joseph Miles Holden and
Timothy J. Scrase*

Vientiane
Transformations of a Lao landscape
*Marc Askew, William S. Logan and
Colin Long*

**State Formation and Radical Democracy
in India**
Manali Desai

Democracy in Occupied Japan
The U.S. occupation and Japanese
politics and society
*Edited by Mark E. Caprio and
Yoneyuki Sugita*

**Globalization, Culture and Society
in Laos**
Boike Rehbein

Transcultural Japan
At the borderlands of race, gender, and
identity
*Edited by David Blake Willis and
Stephen Murphy-Shigematsu*

**Post-Conflict Heritage, Post-Colonial
Tourism**
Culture, politics and development at
Angkor
Tim Winter

Education and Reform in China
Emily Hannum and Albert Park

Writing Okinawa: Narrative Acts of Identity and Resistance
Davinder L. Bhowmik

* Now available in paperback

Critical Asian Scholarship

Critical Asian Scholarship is a series intended to showcase the most important individual contributions to scholarship in Asian Studies. Each of the volumes presents a leading Asian scholar addressing themes that are central to his or her most significant and lasting contribution to Asian studies. The series is committed to the rich variety of research and writing on Asia, and is not restricted to any particular discipline, theoretical approach or geographical expertise.

Southeast Asia
A testament
George McT. Kahin

Women and the Family in Chinese History
Patricia Buckley Ebrey

China Unbound
Evolving perspectives on the Chinese past
Paul A. Cohen

China's Past, China's Future
Energy, food, environment
Vaclav Smil

The Chinese State in Ming Society
Timothy Brook

China, East Asia and the Global Economy
Regional and historical perspectives
Takeshi Hamashita
Edited by Mark Selden and Linda Grove

Mobile Media in the Asia-Pacific

Gender and the art of being mobile

Larissa Hjorth

Routledge
Taylor & Francis Group

LONDON AND NEW YORK

First published 2009
by Routledge
2 Park Square, Milton Park, Abingdon, Oxon OX14 5RN

Simultaneously published in the USA and Canada
by Routledge
270 Madison Avenue, New York, NY 10016

*Routledge is an imprint of the Taylor & Francis Group,
an informa business*

© 2009 Larissa Hjorth

Typeset in Times by RefineCatch Limited, Bungay, Suffolk
Printed and bound in Great Britain by
MPG Books Ltd, Bodmin

British Library Cataloguing in Publication Data
A catalogue record for this book is available from the British Library

Library of Congress Cataloging-in-Publication Data
Hjorth, Larissa.
Mobile media in the Asia Pacific : gender and the art of being mobile / Larissa
Hjorth.
  p. cm. — (Asia.com)
  1. Cellular telephones—Social aspects—Pacific Area—Case studies.
2. Technology—Social aspects—Pacific Area—Case studies.   3. Consumer
behavior—Sex differences—Pacific Area—Case studies.   4. Postmodernism—
Social aspects—Pacific Area—Case studies.   I. Title.
HE9715.P16H66 2008
303.48′33—dc22
2008015001

ISBN 10: 0–415–43809–8 (hbk)
ISBN 10: 0–203–88989–3 (ebk)

ISBN 13: 978–0–415–43809–4 (hbk)
ISBN 13: 978–0–203–88989–8 (ebk)

Contents

List of figures   xi
Abbreviations and glossary   xv
Acknowledgements   xvi

Introduction: the price of being mobile   1

PART I
Mobile media societies   17

1 Locating the mobile: mobile communication and
   gender in the Asia-Pacific   19

2 Paradigms of mobility: conceptual tenors for studying
   mobility today   48

3 Beyond the 'new rich': consumption, production and
   gender in the region   62

PART II
Mobile media cultures   77

4 Fast-forwarding to the present: the rise of customised
   mobile media in Tokyo   79

5 Engaging rings: the *haendupon* and intimate
   communities in Seoul   119

6 Nostalgic mobility: memory and the mobile phone in
   Hong Kong   151

7 Postal presence: persistence of the postal metaphor in
   Melbourne   188

8 Domesticating cartographies: gendered mobile media
   in the region   226

PART III
Mobile media practices   239

9 Domesticating new media: a discussion on locating
   mobile media   241

10 The big bang: an example of mobile media as new
    media   253

11 On hold: reflections on mobile media in the Asia-
    Pacific   265

Notes   272
Bibliography   275
Index   293

Figures

0.1 *Nagara mobility*. Photo: Hjorth 2004.   16

4.1 A *keitai* sales girl in 'electric city', Akihabara, exemplifies the uniformality that McVeigh discusses as part of Japanese national culture. Photo: Hjorth 2004.   85

4.2 A 2007 graph from CNET Japan featuring the three major telecommunication companies, DoCoMo (blue), KDDI (pink) and SoftBank (yellow). The vertical axis lists numbers of subscribers in the millions, while the horizontal axis lists years from 2008 to the projected 2013. See: http://japan.cnet.com/blog/comm25/2007/02/15/post_dffa/.   86

4.3 Pedestrians with their *keitai* whilst waiting for the train. This picture vividly illustrates Fujimoto's *nagara* mobilism. Photo: Hjorth 2004.   95

4.4 DVD of *Train Man* (*Densha Otoko*).   101

4.5 A *keitai* adorned with a plethora of Hello Kitties. Each Kitty represented a different location in Japan and the respondent spoke about each one, fondly identifying who she was with when she acquired it. In this way, *keitai* customisation extends on objects such as the charm bracelet, in which the owner is reminded of times and places shared by looking at the mementoes. Photo: Hjorth 2004.   103

4.6 An example of SJIS art that features regularly in 2ch. Source: Wikipedia Japan.   104

4.7 An example of a mixi page. Courtesy of Yuji Mori.   110

4.8 Images from left: the end of a solitary dinner; webcam wife.   111

4.9 Clockwise from top left: Friend's funny face, self-portrait as *The Ring* monster, scary cat, friend as 'exorcist', sunset at Aichi World Expo 2005.   112

4.10 The website of the UCC organisation maho-island, featuring the advertisement for *Koizara*, the movie (directed by Natsuki Imai, 2006).   115

5.1 An example of DMB advertising for the TU phone. Note

the use of a traditional Korean colour scheme (*saekdong*) as the colour test screen background. Photo: Hjorth.   125

5.2(a–e)   Examples of Cyworld mini-hompy (featuring the mini-room) – imaginary and actual. The top images are by artist Emil Goh, in which he compared people's mini-hompy with their offline living spaces.   129

5.3   Some examples of students' *haendupon*.   133

5.4   An example of an image from a 'coupled' male student's phone.   137

5.5a, 5.5b   Typical 'everyday' and 'special occasion' images as denoted by 'cute' (photos by female participants).   141

5.5c, 5.5d, 5.5e   Some of the typical 'everyday' and 'special occasions/friendship' images (photos by female participants). Here, the symbolic role of food and its relationship to sharing with friends/family is demonstrated.   141

5.6   Modes of realism or 'reelism' are most obvious in webcam or user-as-media-producer techniques.   142

5.7   An example of a 'feminine' phone owned by one of the female students.   147

5.8   An example of advertising the way in which the multimedia capabilities of the *haendupon* are customised by cute character culture. The symbol of *Hyoja*, Samsung Anycall, in this image has been reterritorialised by the female 'produser'.   149

5.9   An example of hand-made customisation. Photo: Hjorth.   150

6.1   Two examples of mobile phone advertising in Hong Kong in 2004. The advertisement on the left is for a phone by the US company Motorola, which attempted to market its products as Japanese using language plays on 'moto' and imitating Japanese models. Motorola dropped its J-pop references in 2006 in the wake of the burgeoning Korean wave. The image on the right is of a billboard in Causeway Bay (a main shopping area) advertising Samsung (Korea), featuring a Western female model. Identical campaigns for both models were run in Australia. Photo: Hjorth.   154

6.2   Annual SMS traffic volume in China, 2000–2006 (billion messages). Source: MII Annual Statistical Reports on the Telecommunications Industry (cited in Qiu 2008).   160

6.3   Some examples of young people's customisation of mobile phones. The trend for wearing the mobile phone like a lanyard highlights its role as an extension of the user's identity and as a title for others to read. Notice how both phones are customised with cute character attire that transforms them into a teddy bear or stuffed animal. Photo: Hjorth.   163

6.4 An example of nostalgic customisation in the form of Hello Kitty mobile phone customisation. Hello Kitty became very popular in Hong Kong in the late 1970s and 1980s, a time when Hong Kong experienced great upward mobility. Photo: Hjorth. 166

6.5 Mobile phone customisation as a battlefield for cross-cultural consumption (Peko screen saver plus Disney cute objects hanging from the Nokia phone). (Photo: 2005) 172

6.6 Camera phone images taken by male respondent. (Photos: 2005) 178

6.7 Camera phone images taken by male respondent. (Photos: 2005–2006) 178

6.8 Camera phone images taken by female respondent. (Photos: 2006) 180

6.9 Camera phone images taken by female respondents. These images are a good example of user-as-journalist (or paparazzi). (Photos: 2006) 182

6.10 'Communication is borderless.' This retro advertising lightbox found in Causeway Bay in 2004 is a great example of big business around nostalgia – even when dealing with new technologies and future generations. Photo: Hjorth. 186

7.1 The star of mobile cross-platform mini-series, *Girl Friday*. The script, story, visual and aural aesthetics have been worked around the mobile phone as a symbol of contemporary everyday life. Source: Ish Media. 200

7.2 One female respondent's images illustrating genres ranging from paparazzi to tourist postcard. Here the international student utilises the camera phone to take tourist pictures for herself and family/friends at home whilst also providing her a space to perform her identity within an Melbournian context. 215

7.3 Clockwise from top left: images classified by the respondent as personal space, social expected, social paparazzi and everyday poetics. 216

7.4 Images clockwise from top left: two shots of watches, a picture of respondent's toy for a wallpaper, a picture of the ANZ bank in Chinatown taken to win an argument with a friend, a shot of a scooter that the respondent had considered buying (female respondent). 217

7.5 Images clockwise from top left: image of girlfriend, with mates, with a friend, vegemite-brother metaphor, visage in the coffee, cigarette burning. 218

7.6 Images clockwise from top left: gate to house, pet dog, mother, father, sister, hockey game, favourite object (guitar). 218

7.7 Images clockwise from top left: graffiti, cat, Korean food, empty tram, friends playing soccer video game, mini-hompy page. 219

7.8 Images from left (clockwise): mug shot of friend, Melbourne building, eating out with friends, blossoms, 'funny' dog and cat. 220

7.9 These images, which exemplify the classic photographer style, from top left (going clockwise): mother's day cake, mum and dad's anniversary, and three shots of sunsets in Australia. 221

8.1 *Girl Friday*, mobile media movies that feature stories revolving around a mobile phone. The innovative cross-platform content includes regular SMS from *Girl Friday* to the registered user as well as parallel interactive content via Web 2.0 formats. 233

8.2 A *keitai shôsetsu* author at home. Source: CNet (Japan). 237

10.1 *Dotplay* workshop hactivists at work. 259

10.2 *Dotplay's* (hardware and software) art of mobile hacking. 263

Abbreviations and glossary

Abbreviations

2G	Second generation phone (voice calling and texting capabilities)
3G	Third generation (mobile with Internet)
ANT	Actor network theory
ICTs	Information and Communication Technologies
J-pop	Japanese popular culture
K-pop	Korean popular culture
MMS	Multimedia Messaging Service
NICs	Newly Industrialised Countries
PHS	Personal Handy-phone System
SAR	Special Administrative Region
SCOT	Social Construction Of Technology
SMS	Short Messaging Service
SNS	Social Networking Site
UCC	User Created Content

Glossary of Japanese terms

Chaku-mero (ring tones)
Emoji (emoticons)
Gyaru-moji (girl's alphabet)
Heta-moji (awkward alphabet)
Kaomoji (face marks)
Kawaii (cute)
Keitai (short for *keitai denwa*, meaning mobile phone)
Keitai shôsetsu (portable novels)
Kôgyaru (trendy female customer, often in her twenties)

Acknowledgements

This book has grown from a series of fieldtrips to Seoul, Tokyo and Hong Kong. In my fieldtrips I have been kindly assisted by a number of colleagues whose support I gratefully acknowledge. In Japan, I would like to thank Ryutaro Inamoto, Hitomo Toku, Shinji Oyama, Shin Mizukoshi, Itaru Hirano, Toshi Tomita, Machiko Harada and participants from Chiba University. In Seoul, big thanks go to Hyun-suk Seo, Kyungah Ham, Yangah Ham, Hyun Mee Kim, Heewon Kim, Yeran Kim, Young-sun Park, Hyunjin Shin, Shin Dong Kim and all the fantastic students at Hallym University who participated in the interviews and surveys. I would like to thank Tae-jin Yoon and Youngchul Yoon for the wonderful BK research fellowship at Yonsei University.

In Hong Kong, I would like to thank Angel Lin and the participants from Hong Kong City University as well as Robert Clark and Vicki Ho. In Melbourne I would like to thank both University of Melbourne and RMIT University participants, along with my two PhD supervisors, Audrey Yue and Carolyn Stevens. To all my friends who put up with my virtual co-presence – Renata Kokanovic, Olivia Khoo, Paul Quinn, Chantal Faust, Helen Addison-Smith, Sofia Ahlberg, Klare Lanson, Esther Milne, Kate Inabinet, Kate Shaw, Pucca, Kylie Robertson and my comrade. This book is dedicated to all of you.

Special thanks to the great support offered by the Asialink visual art residency, Australia Council for the Arts, Akiyoshidai International Art Village, MIALS fieldwork scholarship and the Alma Hansen scholarship, along with warm thanks to the wonderful people at Routledge – Mark Selden, Stephanie Rogers, Sonja van Leeuwen and Leanne Hinves. I would also like to thank the many friends who are fellow researchers of mobile communication both in the Asia-Pacific and also internationally whose feedback and encouragement kept me inspired and focused, especially Gerard Goggin and Genevieve Bell. And lastly, but far from least, I would also like to thank my mum, dad and my brother, Greg.

A revised and condensed version of Chapter 4 has appeared in *Southern Review Journal*, 38(3), 2006: 23–42 (Hjorth 2006a), and an earlier draft version of Chapter 7 has appeared online in *Fibreculture Journal*, 6 (Hjorth

2005b). A condensed version of Chapter 5 can be found as 'Snapshots of almost contact: case study on South Korea', special issue (ed. G. Goggin) *Continuum*, 21(2): 227–238 (2007a).

This book is dedicated to anyone who has ever wondered what it would be like to be, not in someone's shoes, but, rather, in their mobile phone.

Introduction
The price of being mobile

> With the mobile phone, we are at the centre of post-modernity. The mobile phone embodies many parallel and contradictory dimensions of meaning: utilitarian use with leisure, the facilitation of everyday life versus dependency, freedom and control, richness of interaction or introversion, private practice and public use, social cohesion with separation.
>
> (Kopomaa 2002: n.p.)

Synonymous with global modernity, the mobile phone has become an integral part of contemporary everyday life. As a global medium, the mobile phone is a compelling phenomenon, a testament to the significance of the local in shaping and adapting global commodities and technologies. It is easy to forget that the rise of the mobile phone from an extension of the domestic landline to multimedia device – incorporating MP3 player, video, MMS (Multi Messaging System), SMS (Short Messaging System) and emailing – is relatively nascent. Once a symbol of economic status and class, as epitomised by the businessman in the 1980s (Agar 2003; Robison & Goodman 1996), the mobile phone has grown to encompass various roles and functions, often paradoxically, at an individual, socio-cultural, transnational and global level.

The adaptation and usage of the mobile phone can be read on two levels simultaneously. Mobile phone practices can be viewed as an extension of the user's identity and lifestyle. These practices, in turn, can be viewed as part of localised, socio-cultural rituals of the everyday. The mobile phone literally and symbolically connects users to both existing and emerging forms of identity and identification. Whether we call it a mobile phone, cell phone, *keitai*, handy, *haendupon* or *shouji*, the ubiquity of the object in everyday life is unmistakable. As an indispensable part of the everyday, it provides users with multiple divergent meanings in the coordination of social practices. As a cultural artefact, the mobile phone is a smorgasbord of possibilities – signalling the owner's tastes, values, and constructions of identity such as class, gender and cultural background. It can be a poignant symbol of the various forms of mobility today – technological, cultural, social, political and economic – in a global economy. It can be a cultural index for specific localities. It can provide insight into burgeoning transnational flows, regional resurgence and shifting centres of modernity. It can help to provide acuity into the role of geography and history in a period marked by various forms of mobilities and growing moorings (Hannam et al. 2006). The mobile phone operates upon multiple levels of mobility and immobility – both symbolic and material. Hence, what may appear at first to be a peripheral phenomenon can be seen as playing a significant role in defining, shaping and symbolising the twenty-first century post-modernity.

As symbolic of global ICTs (Information and Communication Technologies), the mobile phone demonstrates that the importance of regional factors cannot be underestimated. The Asia-Pacific is a case in point, with locations such as Tokyo and Seoul gaining global attention for their innovation in, and output of, mobile technologies and practices. Behind the multiplicity of contesting identities in the disparate region, technologies have had an increasingly central influence on production and consumption patterns since the 1997 economic crisis. For NICs (Newly Industrialised Countries) such as South Korea, the production of technological hardware and software was pivotal in recovering from financial instability, reflected in the deployment of techno-nationalist strategies that nurtured local industry, established infrastructure such as broadband and ensured a strong platform for the exporting of technologies globally. Continuing with the example of South Korea, it is impossible to separate the image of the nation from technological companies such as Samsung, clearly demonstrating a marriage between technology and modernity. The instrumental role of technology in the region's recovery highlights specific socio-cultural and politico-economic factors that are interrelated with emerging transnational modes of globalisation and consumption in the region.

Through the lens of new technologies, and specifically mobile technologies, the region has witnessed a transformation from rapid economic and technological growth into becoming a powerful cultural index globally. This rise in the global currency of 'Asia', and the resurgence of new forms of regionalism and cultural proximity are, as Taiwanese cultural theorist Kuan-Hsing Chen warns, a phenomenon to be wary of, especially since they may signify the re-emergence of old colonial and imperialist agendas (1998: 27). This image and practice of 'consuming Asia' (Chua 2000) reflects not only the divergent models for consumption in the region but also the way the Asia-Pacific is imagined globally. The intensification of transnational capital in the region is concisely diagnosed by Leo Ching as a conflation between consumerism and emerging forms of Asian modernity. As Ching surmises, 'Asia has become a market, and Asianness has become a commodity circulating globally through late capitalism' (2000: 257). The concept of the region as an imperial compression of various cultural identities and histories has come under much scrutiny particularly in the case of the 'shrinking of the Pacific' (R. Wilson 2000: 565) connoted by the Asia-Pacific.

Once a discursive geopolitical rubric, the Asia-Pacific – as a site for contesting local identities and transnational flows of people, media, goods and capital – has come under much radical revision and reconceptualisation

(Wilson 2000; Dirlik 2007; Arrighi 1994, Arrighi et al. 2003). This has led theorists such as Robert Wilson and Arif Dirlik to utilise 'Asia/Pacific' as 'not just an ideological recuperated term', but as an imagined geo-political space that has a double reading as both 'situated yet ambivalent' (Wilson 2000: 567). Specifically, the diverse and perpetually changing interstitials constituting the Asia-Pacific have been repositioned as a contested space for multiple forms of identity. Far from the 1970s' widely held view of the region as being a bloc of satellite NICs (such as Taiwan and South Korea) oscillating around the region's first industrial nation, Japan, we now see a profoundly different picture. In particular, the phone was instrumental (at both a symbolic and material level) in the uneven economic and cultural capital in the region.

In this formation of shifting peripheries and emerging centres such as China (Arrighi 1994; Dirlik 2007), the mobile phone takes on particular significance. This scenario is presciently depicted in Robison and Goodman's *The New Rich in Asia* (1996), which identifies the mobile phone as an index for burgeoning transnational consumption and new narratives of modernity in the region. With the arrival of the mobile phone coinciding with the rise of the Asia-Pacific, the mobile phone became a poignant and 'rich' signifier for analysing local and transnational formations. However, behind the powerful symbol of the mobile phone lies a story about the emergence of localised practices enacted by actual users. To understand the gravity of the symbolic dimensions, we must situate the mobile phone in the very practices it inhabits, namely mobile phone cultures. *Mobile Media in the Asia-Pacific* attempts to capture these mobile media practices, cultures and societies through exploring one of the most pervasive cultural artefacts today. By studying and comparing the diverse mobile cultures in the region, we can gain insight into not only contemporary understandings of communication, but also the impact of various forms of mobility on what it means to be local, or local–global, today.

The mobile phone is a key indicator of the region's accelerated rise into twenty-first-century post-modernity. Moreover, as symbolic of the shift from the mobile phone to mobile media in the region, the young female consumer has attracted much focus as the multimedia transforms user–producer models of consumption and production towards 'produser' paradigms (Bruns & Jacobs 2006). Extending upon Alvin Toffler's (1980) theory that consumers are increasingly being part of the production process in the form of 'prosumers', Axel Bruns utilises the rubric of the 'produser' to address arising forms of creativity and expression within contemporary networked media. In the region, the expansion of mobile media has been concurrent to, and inter-related with, the rise of its symbolic user, the female. For example, in Japan, the rise of twenty-first century *keitai* (*keitai denwa*, or portable phone, abbreviated to *keitai* meaning 'portable') culture has become synonymous with the conspicuous usage by the trendy female consumer, the *kôgyaru*, as she replaces the *oyaji* (salaryman) as a different (and troublesome) national icon (Matsuda 2005: 35); one example in many in which the region's shift into twenty-first-century post-modernity has been marked by the rise of the active female mobile phone user. So much so, that the female user has now become iconic of *keitai* cultures.

Thus, through the lens of gendered mobile media I argue that we can gain new insight into post-modernity – and the various practices of labour, expression, creativity and producers – as the region embarks on the twenty-first century. As I contend, the emerging forms of gendered mobile media – what I call 'cartographies of personalisation' – reflect new modes of gendered performativity in the region that, in turn, echo changing practices of multiple forms of mobility and immobility – images, people, capital and media – within the region. These cartographies of personalisation amplify broader shifts in work and lifestyle patterns – exemplified by the rise of emotional and affective labour. Through the gendered and localised usage of user created content (UCC) of mobile media we can see old and new forms of intimacy – what I call 'imaging communities' – that, in turn, provide a microcosmos of the region's national states that Benedict Anderson aptly defines as 'imagined communities' (1983). So let us begin with locating mobile media in the region.

Mobile @ region: locating the mobile in the region

As Daniel Miller and Heather Horst lucidly note in their ethnography on Jamaican cell phones, 'what one has to study are not things or people but processes' (2006: 7). By studying the mobile phone we can gain insight into many of the cultural processes that constitute contemporary everyday life. It is a poignant lens for gaining insight into twenty-first-century life, evoking the status of icon of the 'new' (Agar 2003) and post-modernity (Kopomaa 2000; Fortunati 2005a, 2005c). This is underscored by the emergence of mobile communication studies analysing the mobile phone as a symbol of youth culture (Ling & Yttri 2002; Ito 2002, 2005a, 2005b; Matsuda 2005a; Ling 2002; Agar 2003), class (Robison & Goodman 1996; Agar 2003; Fortunati & Manganelli 2002), culture (Goggin 2006; Miller & Horst 2006) and individualism (Castells et al. 2007), as well as fashioning identity formation (Fortunati 2005a, 2005c; Law et al. 2006; Katz & Sugiyama 2005), convergence (technological, social, cultural) and new media (Richardson 2007; Hjorth 2007b).

Moreover, the mobile phone has been seen as a focal point for considering contemporary forms of temporality and spatiality (Ling & Yttri 2002), as a repository for global and local tensions (Castells et al. 2007; Katz & Aakhus 2002), as a space for extending earlier social practices and rituals (Taylor & Harper 2002; Ito 2005a, 2005b; Yoon 2003), and as a signifier for mobility and even emerging post-modernity (Kopomaa 2002; Fortunati 2005a, 2005c). Some researchers see the mobile phone mired within contemporary practices of everyday life and important in negotiating sociality and place (Goggin 2006; Ito 2002; Yoon 2003), while others blame the mobile phone for the demise of relationships in modern life, where the virtual overrides the actual (Bauman 2003; Putnam 2000). Others again refute such claims by arguing that intimacy has always been mediated by memories, language and gestures (Morse 1998; Milne 2004). In a period marked by global forms of mobility, the mobile phone – both as a set of cultural practices and as a cultural index – clearly constitutes a lens upon constructions of the local.

The significance of the local in shaping global technologies is highly evident in the Asia-Pacific region's uneven penetration rates and usages of mobile technologies. As well as encompassing the aforementioned global 'centres' for mobile innovation such as Tokyo and Seoul, the region includes other Asian tigers such as Hong Kong, Taiwan and Singapore that demonstrate abnormally high penetration rates of mobile phone ownership. Hong Kong, for example, has penetration rates of 117 per cent (OFTA 2006). The region is a smorgasbord of localised forms of mobile phone customisation. In each country, identity issues such as class, gender, ethnicity and age inform the adaptation. In particular, in the Asia-Pacific it is the role of gender that markedly inflects the socio-cultural adaptations of the technology in different ways.

In Seoul, mobile phones (called *haendupon*) are used by the youth to document, edit and upload their lives onto the dominant Social Networking Sites (SNS), Cyworld's mini-hompy, in a practice of accelerated co-presence between online and offline connectivity and socialising. The role of female adoption of new mobile phone practices such as camera phones is conspicuous (Lee 2005). In Tokyo, the ubiquity of *keitai* practices has seen the rise in the phenomenon of what Sadie Plant called the '*oya yubi sedai*' (thumb generation) (Plant 2002: n.p.); so much so that the *keitai* ring tone industry is booming with yearly revenues surpassing that of karaoke sales (Okada 2005: 55). Moreover, the rise of mobile media in Tokyo has become synonymous with the female consumer as exemplified by the high-school girl pager in the 1990s in which high-school girls appropriated the *oyaji* technology as their own, which was then followed by the *keitai* IT revolution (Matsuda 2005: 35).[1]

China, with over 429.7 billion SMSs sent in 2006 (Qiu 2008: forthcoming), is entering what has been defined as 'the age of the thumb' (*muzhi shi dai*) (Bell 2005: 68). The rise of the mobile phone in the region has become integral in the emerging process of modernity and in the development of new modes of consumerism (Robison & Goodman 1996). In each location – Japan, Hong Kong, Korea and China – the mobile phone acquires a different symbolic status, reflecting already existing cultural practices. The research into the significance and possibilities of the mobile phone as a symbol and a repository for cultural practices, paradoxes and all, seems unabated.

Nevertheless, in the burgeoning of research addressing a diversity of issues from a variety of disciplinary and interdisciplinary, macro or micro, foci, a marked gap is looming – namely a notable absence of longitudinal, qualitative research of the Asia-Pacific. This is despite the fact that the region houses the greatest number of subscribers (Castells et al. 2007: 8) and global innovation centres (such as Seoul and Tokyo) in the world. Why, when the region is examined, is it only researched in terms of specific 'key' locations, such as Tokyo (Ito et al. 2005) or Seoul (Kim 2003)? Why have researchers, when conducting cross-cultural studies in the region, only focused upon quantitative studies, reducing the mobile cultures into a set of statistical comparative practices (Mitomo et al. 2005)?

It is this gap in the research that *Mobile Media in the Asia-Pacific* will explore. Of all the researchers who have published in this area, there is perhaps only one ethnographer, Genevieve Bell (2005), whose wonderfully perspicacious study on the region is exemplary in highlighting the cultural dimensions of the mobile. However, while Bell's study of 'the age of the thumb' has a longitudinal focus it neglects to address interrelated issues such as contextualising the ethnographic with the industrial, to more fully conceptualise the region's uneven mobile phone phenomenon.

Moreover, when researchers discuss 'Asia' or 'Asia-Pacific', why do they do so in a manner that takes the community regional trope as a given and then contextualises it in terms of European or US models of mobility and identity (McLelland 2007)? Another key phenomenon seems to have evaded researchers – apart from local case studies such as those relating to Tokyo (Matsuda 2005; Okada 2005; Fujimoto 2005), Seoul (Lee 2005) and Hong Kong (Lin 2005a) – is the issue of gender. Why is gender only discussed when the mobile phone is being considered as a fashion icon (Katz & Sugiyama 2005)? As I will argue in *Mobile Media in the Asia-Pacific*, the divergent rise of mobile technologies in the region, paralleling its transformation from economic to ideological power, is marked by new gender tropes. With the region's economic – and now cultural – mobility the icon of the female consumer has become central. However, just as the multiple interstitials of the region have shifted, so too has the construction and agency of the symbolic consumer, the young Asian female.

Mobile @ gender: new mobilities and immobilities

In the realm of mobile technologies and consumption in general, the young Asian female has played a central role. In media images of consumption, 'Asian', 'femininity' and 'youth' are conflated. Commodities denoted and demanded particular forms of practice in what has been called gender scripting (Shade 2007). Fashion and 'private' domestic technologies were feminised, whereas 'public' and 'work' related technologies such as computers and cars were more often masculinised. This gender scripting reflected specific modes of feminisation in the region, particularly in Japan (Yoshimi 1999). Much focus has been given to the role of gendered mobile media in relation to fashion (Katz & Sugimoto 2005; Fortunati 2005a), however its impact upon changing modes of women's work and lifestyle patterns has only recently gained attention (Fortunati 2007; Wajcman 2008; Matsuda

2007). In this recent emergence of research, the region and notions of transnational gendered labour has been relatively overlooked, despite the conspicuous partnership between new forms of mobile media, feminised customisation and women's creative role in new media practices.

The rise of mobile technologies in the region paralleled, and influenced, the development of trends that blurred these bifurcated gendered zones. For example, the rise of the female consumer in Japan ran concurrently with the demise of the nation's post-World War II icon, the *oyaji*. Phenomena such as the high-school pager revolution witnessed the pager being appropriated from its original function as a business tool (for the *oyaji*) to a socio-cultural medium in the hands of the young female consumer. This appearance of the conspicuous form of Japanese female mobile phone user, represented by the *kôgyaru*, attracted much moral criticism that saw the *kôgyaru's* use of sexuality come under fire from the Japanese media in the early 2000s (Matsuda 2005).

However, this conflation between gender empowerment and sexual agency suggests that the naturalising of gender as a biological sex – thus relegating the female to the role of procreator rather than creator – was becoming unhinged, not dissimilar to the critiques of gender as a natural category by poststructuralist feminists. As Judith Butler (1991) notes, gender is constructed and naturalised through a set of localised regulations and rituals she defines as performative.[2] Through a revised notion of gender, as with identity, as a set of performances, we can begin to uncover some of the emerging modes of gendered identities, and attendant modes of femininity and masculinity, arising in the transnational interstitials constituting the Asia-Pacific.

The nascent rise of these new gendered subjectivities is only beginning to gain critical attention, most notably in Olivia Khoo's eloquent deconstruction of the Chinese exotic as 'a new mode of representation', which is a 'product of the emergent diasporic Chinese modernities in the Asia Pacific region' (2007: 2). As Khoo observes, much of the discussion around gender in emerging Asian modernities is inflected through an economic rather than cultural focus. Citing Krishna Sen and Maila Stivens's *Gender and Power in Affluent Asia* as an example of the way in which the gendered processes of Asia modernity and globalisation has been framed within economic parameters, Khoo argues that 'economic changes have also influenced Asia's cultural construction' (2007: 18).[3]

Apart from a few important earlier studies of gendered mobile technologies, the link between gender and technology has not featured greatly in the literature. Although there have been many important studies investigating the gendered history of the telephone in Europe (Haddon 1997b; Rakow 1992, 1993), the 'feminisation of the telephone' (Höflich 1996 cited in Roessler & Höflich 2005: 129), and the gendered body politics of mobile virtuality (Fortunati cited in Wajcman & Haddon 2005: 14–15), little research has been conducted in this area in the Asia-Pacific apart from conflations between gender and youth cultures in locations such as Tokyo (Ito et al. 2005).

Key studies such as Ann Moyal's concise discussion in the context of Australia (1992), Lana Rakow's (1992) vivid book *Gender on the Line*, Leslie Haddon's domestic technologies approach to the mobile phone (1997a, 1997b), and Plant's (2002) study on nine cities globally, have sought to apply some of the critical apparatus that feminism can provide. However, the aforementioned studies have not challenged the Western precept of gender.

The situation is exacerbated by the lack of transnational studies conducted in the region on women from a cultural perspective (Khoo 2007). When gender is discussed in terms of culture the focus is often confined to a particular nation, as is the case with Lise Skov and Brian Moeran's (1995) perspicuous collection on women and consumption in Japan. As Mark McLelland's (2007) review on mobile literature recapitulates, it is impossible to find a mobile communication study that engages with the region on its own terms without deferring to European or US precepts about modernity, consumption and identity as so many 'global' studies do; for example, Manuel Castells et al.'s (2007) *Mobile Communication and Society*.

It is in response to this paucity of critical engagement that *Mobile Media in the Asia-Pacific* charts the emergence of gendered and localised consumer identities shaping and domesticating mobile technology in the transnational imaginary of the region. Thus *Mobile Media in the Asia-Pacific* endeavours to open insights into the ways in which gender and locality have shaped, and are shaped by, mobile technologies in the region. *Mobile Media in the Asia-Pacific* is not about technology per se but about modes of production and consumption that have arisen due to the burgeoning significance of mobile technologies. It sees the mobile phone as part of emerging modes of consumption, production, and creative media industries of the twenty-first century. Through the lens of gendered mobile media we can begin to conceptualise the region's post-modernity as the century materialises – in short, what Chris Berry, Fran Martin and Audrey Yue (2003) characterise as *mobile cultures*.

Mobile @ cultures: approaching domestication

From the dawn of this millennium, the mobile phone has attracted academic interest and been regarded as a legitimate research topic by researchers in a broad range of fields including sociology, media studies, IT, economics, politics, cultural studies and anthropology. An extensive body of work has been undertaken on the symbolic dimensions of the mobile phone, especially as a medium through which to understand contemporary actual and metaphoric patterns of communication and meaning. Studies around youth and mobile phones have proliferated, further conflating debates about the new technologies and youth cultures. The mobile phone has been used as a focal point for discussing identity issues of gender, class and ethnicity as well as broader issues such as globalisation and post-modernity.

Mobile Media in the Asia-Pacific aims to redress the conspicuous gap in research into the gendered role of mobile communication and media customisation in the Asia-Pacific. In order to chart the integral role of gender in the mobile media phenomenon, I draw upon a variety of interdisciplinary frameworks including media studies, gender studies, cultural studies, media sociology, virtual ethnography and new media in order to explore the complex evolving terrain of mobile cultures today. One of the dominant and highly successful approaches in the field of studying mobile phone cultures is, undoubtedly, the domestic technologies approach.

One of the ways in which the phenomenon of localisation has been described is through the rubric of domestication. As part of the tradition of approaches that sought to understand the role of media and consumption as a dynamic process, such as SCOT (Social Construction Of Technology) and ANT (Actor Network Theory), the domestication approach – attributed to the late Roger Silverstone – sought to analyse media and technologies as part of ongoing processes involved in cultural practice. The domestication framework perceives domestic technologies such as the radio, TV and the mobile phone as part of an ongoing adaptation and adoption process.[4]

The ongoing influence of the domestic technologies approach attests to its importance as a tool for comprehending the socio-cultural and individualistic symbolic power of commodities, especially communication technologies, as re-enacting older rituals akin to totem and fetish worship. The 'technology' function of artefacts pales in comparison to their symbolic weight and power. As Silverstone and Eric Hirsch observe in their pioneering work into the area, contemporary technological artefacts must be viewed as essentially material objects, capable of great symbolic significance, investment and meaning, while domestic technologies are 'embedded in the structures and dynamics of contemporary consumer culture' (1992: 20).

For the purposes of *Mobile Media in the Asia-Pacific* – with its focus on consumption and localised gender identities – the most useful framework is clearly that of the domestic technologies approach, which studies the symbolic dimensions of technologies in everyday life. In particular, the domestic technologies approach focuses on the meanings that individuals and cultural contexts give to technologies, thereby extending the ways in which users perceive them. A key feature of the domestic technologies approach is that it explores the interaction between the user and the technology as an ongoing process, as well as addressing the ways in which technologies such as the mobile phone can function symbolically as indicators of the user's tastes, values, and cultural and social capital. The mobile phone is not only a symbolic repository for the user's social capital but also signals to others certain unspoken clues about the user's identity and social status.

An earlier version of the domestic technologies approach can be seen in cultural studies scholarship – most notably in *The Story of the Walkman* (1997). In this groundbreaking anthology, Paul du Gay et al. outline a model for consumption consisting of various nodes, which they call 'circuits of culture'. This model for understanding the procedural nature of socio-technologies has been particularly prevalent in helping to read the mobile phone as both a set of practices and a dynamic mode of cultural production. In particular, Gerard Goggin's (2006) wonderfully lucid study on 'cell phone cultures' persuasively argues for utilising the 'circuits of culture' model to analysis the mobile phone. As Goggin argues:

> Cultural treatments of telecommunications are still comparatively rare, but in the case of the cell phone such an investigation is essential. Not only have cell phones developed their own 'little' cultures of consumption . . . we need to grasp and debate the place of cell phones and mobile technologies in our larger cultural settings . . . and what the implications of all this are for understanding culture at the most general level.
>
> (2006: 3)

Mobile @ cartographies: contextualising case study locations

The book's focus, the mobile phone user, means that much of its material is ethnographic – that is, grounded in the study of, and within, mobile practices and cultures. However, it is important to understand how these micro-narratives and practices are mired by political, economic and industry factors. By incorporating a domestic technologies approach into a macro-analysis of political and industrial issues, we can outline a model for cultural production and consumption not dissimilar to Goggin's robust reworking of the 'circuits of culture'.

By focusing on four divergent locations of mobile phone consumption – Tokyo, Seoul, Hong Kong and Melbourne – I investigate some of the local and regional modes of how gender is performed through, and within, mobile technology across the Asia-Pacific. Through the lens of mobile technologies as both a practice and cultural index, we can start to conceive of emerging modes of domestic and social labour, and the way in which women's relationship to technological production and consumption is changing.

Mobile Media in the Asia-Pacific is the result of seven years of ethnographic research in the Asia-Pacific. I began researching mobile phones while living in Tokyo at the beginning of 2000 and witnessing the rise of the 'mobile IT revolution' (Matsuda 2005) with i-mode (branded as a 'mobile with internet'). My curiosity about mobile technologies was motivated by my interest in material cultures, particularly the relationship constructed by globalisation in which place of production and consumption are often circumnavigated, if not completely camouflage. As a child, many of my favourite toys and influential media had been Sanrio's Hello Kitty, Osamu Tetsuka *anime* (animation) such as *Astro Boy* and *Kimba the white Lion*, along with Atari video games such as *Space invaders* and *Donkey Kong*. As I grew up, I became strongly aware that my formative objects and media had not been American or European, but rather, Japanese, Chinese and even

Korean. I became fascinated in connecting the place (and culture) of production with my consumption practices.

Enter the ubiquitous mobile phone, which was one of the dominant modes of global production that soon accompanied great economic, technological and socio-cultural shifts in the region (especially the NICs) and its practices of localised and transnational consumption. In Japan in 2000, the *keitai* became pervasive and ubiquitous as a socio-technology and cultural artefact embedded in everyday life. I began to observe, inquire and document the various ways this ubiquitous technology was transformed into a cultural artefact. From there I began travelling the region to locations such as Taipei, Singapore, Jakarta, Hong Kong and Seoul. Rather than covering all these regions, I decided to focus on four locations – Tokyo, Seoul, Hong Kong and Melbourne – and spend time conducting ethnographic research in each site. The case studies are qualitative and longitudinal, and involved spending at least two three-month fieldtrips in each location between 2000 and 2006. Between 2000 and 2002 I lived in Japan and from 2005 to 2007, I spent my time living between Seoul and Melbourne.

In order to study the mobile phone, I have myself been physically, psychologically, geographically, conceptually and culturally mobile over the last six years. Melbourne has been my base, but in some years I rarely even visited Melbourne. The experience has cast an indelible shadow on how I view both the actual role of mobile phone cultures and practices, as well as how I have reconceptualised discussions about mobility and what Kenichi Fujimoto defines as the socio-geographically bound notion of mobilism (2005). (The epitome of mobile irony was that, near the end of my research, I owned three mobile phones, and all three refused to work. I had the theory but at that moment no avenue for practice.)

The initial fieldwork took the form of travelling around the region, and was conducted from 2000 to 2003. The year 2004 marked the beginning of the focused case studies. I conducted separate case studies in each location, once in 2004, with follow-up interviews and focus groups in 2005 and early 2006. In each location I surveyed and interviewed individual respondents as well as conducting focus groups with a sample study group of 20–25 respondents, aged between 19 and 50 years old, half male and half female. It was important that the demographic did not just consist of young users, who have often gained, and continue to gain, the most attention in mobile communication literature. In order to provide consistency in my sample groups, in each location I limited my sample study to university-related respondents. Although this did not entail that the respondents had tertiary educations – as some of the respondents worked in administration – most had at least an undergraduate degree.

One of the reasons for limiting the range of cultural and economic capital of the sample groups was to ensure that respondents would feel relatively comfortable in speaking in English, which for a majority of respondents was their second language. Even in Melbourne nearly half of my respondents spoke at least two languages, and for some of them English was not the preferred language at home. This linguistic (and thus cultural) factor naturally limits the scope of the study. I was mindful of how being a native English speaker and expecting my respondents to speak in what was often their second language could cause stress and anxiety. One of the ways I attempted to be reflexive to this problem was by spending time living in each of the case study locations and attempting to learn to speak or read the native language at a basic level. The issue of English literacy was less an issue in Melbourne and Hong Kong than it was Tokyo and Seoul. I gave respondents the choice of having a translator present during interviews, as well as allowing respondents to take questions away and answer in their own time and language. A few of the respondents required part translation.[5]

The role of language as a crucial part of one's cultural identity was important, and while I learnt rudimentary skills in the languages, my command was too low to engage with the complexity. My usage of specific terms in their original linguistic context – such as *keitai* – highlights the need to understand how the linguistic is incorporated with socio-cultural economies at both material and symbolic levels. Unquestionably, there is a need for further research into the region's mobile media especially in terms of linguistics – particularly in the case of the SMS. This study aims to focus more broadly upon gendered mobile media literacies in the region.

Interrelated with the role of power around the language issue was inevitably the fact that I am a Western female. I was very aware how being an Anglo Saxon female would alter my respondents' answers and I sought to be reflexive to some of the notions about gender that were a product of a Western context. While I had spent, on average, at least a year, on and off, living in each location apart from Hong Kong (which I visited frequently for periods of three weeks to one month), I was mindful of the need to be sensitive to cultural and socio-linguistic subtleties. In particular, I wanted my respondents to guide me into new ways of thinking about their own and their society's gender performativity.[6] The follow-up sample study groups conducted in 2005 and early 2006 included new respondents, and the shift into mobile media (such as camera phone practices) became apparent. In both Seoul and Melbourne I was fortunate to work with university students for a mobile media seminar in which students researched their own and friends' mobile media practices for three months: this opportunity enabled in-depth discussions about the various forms of mobile media practices. In the case of Hong Kong and Japan, most of my follow-up questions were conducted via email after the initial face-to-face interviews and focus groups.

This study is a pointer to the sort of research that needs to be conducted if we are to gain insight into emerging modes of gendered practices and associated modes of mobility (technological, economic, cultural, social, political) in the region. Ideally, more longitudinal and micro-focused studies, with increased multi-linguistic proficiency, would provide deeper understanding of the multiple mobilities – actual and metaphoric – that are emerging in the region. I hope that this study provides a benchmark for more transnational longitudinal studies, as regional area studies researchers and anthropologists recognise the importance of everyday technologies such as the mobile phone in providing a lens into contemporary forms of locality.

By focusing on qualitative sample studies conducted over a couple of years, I sought to provide a detailed understanding of the shifting individual users' relationship towards the increasing feminisation of mobile technologies in the region. From the outset, two significant and interrelated social phenomena could be noted in terms of gender and technology: firstly the rise of women as 'prosumers' – that is consumers as co-producers (Toffler 1980) and active 'produsers' (Bruns & Jacobs 2006) of new media technologies, and secondly, signs of increasing employment of women in media industries once dominated by men (ILO 2008).

With the rise of mobile media we are seeing new forms of labour, creativity and user–producer/author–audience paradigms that reflect emerging forms of post-modernity and new media in the region. From the sending of a SMS considered as a work of art to the burgeoning phenomenon of millions of readers of *keitai shôsetsu* (mobile phone novels in Japan) that have captured the *zeitgeist* of contemporary media literacies, female subjectivity and forms of expression are being demonstrated throughout the region. The question that motivates this examination into emerging gendered mobile media is how these practices and cultures translate into forms of female empowerment and agency. Is contemporary mobile media a reflection of the newly conscious and powerful woman in the region?

In order to address these questions around the localised engendering of mobile media customisation, this book consists of three sections. The first section is *Mobile media societies*, the second *Mobile media cultures* and the third is *Mobile media practices*. Firstly, this introduction sketches out the motivating questions and context for *Mobile Media in the Asia-Pacific*. Then the first section, entitled *Mobile media societies*, foregrounds the literature and key issues relating to women and mobile media in the Asia-Pacific. In the introductory chapter, I contextualise my methodology and fieldwork, as well as providing an overview of the relevant literature. Part I provides the background for conceptualising gendered mobile media in the region. Through the trope of gendered mobile media, I argue that we can gain new insights into understanding post-modernity in the region. This trope is what I call 'cartographies of personalisation', as symbolised by the young female mobile phone user; this trope operates as a motif for the region's new forms of post-industrial modernity. By exploring the emerging 'imaging communities' of gendered mobile media – and how they form localised and transnational 'cartographies of personalisation' – we can begin to chart twenty-first-century post-modernity in the region. In order to do so, Part I explores the relevant literature and conceptual paradigms.

Chapter 1 surveys the current literature in mobile communication and the rise of mobile media. In this chapter, I explore the literature surrounding mobile media – both globally and regionally – in light of the conspicuous rise of the female mobile media user in the region. In particular, I consider why this phenomenon has been largely ignored in scholarship to date. This is followed by Chapter 2, 'Paradigms of mobility', which focuses on rethinking notions of mobility that have dominated much of the recent scholarship around globalisation, and how this impacts upon reconfigurations of intimacy, social capital and individualism.

Chapter 3, 'Beyond the 'new rich'', discusses the ways in which gender, consumption and modernity in the region have been theorised as distinct from Western or European models. This chapter revisits studies on consumption in the Asia-Pacific while also addressing the theories that have provided new ways of conceptualising contesting modernities in the region. By considering how the role of gender has been configured after the Asian financial crisis of 1997, this chapter focuses on the conspicuous motif of young female as a symbol of – and for – consumption. Reflecting on the rise of feminised, socio-emotive customisation and the increasing deployment of social labour in the use of mobile technologies, I discuss the relationship between feminised mobile media practices and their relationship to female empowerment (and employment) in the communication technologies industry. The chapter asks whether women's creative use of mobile media equates to shifts in their socio-economic position and how this phenomenon, in turn, reflects shifts in the region's construction of post-modernity.

The following section (Part II), *Mobile media cultures*, introduces four case study locations in the region – Tokyo, Seoul, Hong Kong and Melbourne. In Chapter 4, entitled 'Fast-forwarding to the present', we focus upon one of the locations that have gained much attention globally as 'centres' for mobile media innovation, namely Tokyo (Japan). As a site with a long history of innovative post-World War II domestic technologies such as the Sony Walkman, Tokyo regained and reconfigured its national image with such global images as the DoCoMo i-mode 'mobile (*keitai*) IT revolution'. In this phenomenon the role of the female user has been pivotal – providing an evocative symbol for emerging Japanese national cultures and twenty-first-century post-modernity. In Chapter 5, entitled 'Engaging rings', we move to another global centre for mobile media innovation, Seoul (South Korea, henceforth Korea). Like Japan, Korea's techno-nationalism has operated to produce an exemplary model for technological innovation, affording Korea the title of the most 'broadbanded' country in the world (OECD 2006), in which particular technologies such as the mobile phone (*haendupon*) play a pivotal part in the re-imagination of Korea both internally and externally. In the conflations between nationalism, family and new technology (Cho 2000) that is South Korea in the twenty-first century, we are left to question how empowering this fast-tracked post-modernity, *vis-à-vis* gendered mobile media, is in terms of women and female performativity in Korea.

Chapter 6, 'Nostalgic mobility', shifts away from the two 'centres' whose highly regulated local industries have ensured innovation, to a market bombarded by an array of global brands and service providers: Hong Kong. This chapter considers some of the ways Hong Kong negotiates between its current Chinese rule and its history as a British colony. Through the particular patterns of gendered mobile media we can gain insights into how Hong Kong is re-territorialising itself within twenty-first-century post-modernity. In Chapter 7, entitled 'Postal presence', *Mobile Media in the Asia-Pacific* migrates to 'the west in Asia' (Rao 2004) specifically to the 'fashion capital' of Australia, Melbourne. In Melbourne we see a culture obsessed with texting in a market saturated with images from the Asia-Pacific. Through the trope of the postal metaphor we see how SMS and camera phone practices are shaped by unmistakably gendered modes of mobile media performativity that reflect the place of Melbourne in the region's gendered post-modernity.

In Chapter 8 ('Domesticating cartographies'), the concluding chapter of Part II, *Mobile Media in the Asia-Pacific* reflects on the four aforementioned diverse locations. The chapter asks: how can we define these emerging forms of gendered mobile phone practices and cultures I call cartography of personalisation, and how do they reflect arising modes of mobility (and immobility) in the Asia-Pacific? How, in turn, is the symbol of the mobile phone revealing formations of transnational communities? Does the female consumer, as both an image and a mode of performativity, expose new gendered practices and power relations in the region? Finally, how does the female mobile phone user represent localised and transnational forms of post-industrialism and post-modernity in the region?

The third and final section, *Mobile media practices*, moves onto some of the emerging modes of new media, specifically mobile gaming, in two case studies. In Chapter 9, entitled 'Domesticating new media', we explore the role of convergence and media practice in the region. This chapter considers the role of mobile media as a marriage between new media and domestic technology traditions. Chapter 10, 'The big bang', offers a case study on Korean mobile media group INP, whose projects can provide insight into this burgeoning phenomenon of mobile media as new media. In this section I provide examples of the ways in which mobile media is deployed in creative ways, through convergent media tactics and burgeoning creative industries such as gaming. These two chapters are meant to provide two examples within a myriad of ways in which the region's deployment of mobile media can point to various creative and exploitative futures in which gender will continue to feature prominently. I conclude with Chapter 11, 'On hold', in which I meditate on the partnership between mobile media and new media in the Asia-Pacific and how the various *imaging communities* reflect gendered and localised intimacies that I call 'cartographies of personalisation'.

Throughout this book I will demonstrate how, through the lens of gendered mobile media, we can gain insight into new forms of mobility and immobility in various locations in the region. This book examines how gendered mobile media consumption has enabled new 'cartographies of personalisation' in the Asia-Pacific. 'Cartographies of personalisation' have both

Figure 0.1 Nagara mobility. Photo: Hjorth 2004.

symbolic and material manifestations – they can be part of the customisation of the outside of the phone with cute charm given by a best friend, to inside the phone in the form of the multimedia practices of SMS and camera phone imagery. In each location, the various specificities such as technological infrastructure, governmental and industry regulation, socio-economic demographies as well as socio-cultural and linguistic particulars all inform the 'cartographies of personalisation'. In particular, the various forms of UCC mobile media – new micro 'imaging communities' – that are being produced can be seen to characterise new forms of intimacies that are not just an example of social labour exploitation but rather new forms of creativity. As one of the most ubiquitous lenses for twenty-first-century production and consumption in the Asia-Pacific, mobile media can give us insight into residual and emerging local, national, and transnational lifestyle patterns. This is the art of being mobile.

Part I

Mobile media societies

1   Locating the mobile

Mobile communication and gender in the Asia-Pacific

> In recent years the imagination of the West, and indeed, of the East as well, has been captured by the dramatic emergence in East and Southeast Asia of a new middle class and a new bourgeoisie. On the television screens and in the press of Westerns countries, the images formerly associated with affluence, power and privilege in Asia – the generals, the princes and the party apparatchiks – however outmoded in reality, are being increasingly replaced by more recognisable symbols of modernity. Western viewers are now familiar with images of frustrated commuters in Bangkok and Hong Kong traffic jams; Chinese and Indonesian capitalist entrepreneurs signing deals with Western companies; white-coated Malaysian or Taiwanese computer programmers and other technical experts at work in electronics plants; and, above all, crowds of Asian consumers at McDonald's or with the ubiquitous mobile phone in hand.
>
> (Robison & Goodman 1996: 1)

At the dawn of the twenty-first century, the Asia-Pacific provides a compelling model for analysing emerging forms of mobility, globalisation and post-modernity. The region is a powerful player in the circulation of mobile technologies – both materially and symbolically – and in shaping the burgeoning lifestyle patterns associated with them. The cultural and economic power of global mobile technologies in the region can no longer be sublimated under the symbol of Japan as the production epicentre of portable technologies such as the Sony Walkman. Concurrent to the rise of mobile technologies globally, the region has grown to become both a powerful economy and a conveyer of soft cultural capital. Through various forms of innovative mobile technology in locations such as Tokyo, Seoul, the Shanghai–Suzhou corridor and the potentialities of colossal new markets – particularly China – the Asia-Pacific now plays an important role in global production and consumption circuits. In sum, the region's formidable economic power has now transformed into a rising cultural currency globally.

The multiple forms of cultural capital that the Asia-Pacific commands worldwide are undisputed, particularly the rise of mobile phone cultures as part of the region's techno-cultural capital. This phenomenon parallels the unshakable position mobile phone cultures occupy globally. Demonstrating the world's highest mobile phone subscription rates (Mitomo et al. 2005; Castells et al. 2007) and housing key centres for globally innovative production (Tokyo and Seoul), the region is unquestionably a centre in mobile capital; in this positioning we can see the deep interconnections between mobile phone consumption and production patterns. The question this raises is, given how central mobile phone consumption and production has been in the rise of the region as arguably this century's new global power (Arrighi et al. 2003), to what extent is the transnational imaginary vested in, and represented by, the cultural index of the mobile phone? And how might studying the mobile phone in such a context provide insight into burgeoning imaginaries of the region both internally and on a global scale?

Returning to Robison and Goodman's use of the mobile phone as a symbol of the region's emerging modernity of *The New Rich* (1996) can provide us with great acuity. Apart from gracing the collection's front cover, the mobile phone actually only gets two mentions, both in the context of presenting two different paradigms for conceptualising consumption and modernity. McDonald's represents a conflation between Westernisation (or, more specifically, Americanisation) and consumption, in which globalisation is conceived of as a homogeneous product. By contrast, the adoption of the mobile phone suggests, partly because of its emerging 'class' status and lifestyle, a more localised appropriation of consumption and modernity.

But things were not that simple, as was exemplified by the following year's economic crisis in the region. Consumption and modernity post-economic crisis of 1997 became a carefully orchestrated venture, with each 'nation' attempting governmental, industrial and socio-cultural revisions. In the decade following 1997, the region has grown unevenly. This shift was no easy feat and involved socio-cultural, ideological, economic and technological changes that marked emergences of new forms of modernity. Following 1997, Singaporean cultural theorist Chua Beng-Huat (2000) observes that many locations in the region revitalised consumption as no longer a subset of production or a discursive form of Westernisation. This reformation was important if the region was to quickly recover from the crisis that saw some countries, including Korea, having to be financially rescued by the International Monetary Fund (IMF).

Once a symbol for emerging class and leisure cultures in the region – as identified by the opening quote from Robison and Goodman – the mobile phone has come to encompass many social, cultural and economic dimensions. These dimensions are multiple, divergent and always evolving, like the region itself. To explore mobile technologies is to investigate the ongoing significance of localisation practices in negotiating the influences of global trends. By focusing on arising gendered consumption practices in mobile technologies, and by recognising that these technologies can operate as conduits for localised practices in the face of globalisation, this book aims to provide insights into the ways in which the region is reconfiguring itself both internally and trans-culturally within globalisation.

Indeed, as pioneers in mobile communications globally, the region's various production centres have seen this technological development, and the politico-economic power it secures, translated into forms of cultural capital both within and outside the Asia-Pacific. This is exemplified in the case of Korea, where the advancement of technological innovation has been instrumental in the rise of Korea's transcultural capital in the form of the Korean wave (*Hallyu*). Once a centre for the production of domestic-technology hardware (manufactured by companies such as Samsung and LG), Korea has now become a major exporter of cultural products in the form of films, TV dramas and online games.

As well as housing global leaders in the development, innovation and manufacturing of mobile technologies, the region has also exemplified the emergence of paradigms around user agency such as the rise of 'produsers' (Bruns & Jacob 2006) through UCC. From the aforementioned example of the Japanese high-school pager revolution in the early 1990s (Ito 2005; Hjorth 2003a, 2003b) to the camera phone empowerment pioneered by women in Seoul (D.H. Lee 2005), the region is awash with the rise of mobile phone users, and agency, inextricably linked to female users. This phenomenon has resulted in the region's particular domesticating of mobile technologies being implicitly tied to gender practices of consumption. Hence, to explore mobile consumption in the region is to investigate the rise of gender-inflected technologies in burgeoning consumer identities.

As a key icon of mobile phone consumption and multiple forms of 'mobility' (social, economic, physical and cultural), the construction and representation of the young Asian female operates across various levels – national, transnational, governmental, social, cultural and economic. The rise of the mobile phone has been accompanied by increased subversive appropriation of the technology by the active female user (Hjorth 2003a, 2003b; Matsuda 2005; Fujimoto 2005; Bell 2005). Parallels can be made with other domestic technologies, which illustrate the instrumental role of gender and power in inscribing technology with the socio-cultural.

Just as the landline telephone, originally designated as a business tool for men (Martin 1991a, 1991b; Moyal 1992) was appropriated by female consumers in the domestic sphere and transformed into a social medium, the mobile phone followed a similar trajectory. Initially it was women who played a major role as operators, while men – as symbols of business – were the main consumers. Unlike domestic landline telephones, which were fixed to the domestic sphere and thus reinforced gendered spatial division and reproductive (unpaid, domestic) labour (Fortunati 2002a, 2002b, 2008; Massey 1995), the mobile phone saw the 'domestic' going out into the public sphere, thereby partially breaking down the various gendered divisions of public/private, intimate/anonymous and paid/unpaid labour. In this disintegration of work/leisure distinctions, the mobile phone has often just amplified existing forms of post-industrialism.

However, rather than mobile media resulting in the erosion of gender divisions, notions of social and reproductive labour become increasingly pre-carious – proffering both empowerment and exploitation. In the case of developing nation contexts such as the Philippines, domestic landline usage was never the norm; thus the mobile phone stepped in to facilitate the domin-ant economy of the country – Filipino women working abroad. As Deirdre McKay (2007) observes in the case of Filipino houseworkers in Hong Kong, the mobile phone helps to provide particular forms of intimacy that were previously not available. Hence the charting of telephonic genealogies within the region is heterogeneous to say the least, mired by gendered work and lifestyle practices. In these multiple and uneven forms of emerging mobility and post-modernities, signposted by the iconic rise of mobile technologies, it is the symbolic prime consumer – the young Asian female – who becomes embroiled in the shifting of consumption, mobility, gender and 'Asia'.

Moreover, as Sen and Stivens (1998) observe in their study on gender and power in 'affluent Asia', gender, modernity and globalisation are interwoven in the construction of class. This is most notable in the symbolic role of women as signifying consumption and lifestyle. Through the lens of political economy, Sen and Stivens demonstrate the central role of gender in the mod-ernising and globalisation of Asia. Thus it is not surprising that the symbol of the Asian female consumer and the mobile phone became truncated signifiers for divergent forms of consumption, mobility and modernity in the Asia-Pacific.

For Thanh-Dam Truong (1999) the pivotal role of the female – as both consumer and producer – in the rise of industrialisation with the 'Asian miracle' of the region post-1997 cannot be underestimated. As Truong observes, the rise of industrialisation is also the product of conflicts around gender, and specifically around the role of gendered classes and forms of labour. Far from post-industrialism liberating women from simplistically being defined in terms of sexuality and economy, women's roles are further naturalised and globalised (in the case of women from developing countries such as the Philippines) into positions of care and reproductive labour.

The connection of the mobile phone with feminised forms of labour can be found throughout the world. For French sociologist Chantal du Gournay (2002), the mobile phone constructs new forms of intimacy in which gender features prominently. As du Gournay notes, these forms of intimacy and femi-nised labour are particularly prevalent in the public through the rise of social reproductive labour; so much so that the mobile phone becomes a fetish of the hyperfeminine. According to Leopoldina Fortunati, since the mobile phone is one of *the* most intimate items in everyday life, it reflects particular gendered performativities (Fortunati 2002b). For feminist sociologist Judy Wajcman, the association with 'affective or emotional work is part of the unequally distri-buted gender division of labour' (Wajcman et al.: forthcoming). However, in the context of the region and the uneven power relations between women in de-veloped and developing countries, the feminisation of labour is most apparent.

The 'feminisation' (Truong 1999: 134) of industrial relations, and the rise of feminised modes of mobile cultures and consumer practices, is inextricably linked. This broad engendering of the region's production and consumption practices is linked to general symptoms of globalisation such as the outsourcing of care cultures. In the various forms of mobility – people, ideas, labour and capital – it is undoubtedly women who are implicated the most (Hochschild 2000; Parreñas 2001). Unquestionably, the role of the mobile phone as a technology of propinquity is both instrumental in, and symbolic of, the transnational flows of gendered modes of labour and consumption.

This dynamic could be characterised as a growing dialectic between arising forms of *materialism* and *maternalism* that are intertwined with practices of intimacy as well as, on a broader level, reconstructions of nation-state in the face of global mobility and transnationalisation. As Brian McVeigh (2004) observes in the case of Japan, the role of women as 'good wives and wise mothers' (*ryôsai kembo*) is characteristic of the maternal materialism firmly entrenched in Japan's various levels of nationalism. Through the lens of gendered mobile media, we can gain insight into emerging forms of post-modernity on both a micro and macro level via symbolic and material practices. On the micro level, there are the emerging gendered practices of intimacy, labour and community, and, on a macro level, formations of national culture, transnationalism and post-industrialism in the region.

Thus the mobile phone is a poignant symbol and set of material practices within the region's various contesting postmodernities. Yet despite the region's pivotal role in the global production and consumption of mobile technologies, the region has been under-explored (McLelland 2007). Moreover, the crucial role of the young Asian female as synonymous with the rise of mobile media and UCC practices has also been overlooked. In order to address these gaps in the research about mobile media, this chapter will firstly outline the rise of recent field of inquiry called mobile communication. It will then chart the current literature on the region and the concurrent gaps in research. In the last two sections of the chapter, I delineate the emergent rubric of mobile media and its relationship to new media practices.

One of the central intentions of this book, as I will demonstrate in Part I, is to address the lack of theoretical and empirical engagement with the pervasive symbol of the young female mobile media consumer and how this reflects post-modernity in the region. Despite mobile media increasingly being a repository for burgeoning localised and gendered practices of intimacy, labour and community and how this, in turn, reflects shifting paradigms of national culture, transnationality and post-modernity in the Asia-Pacific, the area has continued to be under-investigated.

Moreover, through the interrelated rise of UCC practices and mobile media, we can explore some of the evolving paradigms for labour and creativity that suggest new – but also rehearsing and adapting older or 'remediated' (Bolter & Grusin 1999) – media tactics in which women figure prominently. Via the rubric of gendered mobile media, this book investigates arising patterns of intimacy, creativity, labour and community encompassed under the trope of 'cartographies of personalisation'. In order to do so, we firstly need to locate the place of the region's mobile media in the context of global mobile communication research.

The place of the mobile: current literature in the field

> ... the mobile phone is far too much of a newborn creature to have a storied history, or even much of a reputation in social science research. Its advent and rapid evolution have bypassed most researchers who are deeply engaged in their own research pursuits, but few if any social scientists would fail to recognise the impact this technology has had on all of us and on aspects of our behaviour.
>
> (Beaton & Wajcman 2004: 2)

As John Beaton and Wajcman observe, the social impact of the nascent rise in mobile communication cannot be ignored. In their important study of Australian mobile telephony, they note the transformation and diffusion of boundaries between traditional private and public spheres (ibid.: 9), a trend that sees mobile telephony penetrating 'new geographic spaces that enable the consumption and communication process to be applied in new social, cultural and psychological spaces' (ibid.: 12). In 'Intimate connections: the impact of the mobile phone on work–life boundaries' (2008), Wajcman et al. note the mobile phone 'characterises modern times and life in the fast lane' and has become iconic of 'work–life balance' – or lack thereof – in contemporary life (ibid.).

Wajcman et al. observe that manipulating 'the boundary between work and life was one of the principal ways that many people controlled their time' (ibid). These boundaries of time and space are determined, in part, by 'debates about work/life boundaries' that are imbued by traditional gendered divides 'between the separate spheres for market work (male) and domestic work (female) wrought by industrialisation' (ibid). Thus the mobile phone is deeply implicated in debates around various forms of mobility and immobility that cut across gender, labour, technology and capital within contemporary globalisation.

This is, in part, due to the multiple dimensions of mobile communication as metaphor, icon, culture and practice. As a consequence, it lends itself to interdisciplinary and transdisciplinary analysis. At mobile communication conferences, the rooms are filled with sociologists, media theorists, anthropologists, philosophers, new media artists, economists and IT researchers. The mobile phone can be read for its social, technological, economic, and creative properties. And yet within this burgeoning area gaps remain – most notably the role of the mobile phone as a cultural index (Goggin 2006a), the implications of the mobile phone within the changing modernity of the Asia-Pacific (McLelland 2007), and the way in which the rise of the mobile phone – as a symbol and practice – is imbued by gendered genealogies (Hjorth 2003a).

Far from being merely a functional technology, the mobile phone is a cultural artefact overflowing with meanings inflected by cultural, social, economic and even political (techno-nationalist) factors. It operates simultaneously upon individual, social and cultural levels connecting people both literally and symbolically. As the mobile phone has grown into a multimedia device in the form of mobile media, it has come to be a repository for new forms of social, economic and cultural capital. This diversity and polyvocality is exemplified in the Asia-Pacific, which not only has the highest subscription rates (Mitomo et al. 2005; Castells et al. 2007), but also exemplifies the most varied modes of localised consumption and production practices (Bell 2005).

Yet despite the region's significant role in both producing and consuming mobile technologies, it has gained little focus in comparison to research on mobile communication in Europe. Of the handful of researchers specialising on the region, only a few such as anthropologist Bell have explored the role of mobile communication in the region (2005), with most focusing on specific locations such as Tokyo (Ito 2002, 2005a), Seoul (Kim 2003; Yoon 2003) and the Philippines (Pertierra 2006; Ellwood-Clayton 2003). In Bell's ethnographic study in Asia over a three-year period she identified multiple, often competing 'cultures of mobility' that demonstrated that 'what it means to be "mobile" . . . has distinct cultural meanings' (2005: 70). As Bell's study affirms, more transnational and longitudinal studies need to be conducted in the region if we are to gain a sense of the multiple mobilities that are intrinsically linked to the numerous modernities.

Through the lens of mobility we can gain insight into the localised notions of intimacy, co-presence, gender, nationalism, identity, social capital, labour, disapora, and migration in an age of globalisation. It can provide simultaneously both a micro and macro model for understanding the complex, ever-evolving and shifting mobilities that prevail today. As Bell's eloquent study identified, there are gaps that needed to be addressed – most notably the link between socio-cultural and politico-economic mobilities. Research is also needed into the pivotal role of women as the prime consumers and active co-producers in mobile cultures, and into how this influences gender performativity and representations at a national and transnational level.

The uneven adoption in the region has started to gain the attention of mobile communication scholars worldwide (Castells et al. 2007). However locations as such Japan and South Korea – being exemplary models of techno-nationalism (West 2006) – have often fallen prey to new formations of techno-orientalism in which stereotypes around orientalism (Said 1978) and occidentalism are rehearsed through the production and consumption of post-industrial technologies (Nakamura 2003). I will discuss these geo-political imaginaries later in the chapter when I outline the role of production in the region and its reception globally. A recent example of techno-orientalism can be found in the case of Japan and what Douglas McGray characterised as its 'gross national cool' (2002). In his critical appraisal of projections and representations of Japanese globally *vis-à-vis* its commodities, McGray observed that Japan's association as technological centre had been translated into a type of global cultural capital.

Expanding upon Joseph Nye's notion of 'soft power' in which governments utilise non-material resources to exert power over people, countries and markets through persuasion rather than coercion, McGray identifies Japan's 'gross national cool' (GNC) as a heavily orchestrated governmental and industry collaboration to create a type of cultural capital within global markets. Through the continued innovation of 'mobile' technologies that advance a notion of what Tetsuo Kogawa characterises as 'electronic individualism' (1984) from the Walkman onwards, Japan was able, at the beginning of the twenty-first century, to play a central role in the rise of mobile 'personal' (Ito 2005) technologies.

Thus Japan has been able to regain some of its pre-1997 economic power, but the power it now wields is predominantly cultural and ideological rather than economic and technological. The conflation between technology and culture has been highlighted in the case of Euro-American conceptualisations of Japan in terms of techno-orientalism (Morley & Robins 1995; Yoshimi 1999), and in analyses of post-World War II US–Japan bilateral agreements, which have been pivotal in the scripting of a particular US geo-political definition of the region (Arrighi 1998). The transformation from technological and economic prowess into cultural capital both transnationally and globally is particularly prevalent in Japan's reconvening nationalism – a point I will expand upon further in Chapter 4. Thus mobile media is tied to not only burgeoning forms of mobile intimacy, but, as GNC shows, mobile capital (and capitalism).

Mobile @ global: a brief history of the rise of mobile communication studies

Unsurprisingly, the early studies of the mobile phone can be traced to its formation and transformation from the landline. As with the landline, the rise of mobile phone technology was marked by an appropriation of the mobile phone's original intended use as a business tool into an instrument for social and domestic use, particularly by younger women. This transformation from male business tool to vehicle for female social 'gossip' (Martin 1991a, 1991b) and reproductive/social labour has indelibly marked the history of telephony.

Despite this gendered formation, the pivotal role of gender in the domestication of the technology into a cultural and social practice has been marginalised in the literature, with researchers preferring to emphasise the 'youth' aspects and relegating gendered customisation to the realm of fashion (Katz & Sugiyama 2005; Fortunati 2005a, 2005b, 2005c) or motherhood (Matsuda 2007). This is surprising given the often subversive ways in which female users in the region have transformed the technology through innovative social and cultural practices (Fujimoto 2005; Matsuda 2005; Hjorth 2003a, 2003b). Arguably, this 'gender agenda' surrounding technologies is part of embedded constructions of gendered nationalisms in which researchers seem to be unable to discuss women except in stereotypical terms such as the afore-mentioned view of Japanese women as 'good wives and wise mothers' (*ryôsai kembo*) (McVeigh 2004).

The first studies of mobile culture around the early 1990s tended to high-light the implicit role that gender played in the emergence and transform-ation of the business technology into a socio-cultural practice. Moyal's (1992) aforementioned study on gender and the telephone in Australia was not only one of the first studies in Australia, but also an earlier pioneer in what would become mobile communication research. Patricia Gillard's research in Australia in the 1990s (particularly with the Australian govern-ment) was significant in conceptualising new models for studying telecom-munications as a cultural practice. Michele Martin's (1991b) eloquent study explored the transformation of the telephone from business tool to a femi-nised social and cultural artefact. In the same year Wajcman's wonderfully rigorous critique of technology in *Feminism Confronts Technology* (1991) hallmarked the epoch's feminist re-examination of the socio-technological tropes of cyberspace and politics of virtuality (Turkle 1995; Haraway 1991; Stone 1995). This era also saw the rise of the concept of the 'feminisation' of technology/telephony (Brunner 2002) and debates around the gendered body politics of mobile virtuality (Fortunati cited in Wajcman & Haddon 2005).

Most notably, as Fortunati (2002b: 55) observes, reproductive labour became increasingly implicated in the proclivity of mobile telephony to embody parallel movements towards individualisation on the one hand, and social capital, on the other hand (Ling 2004: 176). In the rise of full-time intimacy, the upkeep and maintenance of social capital is no longer the pre-rogative of the female. In developed contexts, both males and females (in various degrees) are caught up in the Web 2.0 leash of full-time social labour through ongoing participation in SNSs (social networking sites) and the 'per-petual contact' of mobile media applications such as SMS: a phenomenon that could easily be encapsulated as an increasing 'feminisation' of work/life practices. However, in developing countries the picture is different, with reproductive and social labour – and thus, in the case of mobile care cultures, the exploitation of 'feminised' labour – being still very much the role of the female (Hochschild 2000, 2003).

There are striking parallels and contrasts between, on the one hand, the rise of the 'feminisation of immigration' (Ehrenreich & Hochschild 2003) whereby most diasporic workers are women from developing countries (such as the Philippines and Vietnam) going to care for families in developed countries and, on the other hand, the financial and social mobility of women in those wealthier families in developed contexts. Both types of mobile women have paid work that results in outsourcing their home caring to others. This blurring of home and work in the buying and selling of emotional labour is affirmed by one of the first European ICT researchers, Dutch research Enid Mante-Meijer, in her co-study with van de Loo (1998) 'Blurring of the life spheres: Flexibility and teleworking'. In the study, the role of women and the precarious and flexible nature of women's work can be viewed as emblematic of broader forms of post-industrial mobility and immobility in which inequalities are exacerbated.

In a similar vein as Martin's study, Lana Rakow's (1992) lucid study investigated some of the ways in which gender has informed conventions around telephonic practices. The issue of reproductive labour and the shifting politics of 'care cultures' that Hochshild details so vividly in her research is presciently outlined in Rakow's and Vija Navarro's (1993) 'Remote mothering and the parallel shift: women meet the cellular telephone'. Here, the role of the telephone as both a product and symbol of particular types of emotional and reproductive labour is emphasised.

Robert Hopper's (1992) *Telephone Conversation* investigated some of the ways in which technologically mediated conversations and modes of co-presence were formed and how these reflected other social activities, while Claude Fischer's *America Calling* (1994) insightfully charted the rise of the telephone in the US to the 1940s, affirming de Sola Pool's early research into communication technology (1980). Haddon's (1997a, 1997b) research on the telephone as a domestic technology firmly embedded mobile communication within a domestic technologies approach. By 1999, American communications scholar James Katz, one of the pioneers both in the US and globally, had begun his formidably prolific research (1999, 2003, with Aakhus 2002).

Despite the fact that during these interesting early years the rise of mobile communication was clearly invested with gendered politics and the socio-cultural economies of the domestic sphere, history repeated itself. Like the landline that started off as a business tool, to be later transformed – feminised – by women into a socio-cultural practice and artefact, the mobile phone replicated the same cycle. So why, second time around, has gender continued to be relegated to a minor field while the exciting, sexy, fun field of 'youth cultures' continues to dominate? Despite the reality of aging populations and mainstream practices, the conflation between youth and new technologies only perpetuates stereotypes about youth subcultures and the wayward role of new technologies.

In the early studies of mobile communication, European scholarship burgeoned, dominated by a focus on youth cultures. In particular, the Scandinavian region, with its early and veracious adoption of practices such as the SMS and then the MMS, gave rise to some of the first pivotal studies such as Ling and Yttri's study on SMS in Norway in the mid 1990s (2002). Another European study, that of Alex Taylor and Richard Harper (2002) on SMS practices as symbolic gift-giving, had a considerable impact on subsequent studies, especially around sharing practices involving camera phone imagery, as noted by Ito and Okabe in Japan (2003, 2005a, 2005b).

The role of mobile phones both as communication devices and as symbolic objects for re-enacting earlier rituals and maintaining social patterns became a prominent theme in debates about 'mobility' causing a demise of place (Ito 2002; Yoon 2003). Studies from the domestic technologies tradition, particularly by British scholars such as Silverstone and Hirsch (1992), Haddon (1997b) and Miller (1987, 1988), drew attention to the role of domestic technologies (the mobile phone being a more recent mode), as extending upon previous media and the attendant social economies of public and private space. Miller's early research paved the way for considering commodities as extensions of individual identity and selfhood, particularly in the 'social economy' of the domestic sphere. For Silverstone and Hirsch, it is not individuals but the context of the household that transforms commodities into meaningful parts of the moral economy. As they avow, 'the moral economy of the household is therefore both an economy of meaning and a meaningful economy' (1992: 18).

Two early publications gained widespread attention by highlighting the symbolic dimensions of the mobile phone as both a practice and a culture – namely Kopomaa's *The City in Your Pocket* (2000) and Plant's 'On the mobile' (2002), commissioned by Motorola. Kopomaa's book articulated some of the proliferating paradoxes of the everyday, which were exemplified by the symbol of 'post-modernity', the mobile phone. He pointed out that conventions around public and private, work and life had already blurred and that the social index of the mobile phone encapsulated this contemporary milieu. Plant's important and well-cited study was one of the first of many that attempted to engage with the mobile phone as a global social practice. Surveying nine locations around the world, she identifies types of public etiquette that are associated with the use of mobile phones. Utilising the analogy of bird types, Plant explored the various roles of mobile phone performativity in modes of public intimacy – particularly among strangers. In this study, she clearly demonstrates the way in which gender, age and locality informed types of mobile practices.

By 2002 mobile communication studies had begun to proliferate, most notably in the form of anthologies. These anthologies, such as Katz and Aakhus's groundbreaking *Perpetual Contact* (2002) and Hungarian philosopher Kristof Nyíri's wonderfully prescient series of philosophical conferences and attendant publications (such as 'place' [2005], 'mobile learning' [2003b] and 'mobile democracy' [2002]) were important in formatting some of the key rubrics for mobile communication. The year 2002 also saw two other key anthologies: Brown et al.'s *Wireless World: Social, Cultural and Interactional Issues in Mobile Communications and Computing* (2002), and Lindgren et al.'s *Beyond Mobile: People, Communications and Marketing in the Mobilised World* (2002). In these, the possibilities of the mobile phone as a lens for conceptualising broader social phenomena became apparent.

The same year saw Howard Rheingold's popular *Smart Mobs: the Next Social Revolution*, which extended, via anecdotes and secondary sourced material, the global sampling that Plant had conducted the previous year. 2002 also saw the beginnings of mobile communication research in Asia, typified by the then small but annually expanding conferences held in Seoul by Shin Dong Kim. By 2004, the Korean conference *Mobile Communication and Social Change*, organised by Korea Telecommunications giant SK, hosted over 80 international delegates – indicative of the growing interest in the mobile communication field in Asia, after much emphasis on Europe and the US.

The global popularity of the mobile phone became apparent in 2003 with the release of Jon Agar's pithy and enjoyable *Constant Touch: a Global History of the Mobile Phone* (2003). The potentialities of the mobile phone as both social practice and artefact were epitomised by the Bruno Latour (1987)-inspired anthology by Fortunati, Katz and Riccini entitled *Mediating the Human Body: Technology, Communication, and Fashion* (2003). This deployment of the social readings of technologies by Latour's Actor Network Theory (ANT) was also apparent in Katz's anthology *Machines That Become Us: the Social Context of Personal Communication Technology* (2003). The year 2004 saw such conferences as *The Life of Mobile Data* (organised by Nicola Green at Surrey University in London), featuring keynote speaker David Lyon, which shed some light on some of the darker aspects of the mobile phone phenomenon, such as surveillance. The same year saw one of the key scholars in mobile communication, American–Norwegian teen expert, Rich Ling, release his global survey of the field *The Mobile Connection*.

The following year, 2005, saw a deluge of mobile communication anthologies including Hamill and Lasen's (eds) *Mobile World: Past, Present and Future*, Ling and Pedersen's (eds) *Mobile Communications*, Richard Harper et al's *The Inside Text*, Peter Glotz and Stefan Bertschi's (eds) *Thumb Culture*, Haddon et al.'s *Everyday Innovators* and S.D. Kim's (ed.) *When Mobile Came*. That year also saw the release of the first anthology to focus specifically on one cultural context (Japan) – namely the wonderfully comprehensive Ito et al. anthology, *Personal, Portable, Pedestrian* (2005).

By 2006, publications such Joachim Höflich and Maren Hartmann's (eds) *Mobile Communication in Everyday Life*, Pui-lam Law, Fortunati and Shanhua Yang's (eds) *New Technologies in Global Societies*, Miller and Horst's *The Cell Phone* and Goggin's *Cell Phone Culture* added weight to the already ever-expanding library of mobile communication. It seemed that the diversity of disciplinary, and interdisciplinary, scholarship was vast and comprehensive. However, as I have noted previously, in this mass of burgeoning research there were significant gaps.

As Goggin (2006a: 4–5) aptly notes, many of the studies to date into mobile communication have focused on the social aspects, exemplified by Ling's *The Mobile Connection* (2004). Sociologists have, until recently, dominated the literature. Although the mobile phone phenomenon is primarily a communicative, social one – despite industry hype about mobile media becoming a vehicle for creative content – it also has other, equally significant dimensions such as cultural and economic. In particular, it has a wealth of possibilities as a cultural artefact, a phenomenon that has been explored by anthropologists such as Miller and Horst in Jamaica (2006) and Pertierra in the Philippines (2002, 2003, 2005a, 2005b, 2006). Goggin convincingly argues for the pivotal role that cultural studies can play in the analysis of the 'under-explored' cultural dimensions of the 'cell phone'. By deploying the 'circuits of culture' approach proposed by du Gay et al. (1997) in *The Story of the Walkman*, Goggin asserts the need to evaluate the socio-cultural within the context of the political economy, in order to comprehend the ever-evolving multi-dimensions of culture. As Goggin avows:

> While there have been quite some studies attentive to cross-cultural context, there has been little work that systematically explores local or international cell phone culture, and its implications for general accounts of culture. In particular, I think there has been a lack of recognition and analysis of how power relations and structures shape cell phone culture. Here international cultural studies have something to contribute with their historical and still current preoccupation with the inescapable constitution of culture through power.
>
> (2006a: 5–6)

Apart from the lack of studies focusing upon the multi-dimensional aspects of the mobile phone as a cultural index for power relations both at a micro, interpersonal level and at a macro, supra-national and transnational level, there are other overt gaps in the field of mobile communication research. In particular, why has the role of mobile phone cultures and practices been relegated to the domains of 'social' activity, without engaging in its role in producing cultural, nation and transnational meanings and implications? This is perhaps where the domestic technologies approach, and ethnographic and sociological approaches in general, have neglected to conceptualise the multi-dimensions of mobile phone cultures identified by Goggin.

Moreover, why have the 'global' studies continued to perpetuate an ungrounded and a-historicised framework for the rise of mobile communication in the Asia-Pacific? How can we reconceptualise notions of mobility, individualism and community beyond the Eurocentric models that are conflated as 'global' studies? When locations of the region are studied, they are either studied in cross-cultural snapshots that reduce complex histories into cultural comparisons; or, alternatively, they are 'separated' into geographic boundaries such as 'Japan' that only further dissociate the culture from its regional and transnational context. Why has the Asia-Pacific remained relatively under-explored, despite its obvious global power both in the production and consumption of mobile technologies?

Mobile @ asia_pacific: regional mobilities

> ... it is fair to say that thus far scholarly study of cell phones has been dominated by a focus on European and North American examples and assumptions. Work on cell phones in other parts of the world – especially Asia – is now emerging, as it is, rather too slowly, in studies of Internet and other new media ... The mobile, as too the Internet, has been mutually implicated in cultural and social change in Asia.
>
> (Goggin 2006a: 13)

As Goggin observes in his survey of mobile literature, 'Asia' has yet to demonstrate its significant and diverse mobile phone phenomena to the world. This, as Goggin notes, is partly to do with the fact that early studies were not written in the *lingua franca* of English. It is also to do with the fact that much of the emerging research on socio-technologies in the region focused just upon Internet, which, in locations such as Japan, emerged pretty much concurrently with the mobile phone. In this sense, such locations have gained global attention as possible examples of the future of convergent technologies in a networked society (Rheingold 2003; Castells et al. 2007). Moreover, the fact that many collections of research on global trends in mobile phone use took Western or Eurocentric notions of individualism and consumption as given, undermined opportunities for understanding the complexity of localised mobile phone cultures and practices in the region.

Mobile communication research in the Asia-Pacific has, like the region itself, evolved unevenly. Unlike those studies in Europe which tend to take a sociological perspective, the first studies in the region to be published in English took the form of anthropological inquiry – most notably Ito's in Japan (2002) and Pertierra's in the Philippines (2002, 2003). Both Ito and Pertierra have gone on to conduct pivotal case studies that have brought a wealth of knowledge and rigour to the literature. In particular they showed ways in which, by analysing the cultural and social implications of mobile phone practices and cultures, one can gain insight into the micro and macro encompassing intimacy, co-presence, individualism, place, lifestyle, community, social capital and even notions of national culture. Like Japan with its model of GNC, emerging NICs such as Korea and China began to reconfigure their technological prowess for global cultural power.

Kim's two studies on the role of the mobile phone in Korea – as a vehicle for political agency (2002) and for reinforcing cultural and national identification with familial hierarchies (2003) – were significant in highlighting the role of technologies within the national. They also showed that despite the proclivity towards existing modes of individualism in locations such as Japan (McVeigh 2003; Kogawa 1984; Okada 2005), in places such as Korea, where the notion of family was significant as a motif for identity and nationalism, the mobile phone participated in pre-existing community building practices. Kyongwon Yoon (2003, 2006) cited the role of the mobile phone as a repository for techno-nationalist 'cyber-Confucius-ism' serving to reinforce these traditional ties, particularly in terms of intergenerational hierarchies.

The aforementioned studies of Korea and Japan highlight how both locations exemplify localised forms of mobile phone practices that are marked by gender and generational differences. In the context of these two very distinct models of gendered generational divisions, the mobile phone is not just a social phenomenon; rather, it is encompassed within the role of technology at a governmental level in the rescripting of national culture. The production and consumption of mobile technologies feature prominently in government policy as noted in the white pages and telecommunication ministry reports. Mobile technologies function significantly on economic and cultural levels both within and outside the nation-state. An example is Samsung and its symbolic role as a technological, but also cultural, icon of Korea.

The linking between this socio-cultural phenomenon and its role in the rescripting of national culture has only been discussed in the case of discrete nationally-defined case studies such as in studies by Ito et al. (2005) and McVeigh (2004) in Japan, and Kim (2002, 2003) and Yoon (2003, 2006) in Korea. Without comparing these burgeoning forms of localised 'mobile' national cultures, we cannot gain insight into the ways in which contemporary formations of post-modernity are being shaped within the region. In particular, studies are needed that supersede imported Western precepts about individualism, intimacy, social capital and community. Such studies could provide more understanding of 'nostalgic' nationalisms (McVeigh 2004; Ma 2000) and transnationalism. The symbol of the female, through the motif of the female mobile phone consumer, provides an inroad into the divergent forms of nationalism and regionalism, as well as into the particular forms of individual and community network formations.

Of late, the field of mobile communication has seen the rise of anthologies and conferences that have attempted to address this glaring omission; but a large gap remains. These include anthologies that attempt to contextualise 'Asia', whether in the context of mass communications (Rao & Mendoza 2004), European (Pertierra 2007) or global communication (Castells et al. 2007; Law et al. 2006). Singaporean mass communications scholar Madanmohan Rao has released two reviews on communication technologies in the context of both Asia and the Asia-Pacific (Rao 2004; Rao & Mendoza 2004) but these collections take a predominantly quantitative political economic view, neglecting the pivotal role these technologies play at a grass-roots level. Moreover the data was collected pre-2003, before mobile phones and the Internet had been fully immersed at the level of everyday practice in the region.

Arguably, the most interesting anthologies on the social and cultural dimensions of technologies in the region were the ones that took a cultural studies approach in critiquing the Anglocentric and English domination of such areas as the Internet. The year 2003 saw the arrival of two anthologies that attempted to address this gap not only by exploring non-Western models for thinking about subjectivity and technology, but also by investigating how these non-Western paradigms played out in transnational flows of the region.

In *Mobile Cultures: New Media in Queer Asia* Berry et al. (2003) addressed the often overlooked but profoundly important role of sexuality and gender performativity that challenged normalised, Western, heterosexual models. Similarly, Nanette Gottlieb and Mark McLelland's (eds) *Japanese Cybercultures* (2003) sought to uncover the multiple forms of subjectivity, individuality and community that were manifesting within the then nascent Internet. In the case of Japan, where the introduction to the Internet was via the *keitai*, the collection provided an early study into mobile media. This was followed by K.C. Ho, Randolph Kluver and Kenneth C.C. Yang's (2003) edited Internet anthology *Asia@com* that explored, through the lens of both cultural and political economy frameworks, the arising and uneven spread of socio-technologies in the region.

In 2004, aforementioned Rao released the survey *News Media and New Media: The Asia-Pacific Internet Handbook*, and his book *Asia Unplugged: The Wireless and Mobile Media Boom in Asia-Pacific* (co-edited with Lunita Mendoza). As McLelland observes in his review of both books, although the titles seemed promising, the books were aimed at journalist and industry specialists rather than academics, and they focused on governmental policies while completely ignoring the socio-cultural dimensions of technologies. In his review, McLelland highlights one of the research gaps in the region's uneven rise from technological prowess to cultural capital. As McLelland concludes:

> Although . . . we now have many interesting and useful articles investigating these different dimensions of mobile communications and their impact, the field still lacks a general overview that attempts to understand these patterns and trends in a regional framework, especially one that does not fall back upon default references to a standardised or normalised pattern of usage in 'the West' . . . [S]uch a review is necessary . . . to think of mobile communications (or indeed any technology) as part of complex socio-cultural systems that are tied into local, regional as well as global flows of people, technology and desire.
>
> (2007: 275)

This 'default' setting – to appropriate Lisa Nakamura's (2002) usage of it in terms of Japan being the West's science-fiction backdrop – is found in so many of the anthologies in which individual wonderful studies on locations such as China or Korea get conceptualised within a Western/Eurocentric model of culture and society. The attempt to readdress this problem surrounding so much of the English language literature is the rubric for McLelland and Goggin's (2008) *Internationalising Internet Studies*, in which they draw from case studies and theoretical tropes that do not oscillate around Western precepts.

Some anthologies have been more successful than others in attempting to reorientate the axis away from 'global' mobile communication as another word for Western/European or Anglophonic frameworks. Pertierra's (ed.) engaging *The Social Construction and Usage of Communication Technologies* (2007), a publication from a conference held in 2004, attempted to address the trope about East/West notions more vigilantly than previous 'global' studies collections such as Katz and Aakhus's (eds) *Perpetual Contact* (2002), which took the Western or Eurocentric notions as a universal given. So too, Law et al.'s collection *New Technologies in Global Societies* (2006), arising from a sociological conference in 2004, seeks to give China as much focus as Italy. This collection has some fascinating studies, most particularly on China as exemplified by Garland Liu and Joel Law's (2006) poignant look at the politics of technology in the sex industry. However, again, the rubric regarding the different cultural histories that underlie the divergent uptake of technologies is not fully explored in the array of different methodologies and fragmented case studies.

Castells et al.'s *Mobile Communication and Society* (2007) provides an ambitious overview. In attempting to surmise the mass of research on global mobile communication, the collection has some wonderful insights – such as Jack Qiu's work on migrant workers in China – despite being overshadowed by Castells's dogmatic insistence on conceptualising the material as part of the rise of the networked society. Unfortunately, owing to the unevenness of the material, the sum of the individual authors' specific case studies (in three regions) do not equate to the global diversity claimed by the book. Some locations such as China are discussed with wonderful depth and reflexivity, while others such as Japan are represented only through secondary literature reviews that demonstrated a cursory understanding of the research without providing insight into the cultural context.

As aforementioned briefly, some of the earlier conferences in Korea by Kim (2003, 2004, 2005) such as *Mobile Communication and Social Change* in 2004, seemed to realign the West/East axis of reference outside the lens of Western precepts. However, the power balance remained largely unchanged, with most of the keynote speakers being Western males (Ling, Katz and Haddon) – apart from one female American Japanese (Ito) – and with a majority of the Asian scholars drawing from distinctively Eurocentric sociological models, such as the work of Erving Goffman.

Moreover, the issue of gendered practices was given little attention, despite the fact that the female consumer dominated the visual imagery commonly associated with mobile phone use in the region. Gendered practices were relegated to the discussion of consumption and fashion, ignoring the extensive work conducted by feminists on debates around work/life, reproductive labour and intimacy (Hochschild 2000, 2003; Fortunati 2002b, 2002a; 2005a), consumerism (Featherstone 1991; Zelizer 2005; Lury 1996; Spigel 1992), and the practice of co-presence (online/offline), virtuality (Fortunati 2005a; Morse 1998) and corporeality (Nakamura 2002; Turkle 1995; Haraway 1991; Plant 1998). Discussions about youth cultures prevailed, further entrenching the conflation between new technology and youth cultures.

In 2005, conferences in the region began to take centre stage, fully engaging with the potential of mobile communications – as both a social and cultural practice – to address issues of Asian modernities. Two such conferences, *Mobile Communication and Asian Modernities*, Part I at City University in Hong Kong (organised by Angel Lin) and Part II at Peking University in Beijing (organised by Peking University and France Telecom), attempted to scratch the surface of this emerging phenomenon. Online journals such as *Fibreculture* published issues about the cultural politics of mobility, with a strong focus on examples from the Asia-Pacific (Murphie et al. 2005).

In 2007, Khoo and I co-edited an issue of *MC journal* on the mobile in the context of the Asia-Pacific, in order to establish some ground for inquiries into this significant but highly under-explored domain. Goggin and I also co-convened an international conference on mobile media, at University of Sydney, in which we tried to readdress some of the gender issues (more female keynotes than males) and reframe what 'global' studies of mobile media entail beyond east/west paradigms. However, such exercises are but a beginning in attempting to address the oversights in the field.

The lack of research is surprising considering the pioneering innovations and adoptions in the region: for example, the handset and software development paraded by Korea's Samsung and LG, and the business ecology success story of Japan's DoCoMo i-mode, which was lauded as the first mobile phone with Internet. The diversity of government and industry regulations, as well as of socio-cultural and linguistic factors, makes the Asia-Pacific a rich model for comprehending some of the complex ways in which mobile communication is localised. The lack of research is due, in part, to the difficulties in addressing the socio-cultural, linguistic, economic and political differences that encompass the region – issues that this study tries to address. Moreover, as the region has demonstrated, the mobile phone is no longer just a phone – it has converged into mobile data. Media convergence has been integral in the practice of mobile intimacy.

Just as the mobile phone has symbolically represented much more than a mere technology, so too has the actual technology transformed into a plethora of multimedia possibilities: what has been defined as 'mobile media'. This is evident in my various case studies, particularly in places like Tokyo where, with the release of i-mode as a gated version of the Internet (Sawhney 2004), it was impossible to distinguish where the telephony ended and the Internet began. Hence, in order to gain a sense of the expanse of mobile phone cultures we need to explore one of its most pervasive phenomena, namely its convergence into mobile media.

Mobile @ media: convergence

> The mobile phone has become the Swiss army knife of consumer electronics, becoming by turn a games machine, emailer, camera, or news browser. Heck, you can even talk to people on them. This feature creep has gone so far it's tempting to think it cannot go much further. But with new technologies on the horizon in Japan, and a market infatuated with the mobile, suggesting that the idea of a phone as a do-everything gadget still has a lot of mileage in it.
>
> (Boyd 2005: 28)

As convergence leaves its mark in the first decade of the twenty-first century, the ultimate exemplar is the mobile device. To speak of mobile media is unquestionably tied to the rise of convergence. Convergence can refer to technological, economic, cultural, media realms. Far from a mere form of communication the mobile phone has become a multimedia device par excellence – a plethora of various applications that operate across aural, textual and visual economies. It is a repository for many performances of localised forms of intimacy, propinquity and co-presence. In the region, the rise of convergent mobile media is interwoven with the growth in personalisation techniques. The personalisation of mobile media, at both an individual and collective level, reflects localised notions of intimacy. Thus to explore mobile media convergence in the region is to investigate expressions of intimacy.

As both Tokyo and Seoul have been centres for technological innovation and high early adopter practices, both locations have attracted some of the first case studies of convergent mobile media such as camera phones (Ito & Okabe 2003; D.H. Lee 2005). The convergence of the mobile phone with multimedia – thus becoming part of new media discourses – has seen it form a discursive space of mobile media in which various histories, genealogies and cultures combine. Thus mobile media has gained much interest in terms of new media debates, particularly those focusing on one of the dominant phenomena of globalisation, convergence.

This has led media studies expert Henry Jenkins to characterise contemporary culture as *Convergence Culture* (2006). Within contemporary culture, we can witness convergence occurring across various levels. As Jenkins observes, this media and technological convergence is occurring simultaneously to socio-cultural divergence. The uptake of convergent mobile media is inextricably linked to the local. For Goggin, 'convergence indicates the merging of media and cultural industries associated with forms of twentieth century media such as radio, television, newspapers, magazines that came to be relatively well established in their cultural bearings' (2006a: 143). Whilst Jenkins and Goggin link convergence to the local and the cultural, they neglect to address the pivotal role personalisation plays in the intimate practice of convergence. Why convergence has been so instrumental in mobile media is undoubtedly linked to its deployment of the personal and intimate.

In the Asia-Pacific – with global pioneers such as Tokyo and Seoul – this notion of convergence had been part of everyday mobile communication from the outset. Personalisation makes convergence intimate. From adoring the outside of the phone with cute customisation to creating UCC, cartographies of personalisation render the media intimate. However, these connections have often been ignored or deemed as expressions of 'youth culture' (Katz & Sugiyama 2005) in global media. Often these locations have gained much global mass media attention as potential futures of mobile media without contextualising the relationship between convergence and intimacy.

But beyond the futuristic stargazing, researchers have been attempting to grapple with the full implications of mobile media as it evolves from communication to creative media. This research saw paradigms being imported from other discipline areas such as new media. By 2007, the interdisciplinary rubric for mobile media was being traced by publications such as Finnish media expert Ilpo Koskinen's chapter on 'Mobile Multimedia' (2007), and conferences such as the *International Mobile Media conference* in Sydney, and Nyíri's (2007) annual mobile communication conference in 2007 entitled *Towards a Philosophy of Telecommunications Convergence.*

The rise of convergent mobile media, rehearsing older media as well as defining new forms of expression, can be viewed as an extension of mobile phone customisation. Just as Jenkins observed that accompanying convergence was its twin, divergence, Castells asserted that in the rise of socio-technologies (what he would call 'networked societies') such as the Internet, increasing standardisation would be met by burgeoning customisation (2001). Once living outside the phone in the form of characters hanging from the device, mobile phone customisation has become increasingly convergent, migrating to the inside of the phone in the form of customised camera phone images, text messages, screen savers, music, etc.

We can view customisation as helping to traverse the co-presence of mobile technologies in everyday life, making intimate and thus incorporating new technology into existing cultural rituals and practices. The customising outside the device can express tastes about the user to strangers in public spaces. What adornments the user may choose can draw on their peer group commonalities, or be a memento of a treasured intimate experience. The growth in customising inside the mobile phone through mobile media such as SMS and camera phone images can be about maintaining intimate co-presence with friends and family. These forms of customised mobile media practice can also be deployed exclusively by the user as a repository for memories, experiences, expressions of creativity and more. Concurrent with this oscillation between inside and outside, internal and external customisation has been discussions of mobile media as a 'democratisation' of multimedia. It is important to realise that these notions of convergence are linked to practices of intimacy.

The rise of theories engaging with complex readings of intimacy that dismiss the face-to-face, heteronormative, Western ideas of propinquity (Berlant 1998, 2000; Jamieson 1999) can be paralleled with paradigm shifts in the reading of the mobile phone as a symbol of intimacy and co-presence. For example, since the phone is the most intimate device that accompanies us everywhere (Fortunati 2002a, 2002b), the outside of the phone became a way in which to judge users (Katz & Sugiyama 2005; Hjorth 2003a, 2003b). The first forms of customisation initially adorned the outside of the device in terms of face-plates and cute character phone straps to name but a few techniques. These types of customisation identified something about the owner's identity and circle of intimates (and 'intimate strangers'). Often young partners or best friends bought each other mobile phone accessories so that the recipient was always carrying a memento of their friend or partner around with them. This form of customisation not only operated as a constant reminder of the intimate, but also, being an object outside the often publicly visible phone, it signalled a type of public intimacy trenchantly addressed by Berlant (1998, 2000). Public intimacy is just one example of cultural convergence.

It is in this space that we see an awkward transition in the history of the medium from its beginnings as a social, communicative device to its industry-hyped possibility as a creative (and commercial) venture. As an extension of Toffler's (1980) notion of 'prosumers', whereby he saw the diffusion between consumers and producers as a symptom of increasing conflations between work and social life, current proclamations about the potential of mobile media as an artistic and political tool for the people are debateable. In everyday life, practices of the 'prosumer' can be found everywhere, but it is within the convergent space of mobile media that paradigms of the users appear. As Katz and Sugiyama note, 'users' are no longer just passive consumers; they are prosumers, 'produsers' (Bruns & Jacob 2006) or 'co-creators' (Katz & Sugiyama 2005: 79). In short, the rise of mobile media is characterised by the ascent in the active user. The transformation of the everyday user into photo-journalists was highlighted by the London subway terrorists attacks in 2005, and has continued in the rise of netizen media such as online *OhMyNews* in Korea. As Castells et al. observe, 'communication can be both instrumental and expressive' (2007: 153).

With tools such as texting, emailing, camera phone imagery, video and sound, the mobile phone provides many vehicles for self-expression. These forms of expression play across various levels – individual, social and cultural. Such practices as texting can 'express social inequalities' (Pertierra 2006: 100) concurrent to, as Pertierra notes, creating 'an amplification of inner subjectivity' (2006: 101). In the face of globalisation and its 'shaping the tension and dialectic between society and culture', we can also see 'the rise of the sexual subject' through new forms of intimacy and social labour (2006: 101). As Pertierra observes, 'CMICTs (computer-mediated interactive communication technologies) have consequences on notions of individualism and cosmopolitanism' (2006: 59), although 'contrary to expectation, cellphones also encourage authentic relationships' (2006: 59).

The ascent of mobile media has been shaped by such genres as camera phone imagery, SMS, and the coordination of these practices with Internet peer-to-peer sharing. Extending practices of analogue photography through its so-called democratising of photographic media, camera phones are affording users the ability to document, re-present and perform the everyday. These emerging practices are underscored by the 'exchange' and gift-giving economy of the mobile phone (Taylor & Harper 2002) that see new forms of sharing and distribution through various contextual frameworks from MMS, blogs, virtual community sites to actual face-to-face digital storytelling. Thus the content of mobile media practices such as camera phones are informed by what Ito and Okabe (2003, 2005) have defined as the 'three Ss' – saving, storing and sharing. For Ito and Okabe, the meaningful aspects of camera phone practices are interwoven with the politics of networks/contexts, distribution and interactivity, and the 'three Ss' are central in determining possible interpretations and audiences.

On the one hand, these new forms of visuality, textuality and aurality have resulted in what Koskinen describes as the grammar, aesthetics and logic of mobile media banality (2007). It is this banality that legitimates the user's content in the practice of sharing with others. In an age of highly edited media images, this provides a refreshing space for everyday users to engage with constructions of presence. On the other hand, mobile media deploys the logic of new media practices. In this way, we could say that in order to engage with these practices we need to domesticate new media. The often-rarefied discourse of new media has the potential to become the practice of the everyday user.

Through the lens of mobile media we can gain insight into various forms of contested localities in the region. One of the compelling features of convergent mobile media, as distinct from other new media, is the role of locality. Emerging from a tradition of domestic technologies, mobile media re-enacts the performances around redefinitions of public and private space. In particular, the aesthetics, and thus politics, of mobile media are very much predicated on its formation as a domestic technology. To contextualise these various forms of convergent media, I will discuss the twin histories of mobile media in terms of the parallels and differences between domestic technologies and new media approaches and their ability to discuss some of the efflorescence textualities, visualities and haptic practices that are evolving around this increasingly convergent, but at the same time divergent, socio-technology.

Mobile @ new_media: domesticating new media

As mobile communication and media industries converge, the pervasive futurist rhetoric becomes stifling in its optimism. And yet, if the twin histories of new media and mobile communication have taught us anything, the 'new' is always mediated and remediated. At the core of all communication and cultural practices is the role of intimacy. As Jay Bolter and Richard Grusin (1999) note, 'remediation' is a reworking of Marshall McLuhan's (1964) argument that the content of new technologies is that of previous technologies. That is, new technologies remediate old technologies in a dynamic, ongoing process. As Terry Flew (2002) observed, there is little that is 'new' about new media; rather, new technologies often rehearse and adapt older practices of communication and representation.

If, as art critic Robert Hughes (1981) characterised, twentieth-century modernism was the 'shock of the new', then could not mobile media, as indicative of twenty-first-century new media, be defined as the 'banality of the new'? Certainly, Koskinen's prescient work on the banality of mobile media as multimedia suggests so. However it is important to realise that the politics of banality is one deeply entrenched in the practice of the everyday. As Meaghan Morris (1988) notes, the politics of what is conceived as banal partakes in power relations of normalisation and naturalisation that should not be overlooked. Because mobile media is everyday media it is unequivocally embedded in debates about the normalising role of the everyday. It is a banal and everyday media, yet at the same time it is significantly encoded into remediated and sublime practices of intimacy.

As Timo Kopomaa (2000) observes, today's mobile media can be seen as an extension of nineteenth- and twentieth-century mobile media such as the wristwatch. Technologies such as mobile media re-enact earlier co-present practices and interstitials of intimacy: for example SMS re-enacts nineteenth-century letter writing traditions (Hjorth 2005b). As Milne (2004) observes, new forms of telepresence such as email can be linked to earlier practices of co-present intimacy such as visiting cards. For Lynn Jamieson, part of the problem in discussing contemporary forms of intimacy is that it often assumes the western, heteronormative and face-to-face model as a given precept (1999). Rather, the intimate co-presence enacted by mobile technologies should be viewed as part of a lineage of technologies of propinquity (Milne 2004; Hjorth 2005a). The re-orientation of intimacy as part of broader shifts towards a blurring between public and private spaces is exemplified by confessions and disclosures of private matters through various forms of mobile performativity by intimate strangers (Plant 2002). Once associated with a sense of the private, practices of intimacy have increasingly become public.

As Berlant observes, intimacy has taken on new geographies and forms of mobility, most notably as a kind of 'publicness' (1998: 281) that is epitomised by the mobile phone (Fortunati 2002b: 48; du Gournay 2002). Here we see both the gestures and spaces of intimacy draw from a history of 'emotional grammars' (Beatty 2005), as well as being an increasingly commodified notion that is performed. In the case of mobile cultures, we can see this phenomenon as part of burgeoning emotional grammars and 'geographies of intimacy' (Margaroni and Yiannopoulou 2005) across visual, textual, aural and haptic registers. These geographies of intimacy are indelibly shaped by the role of the local and the vernacular, as well as being part of a broader global trend towards personalisation that are exemplified by the rise of Web 2.0 SNS and mobile media full-time intimacy. It is important to recognise that geographies of intimacy have always involved the co-presence of the virtual and the actual.

Like Jamieson, Milne dispels the myth of a universal, unmediated form of intimacy by evidencing the long history of telepresence in the British eighteenth and nineteenth centuries in the formation of intimacy as we understand it today. Specifically, Milne exposes the fallible premise of the notion of unmediated intimacy, mirroring Jamieson's critique of intimacy as inter-related with face-to-face contact. As Milne observes, the history of epistolary is marked by a genealogy of mediated forms of intimacy. It is important not to fetishise the newness of mobile media, but rather to view it as transforming existing practices.

In the case of the Internet, as Margaret Morse (1998) concisely notes, we must remember that all forms of intimacy are mediated – by language, gestures, and memories. The emerging forms of visual, textual and haptic mobile genres such as SMS and camera phone practices – re-enacting earlier rituals such as epistolary propinquity (Hjorth 2005b; Milne 2004) and gift-giving customs (Taylor & Harper 2002) – only serve to highlight the remediated nature of the rise of mobile media. Thus, in order to understand mobile media, we must view it as a synergy between new media's remediation paradigm and mobile communication domestic technologies approaches.

The study of new media through the lens of remediation echoes a similar philosophical stance to that of the domestic technologies approach. As an influential theorist in the field of media-archaeology, Erkki Huhtamo has argued, the cyclical phenomena of media tend to transcend historical contexts, often placating a process of paradoxical re-enactment and re-enchantment with what is deemed as 'new' (1997). On the one hand, the project of examining mobile media entails observing the remediated nature of new technologies and thus conceptualising them in terms of media archaeologies (Huhtamo 1997). On the other hand, mobile media's re-enactment of earlier technologies is indicative of its domestic technologies tradition that extends and rehearses the processes of precursors such as radio and TV. For Huhtamo, media archaeology approaches are 'a way of studying recurring cyclical phenomena that (re) appear and disappear over and over again in media history, somehow seeming to transcend specific historical contexts' (1997: 222). As Jussi Parikka and Jaakko Suominen note, the procedural nature of media archaeology approaches means 'new media is always situated within continuous histories of media production, distribution and usage – as part of a longer duration of experience' (2006, n.p.).

Both the domestic technologies and new media traditions emphasise the cyclic and dynamic process of media technologies that cannot be simplistically divided between old and new. Rather, the cartography of mobile media is one imbued by paradoxes. In the case of camera phone practices – whether involving still or moving images – mobile media demonstrates two distinctive paradoxes, that of the (analogue) *reel* in the real, and the inherent poetics of *delay* in the practice of immediacy. The role of delay, whether intentional or unintentional, is an integral component of mobile media practices such as SMS and camera phone imagery. In location-aware gaming (mobile games that involve technologies such as GPS – geographic positioning systems) there are always elements of online and offline disjuncture that add to, rather than subtract from, game play.

One of the most compelling examples of the real/reel phenomenon, where the tactile process of the analogue is fully *felt* – both metaphorically and literally – is the rise of screen cultures in mobile media. In particular, the rise of such mobile media devices as iphone, LG Prada, and Samsung Armani phone – to name a few – all incorporate one key feature, *haptic screens*. Here the reel/real paradox is rehearsed in the haptic versus visual, in which the haptic is undoubtedly the more meaningful factor that 'domesticates' the device into the user's everyday life. Much of the analogue reel spectres are more about the tactical experience of image processing; and while these processes have been deleted in the rise of the digital, the legacy of the haptic – which has moved from the filmic developing process to the politics of the touch screen – continues unabated. However, in the language set of twentieth-century media cultures, much discussion was given to visuality rather than the increasing role of the haptic. As Lev Manovich (2003) identifies, contemporary new media and digital practice are consumed by fetishising the real through the lens of the reel – that is, through the *texture of the analogue*.

This aura of the reel also has its feelers in the rise of intimate 'publicness' that Berlant identified in contemporary disclosure politics; publicness that can be seen in the pervasiveness of reality TV programs and the general proclivity towards the 'interior' as 'the new exterior' (Sukhder Sandhi cited in Margaroni & Yiannopoulou 2005: 222). Thus emotion becomes another commodity, another form of economy in a capitalist culture lauding the celebrity. The banality of mobile media is its power, conveying a sense of the familiar, a moment in the real, a gesture of intimacy. One of the dominant features of mobile media is how it further fetishises the analogue by way of its obsession with modes of realism. This is one of the striking features of camera phone images: they are, more often than not, seemingly banal in their subject matter and composition.

In an age of highly edited and photoshopped images circulating various sites of public confessions by intimate strangers, the camera phone image provides an antidote. As Hille Koskela notes, 'by revealing their intimate lives, people are liberated from shame and the "need" to hide, which leads to something called "empowering exhibitionism"' (2004: 199). This has led Koskinen to succinctly define camera phone imagery as 'the aesthetics of banality' (2007). This aesthetic is about a sense of intimacy, grasping a moment, sharing of a thought. As Koskinen (2007) notes, the banality of images renders them reliable and thus worthy of sending to others. Koskinen also argues that mobile multimedia, unlike mobile telephony, 're-territorialises' experiences and communication (2007: 48–60; Scifo 2005).

Building on discourses of analogue photographic practices and the proclaimed democratising of photographic media, camera phones are affording users with new ways to document, re-present and perform the everyday. Camera phone practices enact what Anita Wilhelm et al. observe as the 'power of now' (2004). Co-presence shifts from a noun to a verb; as Joichi Ito (2004) observes in the case of contemporary networked media such as Web 2.0, we 'do' presence. Technologies such as camera phones speed up the process of recording, editing, deleting or disseminating experiences; so much so that often the event can only be experienced as a mediated memory. The role of mobile technologies in shaping and reflecting emerging forms of temporal and spatial dynamics has been characterised by Ling as 'micro-coordination' (2004). In the case of camera phone practices, I have called this phenomenon 'fast-forwarding present' (Hjorth 2006a, 2006c) – that is, the compulsion to document and share every detail of everyday life to the point where one could become a voyeur of one's own life, only living the present as a spectre of the past.

For Koskinen, mobile media's commitment to realism – as evidenced by camera phone practices and MMS – is situated around 'the mundane as a problem' (2007: 50). As Koskinen says, it is the demonstration of the banal that is important in defining the sender as 'ordinary' and thus *reliable*. This can be seen as a furthering of the politics of the everyday and personal as legitimated by Kodak snapshots (Gye 2007). Certain archetypical images of homes and family occasions were an important part of presenting a family as 'normal' both to themselves and others. In this way, mobile media oscillates between the *real* and the *reel*. On the one hand, it rehearses vernacular analogue photography and the normalcies about family (Gye 2007). On the other hand, the ways in which it is shared and distributed can see it functioning as a continuum of the earlier co-present media such as postcards. As with the postcard, there are archetypical 'tourist' themes and images of the 'reel' (echoing analogue genres and images) that are often meaningless to the bystander. However, if, like the postcard, the sender personalises and contextualises it, then the images represent a meaningful exchange, encapsulating the practice of intimacy as something perpetually evolving, changing and 'transferring' (Margaroni & Yiannopoulou 2005: 222).

As Margaroni and Yiannopoulou (2005) observe, the 'intimate turn' in the humanities sees a reassessment of the value of emotions and localised and contingent forms of intimacy that are part of contemporary global mobility. It is a reinvestment of place and locality in the face of multiple forms of mobility. The intimate, they note in citing renowned psychoanalyst Julia Kristeva (2002: 43), can be a site of 'political, social or personal re-volt' (2005: 225) that promises to shake the foundations of universal truths and values. Here we see that the intimate, like the banal, is embedded in the complex weave of the politics of everyday life. It helps create what Raymond Williams (1974) called 'structures of feeling' so pivotal in the patterns of everyday life. Through the *pièce de resistance* of intimacy, the mobile phone

– as both a symbol and repository for emotional grammars – we can gain insight into multiple forms of subjectivity and agency.

This re-enacting of the *reel* (ode to the analogue), through the analogy of the postcard, is significant in camera phone imagery's circuit of culture around 'sharing, saving and storing' (Ito & Okabe 2003). One of the significant features of the postcard is that it bears the markings of place, in the form of the stamp and also the damage that occurs whilst in transit. The postcard is not an exclusively visual experience, and to vindicate it to the realm of aesthetics misses the point. It is also a haptic phenomenon. It is the tactility of the paper – the holding in the hand, the stroking of the textures – that makes receiving postcards a unique encounter, an experience in which the recipient is literally touched by the sender's message.

As Barbara Scifo notes, camera phone practices operate on 'two different levels of experience' – 'on an individual level' and 'on the socialisation level' (2005: 365). I would add to this process a *meta-social* level, a concept whereby users become hyper-reflexive to the ways in which strangers can possibly de-contextualise their mobile customisation. The rise of increased forms of distribution also augments the various contexts for interpretation and mis-interpretation. For example, in the case of camera phone practices, female users often re-appropriate the depictions of female imagery seen in media (D.H. Lee 2005) in a form of gender performativity (Butler 1991) that could be read without irony and parody by strangers. Hence, the important issue here in relation to mobile media *content* is the choice of viewing/sharing *context*. We will look at some examples in the case study chapters. In the instance of Korea, D.H. Lee's (2005) ethnographic work into gendered camera phone use has been instrumental in contesting the argument that camera phone practices merely reinforce narcissist and exhibitionist forms of 'public' intimacy (Palmer 2005).

Daniel Palmer persuasively argues that the corporatising of camera phone practices, exemplified by Nokia patenting the 'Kodak moment' (Gye 2007), has led to an industry promoting exhibitionism. However, while the rise in 'individualism' – or 'networked individualism' (Castells 2001) – might be a global phenomenon, what it means to be individual, as with what it means to be intimate, is very much linked to the local. The flaw in Palmer's argument could be that he defers to 'universal' norms around western, heteronormative ideals about intimacy and individualism. Certainly, as D.H. Lee's (2005) study in Seoul demonstrates, young women are finding camera phone practices useful not only in developing agency around self-expression, emotional grammars, representation and self-actualisation, but also in initiating artistic pursuits such as studying professional photography. Here we see the notion of remediation at play, with the practice of camera phone imagery and genres re-enacting analogue forms of representation.

Thus mobile media reality is about its ability to be 'reel' – that is, its capacity to make sense of the everyday in the user's life. This reel, as we will see in the case studies, is less about ocular-centrism and the continuing tyranny of the visual – rather it is about the banal haptics of the everyday. Unlike the twentieth century 'reel' – in the form of the aural modes of address embroiled in 'screen-ness' such as cinema and TV – the mobile reel, and thus possible creative worlds and realities, is governed by the haptic (Richardson 2007). This is due to the intimate (Fortunati 2002a, 2002b) and personal (Ito et al. 2005) nature of the mobile, symbolised by the fact that it is permanently attached to the user – in their pocket or held in the hand as well as metaphorically encapsulating various forms of self-presentation *vis-à-vis* customisation.

For Chris Chesher, mobile media is not an engagement of gaze, nor the glance, but rather akin to what he characterises as the 'glaze' (2004). Drawing on console games cultures, Chesher identifies three types of glaze spaces: the glazed over, sticky and identity-reflective. According Chesher, these three 'dimensions' of the glaze move beyond a visual economy, deploying the filters of the other senses such as aural and haptic. This formation of the glaze – a combination of aural and haptic into a 'hapral' – was apparent in my observations of spectators engaging with visual mobile media in which they often hold their ear towards the mobile phone when viewing, as if to listen to the pictures. The movies that were most popular were not those that featured vivid visuals, but, rather, those that featured compelling aural narratives. One of the features that are becoming increasingly apparent from studies of camera phone ethnographies in locations such as Tokyo (Ito & Okabe 2003, 2005a) and Seoul (D.H. Lee 2005; Hjorth 2006c, 2007a, 2007b) is the rise of other senses such as aural and haptic in the contextualisation and making 'reel' these images.

In Ingrid Richardson's compelling argument about mobile media she calls on the need to harness the importance of the haptic (2007). Conducting a small ethno-phenomenological study on the use of phone-game hybrids, Richardson disavows the ocular-centrism prevalent in 'new media screen technologies' to focus on 'the spatial, perceptual and ontic effects of mobile devices as nascent new media forms' (2007: 205). Departing from what Lucas Introna and Fernando Ilharco (2004) characterise as the multiple 'screen-ness' inhabiting contemporary life, Richardson argues, 'this "frontal" relationship which is typical of our engagement with most screens – where the media of cinema, television and computer can be said to discipline the body into a face-to-face interaction – is challenged by the mobile screen' (2007: 210). Richardson avows that mobile media disrupt 'any notion of a disembodied telepresence' deployed by much screen-based media; in turn, we can 'see emergent spatial ontologies of a kind never before experienced in such a collective and interactive fashion' (2007: 214).

However, while Richardson argues for a future in mobile media, particularly location-aware mobile gaming, where the virtual and the actual become seamless, there is much debate arising around the correlation between online and offline identity and how this is tailored by the local. As new technologies promise ever increasing immediacy, one could argue that the future of the 'remediated' mobile media is like its past, and thus dominated by the persistence of the reel and delay in the emerging glaze practices. It is undoubtedly these features that give the mobile its sense of place in the world. Far from the death of geography, place (and boundaries) have never been stronger as the practice of mobile technologies for imaging communities attest. And yet the mobility debate continues to persist.

2 Paradigms of mobility

Conceptual tenors for studying mobility today

The study of the mobile phone, as one of the most evocative signifiers for contemporary debates around mobility today, raises potentially far-reaching questions. One of the dominant questions arising in the social sciences is the role of mobility – what it means to be 'mobile' in an age of globalisation, and the impact mobility is having on work, life, locality, and nationalism. The paradigm is characterised by British sociologist John Urry as the 'mobility turn' (Urry 2003). For some, contemporary globalisation is marked by mobility (Urry 2003; Castells et al. 2007), for others it is characterised by immobility and the re-creation of borders and enclaves (Turner 2007).

Although this book is not focused squarely upon the 'mobility turn' as such, to discuss mobile media and its symbolic and cultural dimensions necessitates some elaboration upon the changing paradigms of globalisation. This is particularly relevant to the study of gendered mobile phone practices and cultures in the Asia-Pacific, since gender, age, class, ethnicity and locality unequivocally determine types of mobility and immobility. Moreover, in order to conceptualise the particular localised forms of gendered mobile media as part of the region's 'cartographies of personalisation' – we need to comprehend shifting definitions of what it means to be co-present and intimate in relation to localised notions of social capital and community.

Mobile @ place: waiting for immediacy

It is impossible to discuss the mobile phone without becoming embroiled in debates about the 'mobility turn' or the 'sociology beyond society' debate. 'Mobility' – as a concept with multiple meanings and implications – has become a pivotal rubric for the conceptualisation of globalisation, transnationalism and cosmopolitanism. Within the spatial turn, sociology, human geography and social anthropology merged. This resulted in broadly two philosophical tropes, one that viewed globalisation as diminishing and eroding nation-states, and one that viewed globalisation as part of the broader processes of modernity that were producing new forms of national boundaries. The first approach saw the likes of Urry ('mobile society', 2000b) and Castells ('networked society' 1996) arguing that mobility is a key feature, force and product of contemporary global flows. The other camp saw the increase in immobility and enclaves (Shamir 2005; Turner 2007). In between these arguments is the model put forth by Kevin Hannam, Mimi Sheller and Urry as 'Mobilities, immobilities and moorings' (2006). Indeed, mooring aptly describes the place of mobile media in an age of mobile capitalisms.

Within the contemporary milieu of the social sciences, particularly sociology and urban geography, the 'mobility turn' has undoubtedly become all-pervasive (Urry 2003). This mobility turn is interrelated with the rising ubiquity of global ICTs. In anthropology, the last two decades have seen a shift in ideologies concerning the relationship between space and place and the role of material culture in producing and reproducing social life (Miller 1987). As Igor Kopytoff (1986) perspicuously noted, objects, like people, have biographies. For Judy Attfield (2000), everyday objects reveal much about the spaces they inhabit and the people that they meet, whilst traversing space, time and body. An integral part of contemporary everyday life is the multiple levels in which forms of mobility and immobility resonate in a sense of place.

As a phenomenon and a symbol, the notion of mobility encapsulates various debates around place, locality, globalisation, cosmopolitanism, and diaspora. Mobility, and its obverse immobility, have become poignant concepts in an age of global flows in which people, ideas and capital move – some by choice and some by lack of choice. Issues about mobility are, and have always been, about power. They are also, on a micro level, embroiled in debates about the balances between work and life and what it means to be intimate and co-present. In short, they are about the various ways in which we configure contemporary forms of labour: physical, social, electronic, economic, political, and ideological. Critics of globalisation have highlighted that while capital is free to move globally, this mobility does not translate to such subjects as low-income citizens. Hence, although the mobility of capital may be growing, boundaries such as the nation are strengthened – often through the exploitation of economic inequality (Ling 1999; Truong 1999).

As Bell notes, notions of what it means to be 'mobile' are inflected by the local (Bell 2005: 70). If the car was symbolic of the mobility of early post-industrial modernity, then it is the mobile phone that is evocative of post-modernity's mobility. Hence the mobile phone has often become an index for discussing these various shifting clusters of labour within the dynamism between the global and the local. In this section I will discuss some of the encircling debates around mobility, specifically dealing with the politics of space, consumption, individualism and intimacy, before moving on to the reconceptualising of the Asia-Pacific.

One way of tackling contemporary debates around mobility would be to consider how, in an age of interstitial and transnational communities, the local is being re-imagined and practised. In turn, this would lead us to inquire about the ways in which mobility affects a sense of place, and how, simultaneously, locality informs what it means to be mobile. For Aihwa Ong in 'Mutations in citizenship' (2006), we are moving beyond a 'citizen-versus-statelessness' model wherein contemporary globalisation is denoted by zones of 'global assemblages' that represent political entitlements. The space of the 'assemblage' is not a national terrain but a site for political mobilisation by diverse groups. As Ong asserts:

> We can trace mutations in citizenship to global flows and their configuration of new spaces of entangled possibilities. An ever-shifting landscape shaped by the flows of markets, technologies, and populations challenges the notion of citizenship tied to the terrain and imagination of a nation-state. Mobile markets, technologies, and populations interact to shape social spaces in which mutations in citizenship are crystallised.
>
> (ibid: 499)

For Turner, the contemporary milieu characterised by mobility is creating more borders and 'us versus them' feelings rather than eroding them. Deploying Ronen Shamir's (2005) notion of 'mobility regime' that encompasses the paradox of globalisation's proclivity towards mobility creating 'closure, entrapment and containment' (ibid.: 199), Turner argues that Shamir's 'mobility regime' should indeed be redefined as 'immobility regime'. In order to address this immobility regime Turner coins the notion of the 'enclave society'. As Turner observes:

> While there may be an increased global flow of goods and services, there is emerging a parallel 'immobility regime' exercising surveillance and control over migrants, refugees and other aliens. If sociology is to be criticised, it is not because it has neglected globalisation; it is because it has neglected the rise of global security systems whose stated aim is to protect residential populations against the perceived risk of mobile populations.
>
> (Turner 2007: 289)

Turner's notion of the enclave society resonates with recent work by Nikos Papastergiadis. According to Papastergiadis in 'Mobility and the nation: skins, machines and complex systems' (2005), current sociological theories on migration view the 'nation-state as a bounded system' (ibid.: 1). Papastergiadis argues that theories of migration, predicated on ideas of nation-state 'as a unified and exclusionary social system' (ibid.: 1), abide to a binary mode of 'belonging' and 'movement'. Drawing our attention to the new body-machine (Seltzer 1992) as a political repository for prevailing and relentless binaries of power that uphold bounded notions of nation-state, Papastergiadis argues that it is indeed these rigid, simplistic and oppositional models that generate and perpetuate 'enduring suspicion towards difference, and creates a propensity to equate strangers with enemies' (ibid.: 3).

For Dirlik in 'Asia Pacific studies in an age of global modernity' (2005), mobility has not erased the construct of nation, just the reverse. Deploying the example of diaspora, Dirlik avows, 'rather than deconstruct nations and nationalism . . . diasporas may also serve to further project into transnational spaces the powers of capital and the nation-state' (ibid.: 166). This sentiment is echoed in Khoo's study on the 'Chinese exotic' as an index for emergent diasporic Chinese modernities in the Asia-Pacific. Although this new mode of Asian femininity challenges traditional mainland Chinese tropes of gender, nation and culture, it generates ideologically discursive formations of culture that extend beyond national geographic boundaries yet undoubtedly inform and reconstitute what it means to be 'Chinese'. This diaspora was further complicated by the concept of 'Confucius revivalism' (Dirlik 1995).

Writing at the height of 'Confucius revivalism' in the region, Dirlik (1999a) eloquently condemned a growing dissociation between culture and history in the politics of East Asian identity. He argued that the quest for 'essential' cultural values not only undermined the diversity of the region, but also, due to its ahistorical tenor, perpetuated a type of self-Orientalism. By disavowing history, the 'Orientalism' projected by the 'Occident' became internalised and naturalised, thus reproducing the violence particular of representational tropes. As Dirlik surmises:

> It is the irony of contemporary anti-Eurocentric movements that they themselves are entrapped in the history and geography of Orientalism. In other words, the very effort to counteract Eurocentrism is bound by the categories of a Eurocentric Orientalism.
>
> (ibid.: 167)

The role of place and space as social and cultural repositories for mobile phone practices has dominated the literature on mobile communication. Rather than announcing the death of place, ethnographies of localised mobile practices describe ways in which place has taken on new significance and importance (Ito 2002; Yoon 2003). The mobile phone is an integral player in what it means to be intimate and have a sense of belonging. As Ito's ethnography in Tokyo (2002) observed, young people used the *keitai* to facilitate and ensure further synchronisation in face-to-face meetings. Yoon's ethnography in Seoul (2003) showed how hand phone (*haendupon*) practices re-enacted traditional socialising rituals and familial relationships. These ethnographies are just two of many studies that demonstrate that mobile phones – as exemplifiers of ICTs – are bound to a sense of place and community. This led Castells et al. to note that mobile networks are augmenting the significance of locality and place (2007: 258).

However, few mobile communication studies link the micro socio-cultural practices of locality to the macro political economy or to techno-nationalism at the level of the nation. Among the exceptions are some of the chapters, such as those by Fujimoto and Misa Matsuda, in Ito et al.'s *Personal Portable and Pedestrian* (2005). Although most of this compilation emphasises the socio-cultural, Fujimoto and Matsuda's chapters identify the political and economic imperatives of nationalism around the scripting of portable technologies. I will discuss this in greater detail in the Japanese case study in Chapter 4.

Despite the discursive role of the mobile phone in scripting constructions of national culture so far being downplayed (in favour of analysing the socio-cultural elements), the mobile phone has often borne the brunt of fears and anxieties around contemporary notions of belonging, dislocation, mobility and defining a sense of place and home. This scenario reflects the need to reconceptualise notions of space and place within twenty-first century postmodernity. As urban geographer Doreen Massey notes, place has always been mediated – by projections, imaginings, representation and the very acts of practicing culture and performing identity. For Massey (2005), space is a realm of possibility – multiple, contingent, heterogeneous and ever-evolving. Massey argues that space needs to be 'uprooted' from static modes of representation and deployed as imaginary trope.

This resonates with Arjun Appadurai's model of globalisation as being a series of 'scapes' in which he explores the tenacious force of regionalism. For Appadurai (2000), locality and region are not fixed geographic boundaries, but rather, mutating and ever-evolving scapes in the disjunctive flows of global objects, media and people. As Appadurai vividly describes, often the analysis of regionality assumes 'a particular configuration of apparent stabilities for permanent associations between space, territory and cultural organisation' (ibid.: 7). Appadurai argues that the 'capability to imagine regions and worlds is now itself a globalised phenomenon', and that regions produce their 'own cartographies of the world' (ibid.: 7).

Through the global circulation and mobility of objects, people, images and discourses, boundaries become unstable. Thus, Appadurai argues the need to conceptualise geographies and cartographies not as fixed to national boundaries but as an ongoing series of socio-cultural, political and economic formations. In order to account for these processes Appadurai utilises the concept of flow to encapsulate the movements of various forms of social and cultural capital that traverse national boundaries and borders. These flows take the form of five 'building blocks' including people (ethnoscapes), money (finanscapes), media (mediascapes), technologies (technoscapes) and ideas (ideoscapes). For Appadurai, these flows operate to construct multiple, 'imagined worlds' that are 'historically situated imaginations of persons and groups' constituting the globe (1990: 7).

For Carsten Sørensen in 'Digital nomads and mobile services', our 'society is transforming itself into a mobile society, where interaction itself is mobilised' (2002: n.p.). Sørensen argues that 'traditional' ways 'of conceptualising the social topology of our interaction with the world' based 'on a notion of a region' have been transformed by the popular metaphor of the network. Positing the 'fluid' metaphor posed by Annemarie Mol and John

Law as a more appropriate model for comprehending decentralised and unordered practices and interactions, Sørensen declares, 'for the 21st century working nomad, there is everywhere to run – and nowhere to hide' (ibid.: n.p.).

The question of what it means to be mobile is intrinsically linked to debates about the relationship between definitions of individualism and social capital that are, in turn, being impacted by consumption and lifestyle. The mobile phone is a poignant symbol for consumer mobility (or immobility) and the role of consumption in the reproduction of lifestyle at the level of the individual and the social. It is also a symbol for the debates around the global phenomenon Ulrich Beck and Elisabeth Beck-Gernsheim (2002) have identified as 'individualization'. In what they perceive as a new form of capitalism, or more aptly capitalisms, new kinds of labour and life are occurring in which it is becoming harder to define nation-states in the flux of global transnationalisation. 'Individualization' is not interchangeable with 'individualism', rather it is a 'structural, sociological transformation of social institutions and the relationship of the individual to society' that characterises a 'second modernity' (ibid.: 276).

For Beck, in an interview with Don Slater and George Ritzer (2001), consumption society is the cosmopolitan society. Beck uses cosmopolitan as a trope of individualisation within globalisation. Rather than opposing nationalism, the cosmopolitan encapsulates a different kind of 'otherness of the other' through the 'art' of inclusive boundaries. Undoubtedly, mobile technologies have embodied this process of individualisation in which formations around public and private spaces have been transformed. The practice of mobile technologies, however, has increasingly involved what Castells et al. see as 'the blurring of the social context of individual practice' (2007: 250) where individualism, rather than mobility, 'becomes the defining trend' (ibid.: 251). If this argument sounds familiar it is perhaps because Williams identified this phenomenon of mobility and individualism in what he defined as 'mobile privatization' (1974). In an essay on 'mobile privatization' Williams characterised the 'unique modern condition' as 'an ugly phrase for an unprecedented condition' (1983: 129). Utilising the metaphor of car traffic, Williams paints a scene of dislocation to characterise contemporary mobile privatised social relations.

Drawing from a revised domestic technologies approach, David Morley sees the redemptive characteristics of commodity cultures and their meanings and values ascribed after purchase; that is, in the ways in which commodities become cultural artefacts in the messy and unruly space of cultural practice. Extending Williams's notion of 'mobile privatization' originally ascribed to technologies such as television and the car, Morley identifies mobile communication as 'mobile privatization' par excellence (2003). As Morley suggests, '[if] the Walkman is one "privatising" technology, then the mobile phone is now perhaps the privatising technology of our age, par excellence' (2007: 220). The mobile phone, according to Morley, has further eschewed the boundary between public and private not by bringing the public into the private – as was the case for television – but by inverting the flow so that the private goes out into the public (2003, 2007).

Whether one defines it through the multiple lenses of modernity, post-modernity, second modernity, globalisation, cosmopolitanism or 'individual-isation', the questions of what constitutes 'society', 'social values' and 'community' have become crucial issues this century. For some, these pro-cesses are part and parcel of what it means to be in the ever-evolving and dynamic space of 'culture'. For others, new paradigms such as individualisa-tion are dramatically changing our world, and not for the good. A key example of this can be found in the work of Robert Putnam's *Bowling Alone* (2000). In it, Putnam is vehement in his critique of the decline of social care and community in the US, where technology operates to mediate and frag-ment social capital. However, at the heart of this debate about declining morality and community values lie assumptions about universalised notions of intimacy and community, naturalised around Western, heteronormative models that idealise face-to-face and discredit 'impure' mediated intimacy, as if intimacy could be unmediated (Jamieson 1999). The debate surrounding individualism and social capital, and their impact on practices of intimacy, is inextricably bound to the discussion of mobile phone cultures.

Mobile @ capital: individualism, social capital and intimacy

> 'Mobile intimacy', the ability to be intimate across distances of time and space, is a global phenomenon. Oddly enough, the sociability of mobile telephony is not homogenous across the world. How the mobile phone is used to extend personal relationships in Asia/Pacific is unique compared to the rest of the world.
>
> (Raiti 2007: n.p.)

The notion of social capital was originally coined by French sociologist Pierre Bourdieu in his important study *Distinctions* (1984 [1979]) when he sketched some new ways for thinking about contemporary taste formations. For Bourdieu, capital was a form of 'knowledge' that helped produce and naturalise taste. Interviewing 1,200 French people from varying class backgrounds about their tastes around art, music and popular culture, Bourdieu deployed the lens of capital to discuss three significant influences – cultural (informed by education and upbringing), social (community and networks) and economic. These factors, along with the individual's own habitus (the regulatory patterns of the everyday), were the contributing factors in determining one's identification with a particular lifestyle niche.

Bourdieu's idea of social capital took on new signification when it was reworked by James Colman (1988) to imply a more ego-centred concept, echoing Beck's argument about globalisation increasing processes of indi-vidualisation. It was the aforementioned Putnam (2000), in his savage exposé

on the declining role of community and social welfare in the US, who characterised social capital as societal orientated activity based upon notions of
trust and reciprocity. This issue of interrogating patterns of intimacy as a way
into comprehending contemporary global (mobile) life resonates with the
work of many feminists exploring the role of global care cultures.

Hochschild (1983, 2003), for example, identifies the social patterns that
are blurring work and life through the globalised forms of care chains
that link the 'emotional surplus value' of one (developing) family to another
(developed) family. McKay (2007) observes that in order to investigate 'the
emotional dynamics and material structures that characterise a socially
embedded transnationalism' (ibid.: 175), we need to utilise new models for
viewing migrants as diverse agents rather than as heterogeneously passive
victims of globalisation. She also observes that we need to comprehend
notions of intimacy as part of the practice of global post-structuralism, and
debunk the conventional Western and romanticised notions of face-to-face
intimacy being 'real' and mediated versions as having less legitimacy.

The exploration of contemporary life through the lens of gender and
intimacy is eloquently detailed in Vviana Zelizer's (2005) *The Purchase of
Intimacy*, where she describes the intertwining of work and intimate life and
how these structure, and are structured by, the role of family, gender, and
ethnicity. Here, questions about mobility and mobilisation – informing what
it means to practise co-presence, intimacy and social capital – are inextricably
linked to the increased propensity of ICTs, and the attendant 'feminisation of
technology', to further inscribe insidious forms of reproductive labour. In
particular, researchers have focused on the role of Filipino women as conduits for globalised care labour flows and how these diasporic forms of care
labour maintain transnational forms of intimacies (Parreñas 2001). These
women are, as Parreñas so trenchantly states, 'servants of globalisation'. By
looking at the role of intimacy in terms of gendered modes of mobility, we
see the simplism of Putnam's argument and the idealised values that it
upholds.

It is important to recognise the often loosely defined, localised, intertwined
and ever-changing nature of tropes such as 'social capital' and 'individualism'. Moreover, as Beck rightly identifies, it is essential to distinguish between
global processes of individualisation and constructions of individualism. For
Ling (2004), whose interest in predominantly youth cultures in Norway
and the US steers him away from issues of gender, power and labour, the
divisions between social capital and individualism are best conceptualised as
a dualism. He avers:

> One area where the effects of the mobile telephone – and indeed of all
> forms of electronic communication – is being played out is in the area of
> what has been called *social capital* and its opposite twin, *individualisation*.
> This discussion is a continuation of the traditional sociology project in
> which the interaction between a technological innovation and the

workings of society is examined. If the mobile telephone contributes to individualisation, it follows that the device also plays into out experiences of social capital.

(Ling 2004: 177)

One of the most eloquent writers on the changing condition of societal milieu, 'social capital' and social patterns is sociologist Zygmunt Bauman. Bauman developed the term 'liquid modernity' (2000) to describe the transformation of once rigid social structures into fluid and contingent arrangements and renegotiations. As Bauman notes, the vehicle for this liquid modernity is 'the smaller, the lighter, the more portable' (ibid.: 14). Thus it is not surprising that the mobile phone has featured in his later work where he has elaborated upon this 'liquid' characteristics in the form of 'liquid love' (2003), a state in which paradox governs our practices of intimacy in an epoch marked by technological co-presence. Here, according to Bauman, we see connectivity struggling, and competing, with the practice of actual contact.

Bauman's scrutiny of mediated social spaces, and their impact on notions of responsibility, is aptly summarised in his earlier work on 'postmodern ethics' (1993). For Bauman, the new forms of spaces evolving within liquid modernity see people adorning themselves with the attire of 'postmodern tourists', travelling in and out of people's lives without moral engagement or responsibility. The demise of moral and social responsibilities features in the work of both Bauman and Putnam, which argue that commodities such as the mobile phone only exacerbate disjunctive co-present practices revolving around the growing significance of the individual and dwindling significance of community.

This view resonates with Castells et al.'s argument that it is indeed 'individualism, not mobility, that is the defining trend of the mobile society' (Castells et al. 2007: 251). In the vein of Bauman's liquid modernity, and specifically liquid love, Castells et al. argue that mobile communication 'dematerialises social structures and transforms them into individually centred networks of interaction' (ibid.: 251). However, far from distilling social relations into liquid and eroding time and spatial boundaries, mobile technologies 'enhance the presence of a culture' (ibid.: 258) and must be understood as a social practice that extends existing social networks (ibid.: 246).

Here Anthony Giddens's discussion about consumption and intimacy is significant, particularly as for its neo-liberal focus upon individualism and agency. For Giddens, writing prior to the onslaught of mobile media, onto-logical security, the sense of 'being in the world' is 'an emotional, rather than a cognitive phenomenon' (1991: 2). This resonates with the work of Miller (1987) on commodities being extensions of individual's identity and self-hood, and the domestic technologies approach viewing commodities as meaningful parts of the moral economy. In Giddens's *Transformation of*

Intimacy (1992), he identifies a notion of 'third way politics' in which new forms of citizenship and agency have evolved in the sphere of consumption.

For Giddens, the erosion of the nineteenth century Romantic passion has given way to a practice he characterises as 'pure relationships'. Rather than passion, they are based on talk and 'emotional communication', formed around 'interpersonal equality' (ibid.: 130) and 'only maintained while both parties are satisfied' (ibid.: 58). I would argue, however, that Giddens's blanket notion of 'pure relationship' – as if somehow intimacy were able to become equal despite cultural, social, economic, religious and political differences – is at best idealistic, at worse naïve (Jamieson 1999). For example, how does locality inform what it means to be intimate? How do gender, ethnicity, class and age inform this practice? Moreover, how does mobile media complicate conceptualisations of intimacy?

The issue of intimacy as a localised practice is highlighted in ethnographies of the region. In her ethnographies in the Asia-Pacific, Bell found a variety of socio-cultural definitions of home and how users related to their mobile phone. Examining the way in which domestic technologies such as the mobile phone and the computer functioned in both public and private spaces in Indonesia, South Korea, Vietnam, China and Japan, she demonstrated that they are inflected by various forces, most notably locality and gender. For Pertierra, it is through the lens of intimacy that we can gain insight into the various symbolic dimensions of the mobile phone in everyday Filipino life (2005a, 2005b). This idea is extended in the wonderfully eloquent discussion of Randy Solis into the constructions of 'romance' involved in mobile media practices in the Philippines (2007).

Gerard Raiti (2007), in 'Mobile intimacy: theories on the economics of emotion with examples from Asia', utilises Giddens and Bauman in order to explore the power relations embedded within 'emotional economies'. As he insightfully asks, 'if the non-linearity of late modernity has transformed love into a commodity, then how does the commoditisation of love into the realm of the mobile affect how one negotiates intimacy?' (ibid.: n.p.). Raiti's meticulous discussion and critique of Giddens and the commodification of love puts forward his rationale that 'the reallocation of time, space, and place through new media is the bedrock for mobile intimacy' (ibid.: n.p.). However, although Raiti's arguments are eloquent and robust, his treatment of the phenomenal dimensions of the Asia-Pacific does not grasp the full complexity enveloped under the regional rubric. For Michael Hardey, who also deploys Giddens's notions of 'pure relationships' in the world of co-present virtuality, albeit in the case of Internet dating, argues that Giddens's model of commodified and controlled 'pure relationships' is very apparent within the negotiation of online and offline relationships. As Hardey observes:

> The consequent vision of a highly discursive, disembodied late modern intimacy based on talk rather than passion, negotiation rather than commitment, and the advancement of self rather than the development

of the couple suggest that the internet is uniquely placed to facilitate such relationships.

(2002: 574)

Jamieson provides a lively and heated reappraisal of Giddens and his notion of 'pure relationships' in the context of the role of intimacy as a gendered preoccupation (1999). The inequalities in personal life persist, with more creative energy going into sustaining 'a sense of intimacy despite inequality than into a process of transformation' (ibid.: 477). Jamieson's trenchant attack on the assumptions, particularly Western, heteronormative precepts, that underlie definitions of intimacy provides a wonderful opening to engage in complex and contingent models of performative and localised mobile intimacy.

However, Giddens's notion of third-way politics can perhaps shed light on some of the emancipatory and 'democratising' rhetoric surrounding mobile media, most notably in the region. For example, the rise of the mobile phone as a repository for political agency is most prevalent in the well-cited elections of President Roh in Korea (Kim 2003) and the downfall of President Estrada in the Philippines (Pertierra 2005a, 2005b). While one can assert that types of 'individualisation' (Beck, cited in Slater & Ritzer 2001) or 'individualism' (Castells et al. 2007) are seemingly a global phenomenon, they are, just like notions of 'mobility', inflected by the socio-cultural. This has been highlighted in the particular ways mobile intimacy has been approached and conceptualised by anthropologists in locations such as the Philippines (Pertierra 2006; Ellwood-Clayton 2003), Jamaica (Miller & Horst 2006) and Asia (Bell 2005).

Intimacy has for a long time attracted the interest of ethnographers as a way in which to develop an understanding of trust, and, also, as a way to understand the particular values and tastes of a culture. For Berlant (2000), intimacy is a way in which to critique public and private cultures in the US, whereby intimacy is viewed as part of formations of subjectivity and identity. For anthropologist Ara Wilson in her compelling study on *The Intimate Economies of Bangkok* (2004), intimacy operates as a lens to explore the formation of private issues in public spaces and how these relate to, and exceed, gendered or sexual identity. Through the role of intimacy, Wilson provides new insight into how we can conceptualise globalisation and trans-national capitalism at the level of everyday life.

Returning to Bell, she notes that it is important to recognise that such concepts, along with what is public and what is private, are embedded by the socio-cultural context (2005: 70). In differentiating different types of mobility – such as social, geographic, financial – Bell argues, 'it is through these different cultures of mobility that cell phones have been deployed, consumed and sometimes resisted' (ibid.: 70). Thus mobile phones 'are not just objects and technologies; but also a system of ideas' that traverses the spaces of intimacy, family, home and work (ibid.: 90). This symbolic power resonates on many levels that reflect the 'local particularity and cultural difference as dimensions of a larger political economy of value' (ibid.: 90). Unquestionably, the mobile is a complex signifier for many contemporary debates; as a symbol, index, icon, or as a set of cultures and practices that are inflected by the local, mobile media is informed by the local and domestic.

Mobile @ domestic: bringing the mobile, mobility and mobilism back home

The domestic technology of the mobile phone exemplifies that the practice and spatial occupation of what constitutes the 'domestic' can operate upon many levels, just as does what constitutes 'home'. Domestication not only occurs in the actual domestic sphere; the politics of gender, intimacy and sense of privacy so informed by the domestic sphere also migrate outside the physical 'domestic' space. In the case of the mobile phone, although the domestic technology may have *physically* left home it *psychologically* still resides and connects users to a sense of place and home (Urry 2002; Bell 2005). In other words, domestication may have moved out of the home – literally, in the case of the mobile phone – but ideas of locality (Massey 1993) and place are still, if not more, enduring (Ito 2002).

Like other domesticated technologies (Morley 2003), the processes involved are far from simple or finalised, as each specific site locates and adapts to new cultural artefacts in a series of exchanges. We domesticate technologies as much as they domesticate us. In the increasingly conspicuous rise of the 'mobility' phenomenon, the symbolic dimensions of the mobile phone become all-pervasive. In 'Social networks, travel and talk' (2003), Urry considers the role of physical travel in the context of social life and the new ways in which it is being 'networked'. Rather than opposing the virtual with the actual as Bauman and Putnam do, Urry sees a dialectical play between the two interrelated modes of communication, with the significance of the corporeal always fully privileged in what Urry describes as a 'networked sociality' (ibid.).

As highlighted by the debates by Bauman and Urry, the mobile phone is commonly a repository for larger debates about globalisation, co-presence, place, and mobility, and thus often becomes a scapegoat for social problems, particularly in sociological debate. *Mobile Media in the Asia-Pacific* does not attempt to cover the gamut of critical inquiry around the mobility turn, but rather to locate some of the key debates in the context of the region. One of the most prevalent issues here requiring qualification is the confusion and conflation between two parallel but distinct conditions, namely mobility and mobilism.

In his vivid account of the rise of *keitai* cultures in Japan as part of broader socio-historical trends that can be mapped back to the eighteenth century, Fujimoto argues that current *keitai* cultures are characterised by a condition he describes as '*nagara* mobilism'. While *nagara* roughly translates to 'whilst-doing-something-else', mobilism, for Fujimoto, is the 'broader cultural and social dimensions such as malleability, fluctuation and mobilisation' (2005: 80). Unlike mobilism, 'mobility has tended to refer to functional dimensions of portability and freedom from social and geographic constraint' (ibid.: 80). Thus mobilism is tied to socio-geographic factors, whereas mobility infers transcendence, particularly around geographic constraints. How we conceptualise the role of various interrelated components of involuntary and voluntary mobility at a personal, social and transnational level, in a period of globalisation, is subject to localised definitions of place.

The notion that practices of mobility reflect broader socio-cultural practices and values is also apparent in Michael Herzfeld's revising of Benedict Anderson's notion of the nation-state as an 'imagined community' (Anderson 1983), in the form of 'cultural intimacy' (Herzfeld 1997). For Anderson (1983), 'nation' – as we understand it today – was born through the rise of distribution and printing techniques such as the printing press. Thus a sense of belonging and place, an 'imagining' of community, has never involved just face-to-face (ibid.: 18) and always deployed some form of virtuality or co-presence. For example, diaspora does not result in an erosion of the nation-state, since 'imagining' is an important part of the sense of belonging (Dirlik 1999). Mechanisms such as the printing press, railway systems and now global mobile technologies, operate to further instil notions of intimacy, home and cultural proximity. This reading of Anderson's 'imagined community' as involving co-present intimacies that are not bound to face-to-face contact is eloquently deployed by Maria Bakardjieva and Andrew Feenberg (2004). As they note, 'communication media plays a central role in determining the different styles in which communities have been imagined through history' (ibid.: 37).

For Herzfeld, cultural intimacy describes the 'social poetics' of the nation-state and is 'the recognition of those aspects of a cultural identity that are considered a source of external embarrassment but that nevertheless provide insiders with their assurance of common sociality' (1997: 3). It is the negotiation between the personal intimacies and the socio-cultural intimacies that construct a sense of nation-state. The 'imagined community' (Anderson 1983) of the various nations is no longer defined by the co-presence (actual and virtual) role of print press media. Rather, the characteristic of new media communities is that they are, on the one hand, governed by 'mobile privatization' (Williams 1974), and, on the other hand, anchored by 'immobile socialization' (Bakardjieva 2003) in which users negotiate online and offline communities, often through a domesticated, social space.

Through the lens of 'transnationalism', social scientists such as anthropologists, urban geographers and sociologists have been able to rethink the construction of nation and national identity in the realm of perpetual mobility. The lens can reflect upon unabated diaspora in which imagined communities are often forged through transnational ties that combine to create a picture of a nation, of a culture, of a locality. Indeed, Bakardjieva's 'immobile socialization' resonates with Turner's discussion of 'enclave societies' as another paradox in the mobility/immobility debate. As Turner notes:

> Modern societies are in particular characterised by a deep contradiction between the economic need for labour mobility and the state's political need to assert sovereignty ... Globalisation paradoxically produces significant forms of immobility for political regulation of personas alongside the mobility of goods and services ... Walls and other examples of enclavement are produced by a new strain of xenophobia, which strongly counteracts the cosmopolitanism which many sociologists have seen as an almost inevitable outcome of transnationalism.
>
> (2007: 287–301)

The role of localising and personalising 'imagined' notions of home and place have dominated discussions about globalisation and post-colonialism in age of mobility and its Siamese twin, immobility. If we return to Bell's insistence that what constitutes what it means to be at home is determined by the locality, then it is important to re-cast these debates in the context of the particular forms of mobility and mobilism within the Asia-Pacific. Through the symbol of post-modernity (Kopomaa 2002), the mobile phone, we can revise what defines the local, national, regional, and global in a context marked by 'transnational communities of consumers' (Chua 2006). Firstly, we need to reconceptualise the shifting discursive formation encompassed under the trope of Asia-Pacific.

3   Beyond the 'new rich'

Consumption, production and gender in the region

To a great extent, the predominance of consumerism has been exaggerated . . . to signal the rise of Asia's 'new rich' (Robison and Goodman 1996), not only in developed economies such as South Korea, where users can conduct banking, e-signature, and the purchase of small items based on the mobile phone (Lipp 2003), but also in less developed ones such as the Philippines, where the residents of Manila were reported to have experienced a 'mobile mania', especially using texting . . . Like the Filipinos, Chinese youngsters are most active in using such services as SMS owing to its faddish appeal and much lower price than voice telephony.

(Castells et al. 2007: 108)

Having delineated global mobile communication in the context of gender and the region in Chapter 1, followed by Chapter 2's exploration of 'mobility' debates within current globalisation, this chapter brings us back to the region's emerging forms of twenty-first-century post-modernity. As this chapter will highlight, this emerging form of post-modernity is unquestionably shaped by gendered modes of production and consumption. As one of the region's most evocative symbols for post-industrial consumption and production, the mobile phone can provide us with insight into these patterns at both localised and transnational level. This chapter discusses the role of gendered mobile media in the light of broader post-industrial movements in the region; a process in which gendered forms of labour and inequality – or what L.H.M. Ling calls 'hyperfemininity' (1999) – has continued to prosper.

Through the lens of gendered mobile media as a vehicle for hyperfemininity, this chapter considers how hyperfemininity produces tension between, on the one hand, new emotional grammars of propinquity, intimacy, creative expression and female empowerment, and, on the other hand, the increasing exploitation and naturalisation of gendered reproductive and social labour. The former phenomenon can be seen in the burgeoning forms of mobile media as creative industries such as the mobile phone novels (*keitai shôsetsu*) in Japan, whilst the latter is particularly prevalent in uneven post-industrial economies, in which women from countries such as the Philippines become 'servants of globalization' (Parreñas 2001). This tension is at the core of localised mobile media practices – 'imaging communities' – and the ways in which they contribute to the 'cartographies of personalisation' in the region. These cartographies are shaped by, and continue to shape, the region's various modes of mobility.

The region's emerging class and economic mobility, most recently noted in the case of China, has often been symbolised by the mobile phone. As one of the first by-products of the region's post-industrial mobility, economic mobility created new forms of mobile and immobile labour and lifestyle enclaves. These various modes of immobility and mobility – at both the level of the micro 'imaging communities' to the macro 'imagined community' (Anderson 1983) – contribute to contemporary 'cartographies of personalisation'. Behind the conspicuous image of the female mobile media user is a region marked by contesting and changing consumption and production rhythms as it accelerates into twenty-first-century post-modernity.

In the region, some who were post-industrially more developed saw mobile capital in the form of transcending new modes of class and subjectivities. For others it was their labour that became mobile, as was often the case for women in predominantly lower classes (Ong 1999; Truong 1999; Hochschild 2000). In these techniques of arising mobility and immobility, empowerment and disempowerment, the symbolic and actual role of the mobile phone features prominently (McKay 2007). The complex formation of multiple forms of mobility and immobility could explain, in part, why the Asia-Pacific has been so neglected – despite the powerful motif that mobile media provides in representing the region's twenty-first-century post-modernity.

As a construct, it is important to recognise that the Asia-Pacific was a discursive device. Under colonialism, there was little 'regionalism' within what we now understand as the Asia-Pacific; rather there were a series of bilateral ties to the metropolis. This was also true during the US hegemony of the region in the 1950s and 1960s. Today the region is an ever-evolving constellation of economic and political power distributions: a sum of contesting mobilities and modernities that have distinct and yet transnational connections. The rapid economic growth in parts of the region (Japan, Singapore, Taiwan, Korea, Hong Kong, and now China, Vietnam, Malaysia, the Philippines and Indonesia) over the last two to three decades has led to increasing linkages between these nations in creating transnational networks.

As the region's various mobilities have evolved and transformed, so too have the ways in which modernity in the Asia-Pacific has been conceptualised. From the Confucius revivalisms of the late 1990s in search of 'Asian' values, which were criticised for homogenising diversity, self-Orientalisation (Dirlik 1999b) and essentialism (Ching 2000), the region attempted to think beyond Western and Eurocentric tropes through the lens of 'alternative modernities' (Gaonkar 1999; Ong 1999). This is charted in Ong's aforementioned discussion on the 'mutations of citizenship', extending from her earlier discussion of 'global assemblages'. As Ong observes in her introduction:

New connections among citizenship elements and mobile forms suggest that we have moved beyond the idea of citizenship as a protected statute in a nation-state, and as a condition opposed to the condition of statelessness. Binary oppositions between citizenship and statelessness, and between national territoriality and its absence, are not useful for thinking about emergent spaces and novel combinations of globalising and situated variables. For instance, market-driven state practices fragment the national terrain into zones of hypergrowth. These spaces are plugged into transnational networks of markets, technology, and expertise.

(2006: 499)

As Dirlik (2005) trenchantly observes, in this search for identity and cultural modernity in an age of mobility and reformation of nation state, it was important that the region not deny its own history of cultural imperialism and colonial conflicts. According to Appadurai (1990), a more useful model of regionality could take the form of 'Pacific Rim' – echoing the cartography of the international dateline. In such a model, regionality would take the form of temporal, rather than geographic, clusters. This interrogation of regionality is exemplified in the arguments surrounding the Asia-Pacific, as witnessed in debates around the appropriate term to depict the dynamic and contesting spaces it claims to represent. One can find numerous examples such as 'Asia-Pacific', 'Asia-Pacific region', 'Asia/Pacific', 'Asian Pacific' and 'Asia and the Pacific'.

As noted earlier, the Asia-Pacific – as a site for contesting local identities and transnational flows of people, media, objects and capital – has come under much radical revision and reconceptualisation (R. Wilson 2000; Dirlik 2007; Arrighi 1994; Arrighi et al. 2003). Far from the 1970s' widely held view of the region being a sum of satellite NICs oscillating around the region's first industrial nation, Japan, we are painted another picture. Rather than resembling a rigid and highly figurative oil painting cascading with European motifs and colours, the new picture is of an endlessly moving video image in which multiple divergent identities appear and disappear. In this video vignette of post-modernity, in which borders are both eroded and redefined, and transnational flows tread paths well worn, the mobile phone becomes a symbol for locality and gestures of propinquity, labours of love and work, and the realities and fictions of modernity.

As Arrighi observes in his meticulous discussion of capitalism and modernity in *The Long Twentieth Century* (1994), capitalism can be viewed as four cycles from Genoese mercantile capitalism, Dutch finance capitalism, British industrial capitalism and contemporary American capitalism. According to Arrighi, the fifth form of capitalism, synonymous with twenty-first-century modernity, would undoubtedly be East Asian capitalism. In a subsequent essay about the rising power of East Asia globally, Arrighi notes that East Asia has become 'the most dynamic world player in world-scale processes of capital accumulation' (1998: 59). However Arrighi, like Dirlik (1999b), argues that this form of industrialism and power is not nascent. Rather, it is the myopic visions of many Western models of history that seem only to grasp capitalism and industrialism as relatively new forms, neglecting to see that the domination of the 'West' – either in the form of the US and Europe – is part of a much larger process of modernity and mobility.

Extending upon Geoffrey Barraclough's (1967) examination of the twentieth century being defined by a 'revolt against the west' (ibid.: 153–154, cited in Arrighi 2005), Arrighi avows that the second half of the twentieth century will be recorded in terms of the 'economic renaissance rise of East Asia'. Arrighi et al.'s (2003) eloquent *The Resurgence of East Asia* explores and unpacks the ideological terrain of 'East Asia' as distinct from spatial and geographic notions. Their study is dedicated to repositioning East Asia as a political and cultural configuration that has arisen from a China-centred tribute trade system that pre-dates Western intervention. This argument shifts the conceptual frames of reference away from binary East/West, orient/ occident coordinates, and into a more complex and rich understanding of the ways in which various mobilities and modernities have multiple formations.

Alternative terms and definitions have been proffered to describe or delimit the area posited as the Asia-Pacific in an attempt to acknowledge, or subsume, the hierarchies inherent within the region. Dirlik has proffered two terms, 'Asian Pacific' and 'Euro-American Pacific.' He suggests, 'the former refers not just to the region's location, but, more important, to its human constitution; the latter refers to another human component of the region (at least at present) and also to its invention as a regional structure' (1992: 64).

As Wilson and Dirlik note, one way of reconfiguring the trope of the Asia-Pacific would be to restructure the rubric into Asia/Pacific. In their anthology *Asia/Pacific as Space of Cultural Production*, they attempt to challenge the 'hegemonic Euro-American' construction underpinning the Asia-Pacific in order to conceive of new ways of conceptualising regionalism. In particular, they use the configuration 'Asia/Pacific' to discuss the region as a space of cultural production, social migration and transnational innovation, whereby 'the slash would signify linkage yet difference' (1995: 6). As Dirlik asserts, '[an] emphasis on human activity shifts attention from physical area to the construction of geography through human interactions; it also underlines the historicity of the region's formations' (1993: 4).

Writing in a later period marked by Confucius revivalism, Dirlik (1999b) argues that such an activity sought to naturalise cultural constructions and forge cultural proximity whilst ignoring the histories of violent power struggles and imperialisms within the region. He argued that in defining East Asian modernity, the region must not perpetuate problematic regional community 'tropes' such as 'East Asia' and 'Asia-Pacific', and that there is a need to incorporate complex, dynamic understandings of history along with pluralist notions of culture. As Dirlik trenchantly observed, Confucius revivalism, in its attempts to counteract Eurocentric Orientalism, only served to produce reproductions of it (1999b: 167). In a later essay on reconceptualising area studies in the light of contemporary mobility and transnationalism, Dirlik argues:

> How we view the Pacific, and regionalise it, is not just an academic question but a political one as well. They may all refer to more or less the same location, but terms such as East and Southeast Asia, Asia Pacific, Pacific Asia, Pacific Rim and the Pacific have different, and conflicting referents that remain to be sorted out.
>
> (2005: 159)

Dirlik continues this examination in his recent revising of the 'Global South' (2007). In this, Dirlik controversially poses the hypothesis that the twenty-first century saw the axis of power shift from the North to the South. Utilising the case study of China's rising economy, Dirlik sketches a new global order in which the histories of the past become spectres in the future global reach of capitalism. Here we see the overlay between Dirlik and Arrighi's arguments, whereby the region's particular form of modernity becomes synonymous with a new type of global modernity. However, one needs to be wary of unquestioningly celebrating the region's shift from economic to ideological power in the global market, given what this type of pioneering industrialisation of the region has cost in terms of gender and labour inequalities.

Amid the searching for 'Asian' values that can be witnessed in much of *Hallyu* transnationalism, some of the pros and cons of the region's new-found economic and ideological power have been less discussed – in particular the link between the rise of industrialism in the region and the rise of gendered forms of labour and inequality (Truong 1999). Here, the very forms of binary power and inequalities for which the region has criticised the West, have been replicated. This has led many theorists to explore the role of gendered modernity as a way in which to expose the inequalities, both new and persistent, in the practice and politics of the region (Sen & Stivens 1998; Truong 1999).

In the self-Orientalism exemplified by Confucian revivalism (Dirlik 2000), the prevalent role of representations of Asian women, as symbolic forms of Orientalism's 'hyperfemininity' (Khoo 2007) is undisputed. However, the role of technology, and particularly mobile technology complicates this model of localised and transnational gendering of the region. The phenomenal rise of the region's new economic and ideological global power is accompanied by the perpetuation, if not exacerbation, of certain inequalities, notably those associated with the pivotal role of women in the rise of flexible, precarious labour (both paid and unpaid). The integral role of gender in the rise and repackaging of consumption and alternative modernities in the Asia-Pacific cannot be underestimated.

Gendered modernities: the other side of post-industrial mobility in the region

In 'The underbelly of the tiger: gender and the demystification of the Asian miracle', Truong provides a trenchant critique of the 'Asian miracle' as not only a Western construction of otherness, but also a formation that, in its construction of particular 'Asian' values, served to naturalise women with reproductive labour (1999: 133). She highlights the continuing link between sexuality and economy, in which 'regimes of sexual control' forge women's experiences 'as socio-sexual beings' (ibid: 134):

> In many ways the East Asian experiences of female participation in industrialisation confirm the view that pre-existing gender norms have been present in production relations and hence do not simply 'evaporate' once countries reach a certain stage of industrialisation. On the contrary, gender norms in East Asian industrialisation appear to be embedded in practices of families, firms and states which lead to the creation of a four-tier system of industrial work. The two upper tiers uphold production systems and the two lower tiers uphold reproductive systems. Far from being gender-neutral, industrialisation processes in East Asia have deployed the women's labour force in strategic positions to reduce labour costs, which makes rapid domestic capital formation possible.
>
> (ibid.: 135)

Truong notes the rise of the 'feminisation' of industrial relations, in which women are the inevitable losers, highlights that the 'Asian miracle' is a 'fully gendered process' (ibid.: 158). As she observes, the Asian miracle of ANICs 'is best understood through two structures of symbolic domination, i.e. west vs. east, and male vs. female' (ibid.: 158). While this period saw the 'west vs. east' dichotomy disintegrate as East Asian countries gained economic and ideological power globally, the imbalance between male and female remained in force. This led Truong to claim that while 'the east–west battle for recognition may be won . . . the battle for social equality and cultural meaning of industrial progress may not be put to rest' (ibid.: 159). In the emerging forms of patriarchy and capitalism, Truong noted the reproduction of a 'hypermasculinized' model of industrialisation and development that offers few possibilities for women (1999: 147).

In 'Sex machine: global hypermasculinity and images of the Asian woman in modernity,' Ling (1999) argues that this hypermasculinity is not just the product of East Asia but rather of the global economy. For Ling, globalisation carries with it an association of 'capital-intensive, upwardly mobile hypermasculinity', as opposed to a localised and 'socially regressive hyperfemininity' (ibid.: 278). In this light the mobile phone is an ideal vehicle for housing the twins of global industralisation – hypermasculinity and hyperfemininity. The overt hyperfemininity of mobile phone customisation camouflages the role of device in furthering gendered forms of labour and mobility in which women are relegated to casual, precarious labour or to unpaid reproductive labour to compensate for the continuing divisions between the haves and the have-nots. In this context, the mobile phone serves as a vehicle for hyperfemininity through its predominant use as repository for emotional grammars of propinquity and its instrumental role in the entrenchment of 'global care cultures' (Hochschild 2000) and 'servants of globalization' (Parreñas 2001).

For Truong, from 1970 until 1990 she observed two conflicting trends: increasing female participation in the workforce (initially in manufacturing sections and then dispersing into unpaid, welfare sectors) and decreasing male participation in the workforce. These trends corresponded with the rise of industrialisation and with the implementation of cost saving forms of labour – most notably casual, flexible and precarious labour that was then taken up by women. The rise of casual work has, according to Truong, affected Korean women the most. She noted higher levels of female wages and more upward female mobility in Hong Kong and Singapore than in Korea and Taiwan, which would account for the higher levels of transnational employment of Filipino care workers in Hong Kong and Singapore than in the other two countries.

As Truong notes, the operating system of the ANICs (Asian Newly Industrialised Countries) in 1999 could be described as a four-tier structure. The first tier (occupied by men) represented those in protected forms of long-term and stable employment, the second (occupied by women) consisted of casual workers, and the third tier compromised housewives and their attendant reproductive, unpaid labour of caring for children and older and disabled family members. Paid reproductive workers (i.e. sex workers) occupied the fourth tier. As Truong surmises, '[t]he emergence of this four-tier system appears to be connected with the types of symbolic domination narrated by discourse on Asian cultural values and morality in regards to femaleness, i.e. domesticity and self-sacrifice' (1999: 165–166).

According to the International Labour Office (ILO) 'Global Employment Trends for Women' report (2008), between 1997 and 2007, the Asia-Pacific has been one of the dominant regions for growth in female employment. Along with the increase in educated women, it is also predominantly women leading the move away from agriculture sectors and towards service (33.5%) and industry (25.5%) sectors. This shift in female labour markets mirrors the region's retreat from agriculture and towards service and industry sectors – epitomised by the growing role the region plays in manufacturing much of the global mobile technological hardware (mainland China being an example) and ICT outsourcing.

Moreover, just as female paid employment (predominantly in precarious, new media sectors) has increased over this ten-year period (ILO 2008), so too have the new forms of mobile media and social labour accompanied this phenomenon. This parallel and interrelated phenomenon has resulted in the re-working of gender, labour and technology. From social intimacy to creative user content, labour has taken on various immaterial and material guises. These new forms of labour and intimacies can be witnessed within specific modes of mobility.

In the various forms of mobility – people, ideas, labour and capital – it is undoubtedly women who are most adversely affected in the growing inequalities globally. However, it is important not to perpetuate this system, hypermasculine or not, by constructing binaries in which women are 'victims' of this system. As McKay (2007) rightly observes, too much of the discussion about gendered forms of labour and mobility, victimise the care worker – i.e. rendering her a subject without agency. For example, too often the Filipino worker's maintenance of propinquity and intimacy via devices such as SMS or letter writing is interpreted as sacrificing their sense of family and intimacy. Rather, as McKay observes in interviews with Filipino workers in Hong Kong, the Western binary of face-to-face vs. 'mediated' forms in intimacy (face-to-face being the definitive form) neglects to address the localised ways in which intimacy is understood and experienced. There has been much literature about the Filipino love of mediated forms of intimacy; for example, it is common for Filipino people to have text lovers whom they never meet face-to-face (Pertierra 2006; Solis 2007).

It is no easy task to understand the shifting and diverse formations of gendered subjectivities – and attendant forms of intimacy and mobility – in the region. Moreover, as the region has now fully embraced precarious industrial labour reforms across various levels, we see the [dis]empowerment and inequalities of gender becoming more obscured except in the obvious rise in transnational care cultures. Are these symbolic of women's emancipation from reproductive labour (apart from the Filipino that is) and a deconstruction of the 'socio-sexual' categorisation, or are women being relegated to the class of 'kept' women?

To discuss mobility in the region is to identify that debates around consumption, leisure and post-industrialism are deeply embedded within gendered politics. In order to gain understanding of the dynamic cultural production and reproduction it is important to investigate the pivotal role of consumption, and changing views of consumption, in the Asia-Pacific. In doing so, we can gain insight into other forms of localised gendered subjectivities that may provide a different way to view transnational flows of consumption and production in the region.

The sign of the time: changing modes of consumption in the region

The study of consumption in the region has begun to feature prominently in sociological and other academic literature. In particular, the intersection between consumption and media has provided a fruitful lens in which to examine identity and lifestyle in the region. The role of women – as symbols both *of* and *for* consumption – is a prevalent theme. However, many of the studies on gender have focused specifically on one location, such as Japan (Lise Skov & Brian Moeran's insightful anthology *Women, Media and Consumption in Japan* [1995] is exemplary); or Korea, in which Young-Jae Lee (2000), Hae-Joang Cho (2004, 2007) and Misu Na (2001) and Hyun Mee Kim feature dominantly.

Studies that have focused on transnational gender formations have generally done so through a conflation with class or economic precepts – for example, aforementioned Sen and Stivens's *Gender and Power in Affluent Asia* (1998). Ong and Michael Gates Peletz's (1995) wonderful anthology *Bewitching Women, Pious Men: Gender and Body Politics in Southeast Asia* provides a persuasive paradigm for analysing femininity and masculinity as socio-cultural practice. The authors show how gender can provide a way for thinking about constructions of power and knowledge across local, national and transnational formations. However media technologies, given the date of the publication, configure minimally. Moreover, over the last 13 years, the ways in which consumption and modernity have been configured, particularly after the 1997 crisis, have shifted substantially. One anthology that notes the prevalence and importance of gender in media consumption, despite having no reference to gender or women in the title, is Youna Kim's (ed.) *Media Consumption and Everyday Life in Asia* (2008).

As I will discuss in detail in the following section, gender has often been conflated with fashion, and this is particularly the case with gendered mobile media identities. By framing mobile identity under fashion, the role of nation-state and the politics of mobility are couched under an accessory vs. necessity debate. It is important here to recognise the role of fashion as a facet of global commodity processes, and thus relevant to the politics of consumption debate. For example, in Japan *keitai* cultures such as *kawaii* customisation and ring tones must be understood as partaking in localised rituals that have long existed. The role and significance of accessories can be linked to the sumptuary laws in pre-modern Japan wherein dress codes and hairstyles were dictated by law.

In McVeigh's work – whether he is analysing fashion in general or focusing upon 'accessory' culture such as *kawaii* culture in Japan (1996, 2000) – he demonstrates the symbolic role of the 'accessory' and uniform as a necessity in everyday Japanese life. McVeigh's discussion links to Fujimoto's (2005) analysis of *shikôkin* ('recreational consumer products') extending from the symbolic dimensions of tea ceremonies and the role of associated rituals in Japan. Here consumption operates as part of everyday rituals and is linked to the role of social status. Thus 'fashion' and consumption are conceptualised as re-enacting localised rituals (Miller 1987). But, in the case of mobile phone customisation, much of the literature defines (and limits) it to the preoccupation of young females and the fashioning of accessories (Katz & Sugiyama 2005), without understanding it in terms of broader localised and transnational modes of modernity that predate twentieth-century capitalism and globalisation.

In Goodman and Robison's aforementioned *The New Rich* (1996), it is the formation of new economic imperatives and 'lifestyles' that prefigure in the region's new consumer identity. As *The New Rich* so aptly demonstrated, the region's newly found industrialisation came with unprecedented forms of consumer-driven identity and subjectivity. As industrialisation spread across the sectors from production into consumption practices, the region had to realign its way of thinking about consumption and modernity – most notably, in relation to global capitalism and East/West relations. As the region began to experience new wealth, it also began to experiment with new forms of lifestyle. Although this was initially seen as the East mimicking the meaningless and immoral capitalism of the West, the region began to reconceptualise the phenomenon after the 1997 financial crisis.

As Chua (2000) perspicuously observes, one way to understand the region's multiple modernities would be to chart the role of consumption in the region post-1997. After the financial crisis of 1997, government and industry worked to reconstruct consumption as no longer a subsidiary of production or, more importantly, of Westernisation. Through the rise of Asian capital – both economic and cultural – the phenomenon Chua calls 'consuming Asia' (ibid.) has taken on new significance locally and globally. This can be seen in the ways in which transnational communities of online gaming have taken on new phenomenological dimensions, reflecting emerging localities and allegiances.

Writing after the 1997 crisis, Chua noted the growing role of consumption in the region that can no longer be 'subsumed under the mantle of production' (ibid.: 3). Identifying government policies enforcing saving rather than expenditure (epitomised by Taiwan and Singapore), Chua argues that older generations saw much of the conspicuous and egregious consumption by youth cultures as a form of 'Westernisation'. However, after 1997, the region had to sublimate the ' "traditional" morality of savings . . . in order to save capitalism in Asia' (ibid.: 11). In this repression of traditional ideas of morality, Chua argues that Japan again became a technological centre that symbolised well-made products: despite the antagonisms of some neighbouring countries towards Japan as a site for imperialism, the consumption of Japanese goods was deemed more favourable than the consumption of Western commodities. The consumption of J-pop has been just one phenomenon among many that constitutes the transnational consumer communities of the Asia-Pacific in which 'Asian' products dominate.

The formation of the Asia-Pacific region with Asian, rather than Euro-American, products has become part of everyday life. Pan-Asian cinema has become central to the region's communities; it has also become the source of a revisualisation of Hollywood's film industry, with multiple re-makes and odes to the significance of pan-Asian film globally. The recent remake of Hong Kong's *Internal Affairs* (directed by Andrew Lau and Alan Mak, 2003) by Martin Scorsese (*The Departed* 2006) and Quentin Tarantino's *Kill Bill* (2000) series are two of many examples. Chua goes on to assert that the region is now 'a mix of Japanese, Korean and Chinese-language pop cultures' that are part of the 'daily diet of media consumers in East Asia' (2006: 27). In each location, one can find a different 'mix of dedicated consumers' that leads Chua to argue 'consequently, a network of transnational consumer communities, from active fans to occasional consumers, has emerged in the region' (ibid.: 27).

Chua observes 'these transnational consumer communities exist "beneath" the official international relations in a region where traces of colonial histories and Cold War antagonism remain' (ibid.: 27), and argues that there are 'no structural avenues for these pop culture consumer communities to percolate upwards to intervene in the international processes' (ibid.: 27). However, I would argue that through mobile media and the attendant virtual communities and UCC, there are possibilities for these unofficial imaging communities to impact on official imagined communities and transnational synergies.

Chua's notion of 'communities of consumers' is central to rethinking how 'Asia' is consumed both within the region and also globally. Although Chua acknowledges the power transnational commodities have had on gendered usage of technology, he stops short of conceiving the possibilities this has for female consumers and their role as active fans of transnational popular culture. For example, there is the case of middle-aged Hong Kong women learning to use the Internet to join fan club discussions of their favourite Korean dramas. Often this leads to meeting new people offline who are part of the 'communities of consumers'.

The role of mobile media as symbol for contesting values in the region is undisputed. As Bell argues, 'in urban Asia, at least, these cell phones, rather than facilitating an idealised universal communication, actually contribute to the re-inscription of local particularity and cultural difference as dimensions of a larger political economy of value' (2005: 84). Within this contestation of values, gender features both explicitly and implicitly. In the Asia-Pacific, women have been instrumental in the rise of mobile media. While some of the most cited examples such as the Japanese high-school girl PHS (Personal Handy-phone System) phenomenon consisted of female 'youth' cultures, other examples can be found such as that of older females gaining empowerment via camera phones in Korea (D.H. Lee 2005) and SNS in Korea (Hjorth 2007d). But the most evident example of gendered mobile media has been the role of *kawaii* culture in Japan (Hjorth 2003a; McVeigh 2003). The rise of *keitai* culture in Japan was highlighted by predominantly female early-adopters (Hjorth 2003a; Fujimoto 2005).

Within the region, gender has been explored in specific locations such as Tokyo (Ito 2005a; Fujimoto 2005; Matsuda 2005; Hjorth 2003a), Seoul (D.H. Lee 2005; Hjorth & Kim 2005) and Hong Kong (Lin 2005a). Despite the fact that the female consumer has been pivotal in much of the identification and representation of mobile media in the region, there is a curious gap in the current research. What can we learn about the similar and different ways in which female consumption of mobile media operates in the region? In what ways is the overall proclivity towards feminised customisation linked to different modes of gender performativity? Moreover, how are gendered modes of consuming and producing technology changing in the Asia-Pacific? These questions are significant in the analysis of mobile media in the region. How, for example, is the increasing use of, and access to, the Internet via the mobile phone by women – which has seen a shift in the gendered usage of the Internet away from a once male-dominated arena (Kim 2002; D.H. Lee 2005; Y. Lee et al. 2002) – manifest in terms of female agency and representation in production? It is these questions that will drive the proceeding chapters in the construction of *imagining imaging communities*. Before moving each case study, I will briefly outline this mobile media UCC rubric.

Imagining imaging communities: the politics of mobile media in the region

The growth in visual mobile media and the attendant social networking, online communities and avenues for distribution and sharing will undoubtedly change what it means to participate in media cultures in a globalised Asia-Pacific. And as the Internet grows to encompass new forms of context (such as SNS), convergence (mobile Internet) and dissemination modes for UCC, we are left with questions. How are these overtly gendered practices reinforcing or subverting notions of home and place in a period marked by transnational consumer flows? Will mobile media be able to provide an avenue for what Chua (2006) calls 'pop culture consumer communities' to 'intervene in the international processes' and the politics of the national? Can mobile media become a tool to further value and nurture, rather than merely exploit, hyperfemininity? And how will this reflect and transform current gender inequalities around mobility, labour and intimacy?

One of the important components in a country's *imagined community* is the role of the everyday users' depictions – circulated within and outside the socio-cultural context. Often projected, exported, *official* images of a culture differ from the personal, *unofficial*, individual images disseminated amongst small communities. However, an imagined community is a sum of both its official and unofficial depictions. With the rise in mobile media and SNS, the discourse of the unofficial is becoming increasingly important to the ways in which individuals and communities view themselves and other contexts.

As ICTs seem to herald a global move towards mobility – geographic, cultural, social, physical, psychological – the ongoing role of co-presence seems ever pervasive. As new forms of global socio-technologies, mobile media could be this century's printing press; the very vehicle of modernity that Anderson (1983) discussed as integral in constructing what we define as nation-state today. Far from eroding a sense of place and locality, mobile media reinforces the contingency of the local, the fleeting, and co-present.

Much of the research into mobile media in the region has recognised that it is pivotal in what Koskinen (2007) defines as the 'reterritorialisation' of place. Ethnographies such as the aforementioned by Ito in Japan (2002) and Yoon in South Korea (2003) have identified this phenomenon. In an age of so-called 'democratic' media – such as the camera phone – and the rise of distribution systems such as social software, the everyday user can become a 'prosumer' or 'produser'; an active producer of images to be consumed. Does this phenomenon demonstrate new forms of imagined communities best described by the hyperfeminine politics of imaging communities? And how does gender inform these emerging imaging communities?

As I will explore in my different case studies in the forthcoming chapters, we can see some consistencies in modes of gender performativity – especially in terms of how and why gender informs the ways in which users take and share their images. However, in the case of the 'three Ss' identified by Ito and Okabe, we will find sharp contrasts in each location despite the overall similarity in terms of banal subject matter. In Melbourne, as with Hong Kong, respondents tend not to share, but when they do they prefer to share personally (i.e. viewing on the phone in a face-to-face situation) rather than via MMS. In Korea, respondents prefer to share within a social group rather than individually. This is highlighted through their preference for sharing on the virtual community of Cyworld mini-hompy over sending MMS. Japanese respondents are sending camera phone images predominantly individually (via photo e-mail), although interest is growing in distributing them via SNS (such as the Japanese SNS, mixi) as well as saving them as hard copies (on *purikura* machines that print camera phone images onto stickers) and sticking them on theirs, and their friends', *keitai*.

By focusing on one of the most convergent, divergent and under-explored regions, the Asia-Pacific, *Mobile Media in the Asia-Pacific* aims to explore arising and remediated forms of subjectivity and community in terms of mobile media. By focusing on mobile media, it will summarise some of the parallels and disjunctures from a predominantly user-driven context. The aim is to see these practices as part of broader processes of globalisation and post-modernity in the region. Through the rubric of gendered customisation, *Mobile Media in the Asia-Pacific* will investigate emerging patterns of individualism and consumerism that are predicated around the trope of 'cartographies of personalisation'.

I argue that this process is inflected by gender, and most notably by a feminisation of socio-technologies that are affording users the ability to navigate online and offline identity. This feminisation of customisation has a long history in the uptake of technologies in the region, and is not comparable to the European models. These cartographies of personalisation reflect regional modes of intimacy, mobility, labour and postmodernity that are marked and shaped by both literal and metaphoric gender performativity. In the following chapters, I will detail localised forms of mobile media, reflecting upon whether this correlates to actual gender divisions and inequalities in the consumption and production of mobile media in the region.

Mobile Media in the Asia-Pacific looks at the changing role that women users are playing both as users and as emerging 'prosumers', while recognising that the feminised modes of mobile customisation – as a motif for cartographies of personalisation – are no longer the preoccupation of just female users. In exploring the various case study locations (Tokyo, Seoul, Hong Kong and Melbourne) in Part II, *Mobile Media in the Asia-Pacific* will detail some of the emerging cartographies of personalisation that are challenging traditional notions of gendered, and specifically female, use of new technologies. The new imaging communities are part of the unofficial discourse (which inevitably feed into official ideologies) that constitute the various imagined communities of the Asia-Pacific. This phenomenon of cartographies of personalisation not only speaks about women's changing socio-economic role in the region, but also about the region's burgeoning visual and discursive economies that comprise its identity in twenty-first-century postmodernity. Welcome to *Mobile Media in the Asia-Pacific*.

Part II
Mobile media cultures

4   Fast-forwarding to the present
The rise of customised mobile media in Tokyo

Having outlined the context of mobile media and the role of gender within the formation of the region in Part I, it is timely to turn to the location that has gained most attention globally for its innovation of mobile media. Japan's key role in producing technologies and, more specifically, domestic technologies for global markets since 1970 is well-documented to the point of cliché. Behind the global images of techno-savvy youth adorned with the latest technological gadgets in 'electric cities' such as Akihabara, Japan's role in producing and consuming new technologies – from the Sony Walkman, Atari games console, PlayStation and '*keitai* IT revolution' – has been pivotal. For Ito et al. (2005), the market success of these technologies can be best explained by characteristics of new media that they call the three Ps – pedestrian, personal and portable. The significance of these three Ps is that they transform technological gadgets into socio-cultural artefacts by relocating them into the dynamic space of cultural production.

Living in Tokyo at the onset of twenty-first century in 2000, I was unable to avoid the '*keitai* IT revolution' in the form of Japan's telecommunication giant, NTT DoCoMo's release of the i-mode. i-mode was not just a mobile phone (or *keitai denwa*, abbreviated to '*keitai*' meaning 'portable'); it was a key example of twenty-first-century mobile media, converging telephony and the Internet. The relatively slow uptake of the Internet in Japan in the 1990s afforded Japan the opportunity to leapfrog into the twenty-first-century convergence – mobile Internet – with ease. Thus, in the context of Japan, it is impossible to separate the emergence and rise of mobile media from that of the Internet. Suddenly the Internet – or i-mode's walled version of it – was being accessed by tens of millions of Japanese every day. The phenomenon was marked by sharp changes in the gender patterns of mobile use. Whereas in 1995, 90 per cent of *keitai* users were the archetypes – the salaryman (*oyaji*) – within two years the *shôjo* (young female) had become the dominant user. It seemed that Japan, via the mobile Internet 'revolution', had regained its techno-soft cultural power after the economic slump in the 1990s.

Behind these images of the *keitai* success lies a parallel story – the rise of the active and subversive female user. As the *keitai* phenomenon became increasingly pervasive during my time doing fieldwork, it became apparent that the symbolic and literal meanings associated with the *keitai* were concurrent with a steady rise of the visibility of women. The rise of the *keitai* from business tool to social accessory parallels the demise of the national symbol, the *oyaji*, and the growing power of female users – epitomised by the female high-school user (*gyaru*). Consumers, particularly female, have played an integral role in both the upsurge and adaptation of *keitai* cultures; so much so that the emergence of new tropes of female empowered consumers is indivisible from the *keitai* phenomenon, which, in turn, has been inscribed into Japanese national identity. Thus through the lens of the *keitai* phenomenon, we can not only gain insight into new modes of gendered perfomativity but also into burgeoning forms of gendered national culture.

But, we must ask, how do these new images of women – performing various modes of gender – correspond with the actual empowerment of women? How is the phenomenon of subversive *keitai* usage by female users reflective of new modes of Japanese femininity and hyperfemininity? How does the rise of the female user reflect emerging forms of Japanese nationalism? With the increasing deployment of new technologies by female consumers, we need to ask whether these new consumer tropes of 'mobile' female representation are being translated into women's actual economic and political mobility. If the rise of *keitai* cultures in Japan has been synonymous with innovative deployment of the media by women, how do these gendered *keitai* cultures reframe notions of nationalism? If the *oyaji* was an index of Japanese nationalism in the 1970s and 1980s until the downturn of the economy mid-1990s, has the *kôgyaru* (trendy female in her 20s) taken his place? And if so, what type of symbol of the Japanese woman, and her role in the national identity, is being presented here?

Since World War II, Japan's role as a centre for technological innovation has assured a sense of cultural power, particularly in the region. In the wake of Japan's faded imperialism in the Asia-Pacific, NICs such as Taiwan and Hong Kong have acquired an ambivalent awe of Japan's superpower economy and its cultural capital associated with such products as *anime* (animation) and *manga* (comics); while in locations such as Korea the memory of Japanese imperialism still remains fresh. Given the history of Japan's cultural, and more recently technological, imperialism, Japan has often found itself the subject of the aforementioned sci-fi 'default setting' (Nakamura 2002) and various manifestations of 'techno-orientalism'.

With this in mind, before moving onto the case studies, I need to contextualise current debates around Japanese nationalism and the ideological and symbolic role of technology in configuring gendered mobile media. I will follow this with a discussion of the history of the *keitai* in Japan and how this reflects enduring modes of customisation while also signalling new forms of personalisation, particularly in terms of way in which the *kawaii* – and its association with specific types of femininity – transforms in the space of *keitai* cultures. Moving beyond my earlier research conducted in 2000 in which I focused upon just women – especially the young female (*shôjo*) and the urban trendy female in her twenties (*kôgyaru*) (Hjorth 2003a) – in this chapter I draw from my case studies conducted after 2000 in which I focus on a more broad demographic of both male and female users. The aim of broadening this demographic was to extend beyond these familiar tropes of femininity (and, in the case of *kôgyaru*, challenges to traditional notions of femininity) in order to reflect upon the complex ways in which *keitai* and *kawaii* cultures rehearse and subvert traditional notions of femininity and gendered labour practices.

   This chapter will also explore the role of the mobile Internet as a space for gendered performativity and cartographies of personalisation in the form of imaging communities, before turning to the sample study of respondents. In exploring emerging forms of gendered mobile media customisation in one of the global centres for mobile technological innovation and adaptation, I have a key question in mind: how does the agency of female mobile media 'producers' (Bruns & Jacobs 2006) translate in terms of the general empowerment of women in Japan?

National genders: a snapshot of emerging agendas in Japan

To explore mobile media in Japan is to investigate why techno-nationalism has been so pervasive in Japan since World War II, and how this economic and technological success was translated into global soft-cultural capital. The analysis of the socio-technologies is also interrelated to prevailing modes of nationalism, and thus a reading of nationalism is required. Here, McVeigh's (2004) wonderfully insightful, and in some ways controversial, reading of Japanese nationalism is useful. McVeigh manages to expose the Confucius revivalism that was dominant in the region in the late 1990s and early 2000s, whereby, as Dirlik observes (1999a), in this search for 'Asian' values (under the trope of Confucian capitalism) the historical role of Japan in the region came under scrutiny. Dirlik succinctly surmises that it was a question of two interrelated concepts being bifurcated – culture and history. Did 'Asia' want to return to a pre-Western state in which other forms of imperialism, most notably, Japanese, prevailed?

   As Dirlik's history/culture dichotomy illustrates, nationalism operates on multiple discursive levels, and it is at the ideological level that both Dirlik's and Harumi Befu's definitions of nationalism overlap in understanding processes of globalisation. This is best illustrated in McVeigh's intricate analysis of the pervasive multifaceted nature of Japanese 'nationalisms', in which he argues that nationalism has continued to flourish despite a slight decline after World War II – a sentiment echoed in Sandra Wilson's edited collection, *Nation and Nationalism in Japan* (2002). Compartmentalising the various interrelated forms of 'nationalisms' for the sake of clarity, McVeigh argues that one of the most insidious and omnipresent modes of these multiple nationalisms is 'nostalgic nationalism', in which 'tradition' is harnessed to times of greater prosperity. This usage of nostalgia as a mode for instilling a sense of nationalism is not exclusively Japanese (Ma 2000); however, while the politics of nostalgia are transnational, they take many forms that are shaped by the local.

Scholars such as Harry Haroontunian (1970) have identified that Japan's long history of nationalism – which predates eighteenth-century encounters with the West – has always been infused with elements of Confucianism. In Arrighi's detailing of nationalism in the region in 'The rise of East Asia and the withering away of the interstate system' (1996), he challenges the notion that 'European expansion in Asia' saw the demise of the Sino-centric tribute-trade system. Drawing on the work of Japanese historians Takeshi Hamashita (1994, 1995) and Heita Kawakatsu (1994), Arrighi asserts that national identities were formed around multiple interpretations of Sinocentrism. Kawakatsu argues that through the Edo periods (1603–1867) Japan refined seclusion policies into a 'mini-China' and that the post-Meiji epoch 'was not so much a process of catching up with the West, but more a result of centuries-long competition within Asia' (Kawakatsu 1994: 6–7 cited in Arrighi 1996, n.p). The enduring legacy of nationalism is, for McVeigh, deeply etched into the Japan nationalisms of today.

The role of government and economy to enforce a type of conservative individualism within a nationalist rhetoric is particularly apparent in the construction of gender performativity. McVeigh's research on Japanese gendered material cultures, such as *kawaii* (1996b, 2000), and his more recent ethnographic work on college women and constructions of 'femininity' and 'internationalism' (1996a), have led him to argue two broad points: first, the perpetual conservatism of female performativity as 'feminine' and, second, that the rise of *keitai* and Internet media have led to a further disenfranchising of a sense of community, and to an increase in a 'national' form of individualism (2003, 2004).

This conceptualisation of nationalist individualism and of the historical fixture of Japanese women in particular roles is echoed in the work of Kogawa, and most notably in his discussion of 'electronic individualism' (1984). For Kogawa, the individualism provided by electronic media such as Walkman and *keitai* is a product of Japan's long history of individualism. This conceptualisation echoes Ito et al.'s three 'Ps' and Fujimoto's discussion of *nagara* (multi-tasking) mobilism, as briefly introduced in Part I. I will extend upon Fujimoto's (2005) *nagara* mobilism of *keitai* cultures – as part of a genealogy of luxury good cultures, *shikôkin* – later in the chapter.

In light of the preceding observations, discourses around *keitai* cultures would seem to be but another example of McVeigh's 'nostalgia nationalism', particularly as they rehearse and rely on a type of essentialising of social and cultural norms into historical tropes (despite arguing for a social constructivist view). The role of gender is central here as McVeigh argues that the government deployment of *ryôsai kembo* rhetoric serves to perpetuate gendered stereotypes about agency and labour division under the rubric of 'gendered nationalism'. Matsuda's (2007) perspicuous work highlights the relationship between mothers and children, whereby the increasing demand for children of younger age groups to have a *keitai* can be seen as the result of a carefully orchestrated construction of the 'aura of crime'.

Matsuda argues that despite the fact that there has been no discernible increase in crime rates, the 'aura of crime' has compelled mothers to be perpetually 'on call' for their children. For generations, Japanese children have partaken in a routine premature adulthood whereby they travel unattended to various after-school study classes until 10pm. Now, however, the *keitai* – what Matsuda calls 'mom in the pocket' – accompanies children everywhere. In this case, rather than freeing up the mother, the *keitai* further enslaves the female to be on constant call, thus re-enacting McVeigh's 'gendered nationalisms' through insidious 'good wives and wise mothers' ideologies.

So the fact that *keitai* cultures partake in forms of Japanese nationalism is not in question. *Keitai* culture is clearly mired by nationalist discourses around individualism, nostalgia and gender. The main question, therefore, concerns the relationship between the images of empowered female users and the actuality. According to McVeigh's analysis of the *kôgyaru*, females are adopting subversive styles in public that are, once at home, dropped for performing 'good wives and wise mothers' (McVeigh 2003). Is the female agency embodied by 'consumutopia' (2000) just another version of nostalgic nationalism for/of the 'ideal woman' (McVeigh 1996a)? If the power of the cute is 'a key symbol of Japanese society' (1996b: 293) that reflects the 'social world' and 'via communication . . . constructs gendered relations' (ibid.), what picture does *kawaii* mobile media tell us not only of Japanese women – and associated forms of femininity and hyperfemininity – but also of Japanese nationalism both within Japan and the region?

In light of these discussions, the aim of this chapter is to revisit some of these gendered nationalism tropes, particularly within the context of *keitai* cultures. Having outlined the gendered formation, this chapter will consider what contemporary modes of women – both actual and idealised – are circulating in everyday life. I will explore the conflation between *kawaii keitai* cultures, and how this reflects localised modes of customisation. How does *kawaii* culture, a sphere embedded within nostalgic and gendered nationalism, transform as it shifts into new media spaces? And what does this impart about gendered performativity and agency?

As a prominent example of the rise of subversive female subcultural languages in the 1970s (Kinsella 1995), *kawaii* culture in the form of kitten writing has been linked to female and feminine cultures in Japan, particularly the customising of domestic technologies in the form that McVeigh characterised as the 'techno-cute' (2000). 'Techno-cute' entails 'making-friendly' or warming the coldness of new technology. Conspicuous examples include Hello Kitty and *Pokémon*, while subtler examples can be found in

emerging forms of *kaomoji* (emotions expressed through 'face marks' such as
^_^) (Katsuno & Yano 2002).

As the *kawaii* migrates across cartographies of new technology, and
especially as it migrates from hardware to software customisation, it takes on
new forms of representation. An example is one of the first virtual email pets,
PostPet, a *kawaii* pink bear that had a tendency to go awry and write emails
without the consent of the user. Moreover, as these new virtual spaces
afforded multiple types of performativity, the *kawaii* could take on diverse
'avatar' modes. In the case of mobile media, *kawaii* customisation can be seen
as part of the burgeoning grammars of emotions that try to humanise these
spaces.

This phenomenon of a *kawaii* 'sociology of emotions' can be seen as part
of what I define using the portmanteau 'emotology'. By drawing from ethnographic work conducted with actual users, this chapter will attempt to ground
the foregoing arguments from a perspective of everyday life as a site that, as
Dirlik argues, helps to provide opportunities for overcoming the increasingly
ethnic and national divisions in modern life (1999a: 167). Given that emotologies such as *kawaii* cultures are embedded in gendered practices of intimacy
and communication, how empowering are they? Do they fuel the same
debates about a commodification of gendered modes of users? Are users,
through the ever-evolving language of emotology, developing new forms of
gendered representation, identity (individual and community), self-expression
and modes of intimacy?

In this chapter I begin with the history of the rise of mobile technologies
from the pager, PHS (Personal Handy-phone System, a hybrid of the 'personal digital assistant' [PDA] and *keitai*) to the *keitai*, and how this has been
theorised by researchers. I consider the transformation of Japan's economic
and technological prowess into burgeoning forms of transnational cultural
capital through the poignant symbol and practice of the *keitai*. Through
the lens of *keitai* cultures, I consider Japan's changing social and political
ecosystem both internally and regionally.

In particular, I focus on the fundamental role of customisation in the
domestication of mobile technologies, and how through the distinct blend
of emotology that is techno-cute, *kawaii* cultures have been essential in the
success of adapting the *keitai* from a business tool to a socio-technology. I
chart the gendered genealogy of *kawaii* cultures that coincided with the rise
of domestic technologies from the 1970s, and then reflect on how these gendered discourses were transcribed into *keitai* practices (micro) and cultures
(macro).

This contextualisation of *keitai* cultures in the history of Japanese customisation, particularly *kawaii* cultures, will be followed by a discussion of
my follow-up ethnographic case studies, conducted after my initial fieldwork
between 2000 and 2002 (Hjorth 2003a, 2003b), in two different fieldtrips
in July 2004 and December 2005. Whilst my initial case fieldwork did just
focus upon women and especially *shôjo* and *kôgyaru*, my following case

studies aimed to understand the various degrees and modes of femininity and affective labour between the genders. In this case study I explore respondents' attitudes to customisation, and how these practices reflect types of emotional discourses around co-presence, intimacy and 'communities of feelings' (Hochschild 1983). Through these discussions, I consider how the particular emotological logic of *keitai* customisation is imbued upon both an individual and social level, and how this, in turn, reproduces types of gendered performativity.

Moreover, given that most Japanese people access the Internet via the *keitai*, I discuss one of the most dominant SNSs in Japan, mixi. In these spaces, I explore how gendered performativity reflects gendered consumption and production modes of new mobile media. Will the Japanese young female always be relegated to the role of consumer? How much of the phenomenon is orchestrated on behalf of governmental and corporate IT policies in an attempt to relocate Japan as the centre for mobile technologies and symbolically, as a cultural index for re-orientating post-modernity in the region?

Figure 4.1 A *keitai* sales girl in 'electric city', Akihabara, exemplifies the uniformality that McVeigh discusses as part of Japanese national culture. Photo: Hjorth 2004.

Genealogies of mobility: one history of the *keitai* in Japan

> In Japan, thumbs get even more exercise: games are played with the thumbs of
> two hands; messages and calls are made with one or both. Tokyo's *keitai* kids are
> known as *oya yubi sedai*, or the thumb generation: 'It's not only on the *keitai* that
> they use them,' says one man in his early 20s, to whom today's teenagers are
> already remote and alien creatures: 'they even point at things and ring doorbells
> with their thumbs'. These kids are the world's leading textperts.
>
> (Plant 2002: n.p.)

With three major service providers – NTT DoCoMo [i-mode] the largest
with 54 per cent of the market (with over 53 million subscribers), upcoming
challenger KDDI (au and TU-KA) the second largest with 29 per cent of
the market (over 29 million subscribers) and SoftBank (Yahoo! owned,
replacing Vodafone) with 17 million subscribers – and two smaller companies
WILLCOM (number one for PHS) and Emobile (part of Eaccess, *keitai* with
broadband), Japan symbolises a place fully immersed in 3G mobile technologies.[1] Although DoCoMo has dominated the market over the last seven
years, making it synonymous with Japan's *keitai* culture globally, more
recently its reign of supremacy has been challenged by the cheaper KDDI
(au and TU-KA). While au offers cheaper rates than DoCoMo, TU-KA
services the niche market of disabled and elderly users. With the growing
aging population, TU-KA's market is undoubtedly burgeoning.

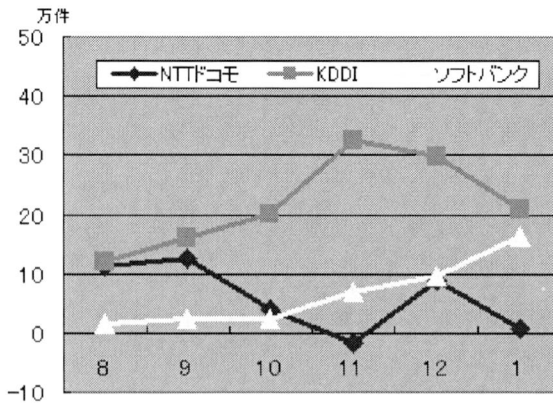

Figure 4.2 A 2007 graph from CNET Japan featuring the three major telecommunication companies, DoCoMo (blue), KDDI (pink) and SoftBank
(yellow). The vertical axis lists numbers of subscribers in millions,
while the horizontal axis lists years from 2008 to the projected 2013.
See: http://japan.cnet.com/blog/comm25/2007/02/15/post_dffa/.

Much has been written about the legacy of DoCoMo as pioneering one the first examples of the mobile Internet globally. By 1999, the streets of Tokyo were graced by *keitai* users accessing the Internet. DoCoMo was the first to exploit the full convergent potential and cross-platform multiplicity of mobile media by implementing the 'mobile wallets' service – phones that have the capacity for electronic credit card-like transactions – as well as Cmode (DoCoMo's collaboration with Itochu Corporation and Coca-Cola Japan), which allows mobile phones to communicate with vending machines. As Beck and Wade note, 'Club Cmode allows users/members to buy tickets and purchase ring tones . . . Cmode units, scattered across the city, are equipped with a printer, sensor, speaker, and are connected to i-mode' (Beck & Wade 2003: 217). Comprehensive publications such as *Personal, Portable, Pedestrian* (Ito et al. 2005) – which are dedicated to the history and contemporary formations of *keitai* cultures and practices in Japan and which allow many non-Japanese speakers access to research previously only available in Japanese – highlight the importance of Japan in the rise of mobile technologies.

One might interpret a snapshot of the rise of the *keitai* in Japan as synonymous with the highly visible group of users, the young female (*shôjo*). However, it was the *kôgyaru* (the urban trendy female in her twenties) who gained notoriety in the late 1990s through her often sexually subversive use of the PHS – and later the *keitai* – to extract capital from the traditionally dominant group, the *oyaji*. Commonly depicted as wearing platformed boots, bleached hair and darkened skin, the *kôgyaru* actively undermined notions of the Japanese female as passive and subjugated by Japanese men. Overtly contrasting any notion of femininity as passive, the *kôgyaru* openly challenged definitions of femininity from their dress to their 'unfeminine' and aggressive language and gestures.

According to Laura Miller (2006), the *shôjo* and the *kôgyaru* represented two very different forms of femininity. Although both deployed *kawaii* and *keitai* cultures in conspicuous ways, the *shôjo* and the *kôgyaru* demonstrated multiple possibilities for femininity, hyperfeminisation and mobile gender performativity. For Miller, *kôgyaru* are 'kogals' and they represent 'a new girl subculture . . . a vehicle for mainstream outrage at the economic and cultural power of youth, especially the subcultural compositions of young women' (ibid.: 30). In particular, the kogal and their various subgroups such as *ganguro* and *yamamba* partake in a 'new aesthetic of the non-cute, or cute infused with an ironic twist' (ibid.: 31). I would like to take Miller's argument one step further, the kogal not only revises *kawaii* and *keitai* cultures on both material and symbolic levels but, in turn, provides novel feminised cartographies of personalisation. The kogal transforms the role of *kawaii* and femininity, adapting it into the realm of irony (ibid.) and playful post-modernism (Allison 2003).

In their open disavowal of Japanese femininity and traditionalism, along with their conspicuous deployment of mobile technologies, the *kôgyaru* was not just subversive but a represented threat to the status quo. It was the way in which the *kôgyaru* usage of the *keitai* reflected her general undermining of Japanese traditional culture – and particularly the role of the female – that signalled a sharp departure from the *keitai shôjo* (symbolising the female high-school pagers revolution). Here the use of the distinctive uniforms by the *shôjo* and *kôgyaru* marked two different roads towards performing femininity – and mobile capital – in Japan. One reproduced traditional forms of femininity (*shôjo*), the other (*kôgyaru*) openly rejected Japanese traditional culture by assertively appropriating non-Japanese forms of fashion (uniform), gestures and language.

Concomitant to this *kôgyaru* phenomenon was the development of *enjo kosai* (subsidised dating with men to maintain their high consumer needs) that rested on technology such as *purikura* (photo stickers or print club) to further her consumer desires. Since then such subcultures as *purikura* have continued to grow, expand and actively participate in the emotology of *kawaii keitai* cultures (Okabe, Ito, Shimizu & Chipchase 2008). According to Miller (2005), the various forms of *purikura* – from *hengao* (funny faces), *kimo-puri* (gross *purikura*), and *yaba-puri* (dangerous *purikura*) – exemplify emerging forms of gender parody within contemporary Japanese girl cultures. However, when the *enjo kosai* mixed with the *kôgyaru* fashion and attitude, the result was an overwhelmingly negative public image of young women out of control. This phenomenon, which came to symbolise the 'youth problem' identified by Matsuda (2005), overshadowed usage by other female user groups and built a very negative image of the typical female *keitai* user.

However, as Matsuda (2005) notes in her analysis of the rise of i-mode in Japan and its exported images of success globally, the negative image of *kôgyaru* was soon replaced by the image of happy and friendly *shôjo*. The uptake of mobile technologies by youth cultures, rather than *oyaji*, initially attracted much criticism, of which the *kôgyaru* bore the brunt. If Japan was to sell its image of being the pioneer of the mobile Internet, it had to change the associated images both locally and globally. That it did so successfully is evidenced by its being lauded globally as a site for the convergent practice of the '*keitai* IT revolution' (Matsuda 2005).

It is impossible to trace the history of mobile technologies in Japan without engaging with the subversive and transgressive usage of the technology by female users (*kôgyaru* and others) and their dominant mode of customising, *kawaii*. The popularity of *kawaii* customisation has seen its migration and translation throughout the region, most notably in locations such as Hong Kong and Taiwan, where Japanese popular culture wielded much cultural capital (Hjorth 2005a). As Ito notes, the history of the *keitai* needs to be understood in terms of its cultural and social dimensions as a socio-cultural object (2005: 1). The history of mobile technologies is one involving the contestation of many identities from class to gender. Hence, in order to situate the role of gendered customisation of the *keitai*, we need to gain a sense of how consumption has operated to extend notions of modernity, socialisation and nationalism within Japan that pre-dated the introduction of Western-style modernity. As noted in Chapter 3, the significance of accessories can be linked to the sumptuary laws in pre-modern Japan, which amongst other things set rules for clothing and hairstyles.

McVeigh has extended this discussion in his study on the uniform's symbolic role in Japanese culture and history, particularly in relation to Japanese women. In McVeigh's 'Cultivating "femininity" and "internationalism"': rituals and routine at a Japanese women's junior college' (1996a) he discusses how Japanese notions of femininity and internationalism are linked with the role of uniforms and accessories in everyday life. He elaborates further in his 1997 follow-up study, 'Wearing ideology: how uniforms discipline minds and bodies in Japan'.

As aforementioned, McVeigh's work connects with Fujimoto's (2005) analysis of *shikôkin* (meaning 'recreational consumer products') such as tea ceremonies, and the symbolic role of customisation as an important component of contextualisation and socialisation in Japan. Here consumption operates as part of everyday rituals and is linked to social status. In 'The third-stage paradigm: territory machine from the girls' pager revolution to mobile aesthetics', Fujimoto evokes Williams's 'mobile privatization' in the form of the distinctly Japanese practice of *nagara* mobilism (2005: 80). For Fujimoto, the 'third-stage paradigm' theory is 'a hypothetical model for mobilizing historical facts for theory building, and is one thread for understanding the backdrop to the girl's pager revolution' (ibid.: 80). In articulating the paradigm in terms of the East Asian context, Fujimoto argues that the increase in *keitai* cultures and new identities must be understood in terms of broader techno-cultural shifts that were symbolised by the shift from military (soldier) to business (salaryman) to socialisation (youth). This phenomenon is, as Fujimoto argues, distinctive to Japan and cannot be compared with European shifts.

Nagara mobilism is re-enacted in the visibility and agency by *kôgyaru* and *shôjo* users against the hegemony of the *oyaji* (ibid.: 80). According to Fujimoto, through the *nagara* mobilism of *keitai* cultures, the conflict between traditional Japan and modern Japan was orchestrated by the shift of attention away from the *oyaji* towards the *kôgyaru*. However, for Fujimoto, this contestation was far from nascent. Rather, Fujimoto argues that this tension between 'two' Japans – the traditional and the contemporary – has its genealogy in the rise of Japan's exportation that can be linked back to the mid-eighteenth century with the mobilisation of trade routes.

Fujimoto also argues that the export of *keitai* culture can be linked to the export of other luxury goods, *shikôhin*. Fujimoto utilises many analogies – from the use of uniform to the use of tea – to differentiate the Japanese *keitai* from Western adaptations of mobile media. The importance of subcultural reappropriation of *keitai* cultures is pivotal to Fujimoto's discussion of Japan's dominance as an global exporter of both hardware and software –

including i-mode, *kawaii* icons, *keitai* straps, wallpapers and ring tones, as well as digital font styles capable of 'unconventional combination[s] of existing characters and symbols named *gyaru-moji* (girl's alphabet) and *heta-moji* (awkward alphabet)', so fundamental to *gyaru's* reappropriation of mobile media (ibid.: 87).

For Fujimoto, the distinctively Japanese use of the *keitai* as more than a mere technology has ensured its export overseas (ibid.: 87). Fujimoto extends the aesthetic paradigm of the *keitai* by invoking the analogy of traditional tea practices and *shikôhin* as foregrounding the rituals and relationship between the *keitai* and the everyday. As Fujimoto notes, just as the *shikôhin* were never seen as mere luxury goods – or as he puts it, refreshing favourites – but were rather embedded in rituals, so too is the *keitai* part of the rituals of the everyday in which customisation plays a certain role.

Drawing on the work of historian Sakae Tsunoyama, who traced the history of tea in the aesthetics of Japanese everyday life and then its export to the west, Fujimoto parallels this genealogy with the consumption and exportation of the *keitai* culture. 'The expanded palette of *shikôhin*' (ibid.: 90) was expanded through eighteenth century exports and souvenirs which, according to Fujimoto, has seen the '*keitai's* replacing *shikôhin* as media' (ibid.: 90). Fujimoto suggests that rather than seeing *keitai* cultures as a distinct break in Japanese traditional cultures – as symbolised by the notion of *shikôhin* – he argues that the emerging phenomenon of *keitai* cultures (in which customisation is intrinsic) could be viewed as *keitai shikôhin*. As Fujimoto notes:

> It is widely recognized that international products such as tobacco, coffee, and tea have been sophisticated informational and media commodities for the past four hundred years, since the emergence of commodity markets in modern Europe. While taking new forms, these basic products have remained resilient through global shifts towards urbanization and informational flows ... *Keitai* are less like books, which tend to be decontextualised, de-localised, and escapist media, and more like *shikôhin*, as objects of recontextualisation, relocalization, and actual media objects.
>
> (ibid.: 90–91)

For Fujimoto, Japanese customising is, like *shikôhin* objects, about contextualisation. This is exemplified by *keitai* cultures as Fujimoto notes:

> Japan has had a tendency to enjoy *shikôhin* not in isolation but within the totality of associated objects, tools, and media. These objects, tools, and media often exist separate from the *shikôhin* themselves and take on their own independent gadget identities in the form of *kawaii* (little, pretty, cute) stationary, fashion accessories, and character goods.
>
> (ibid.: 91)

The role of *kawaii* customisation of the *keitai* to fulfil the sensory experience can be seen as an extension of the legacy of *shikôhin* in Japanese culture. Thus *kawaii keitai* cultures are far from nascent or contingent but, rather, part of ontological processes particular to Japanese culture. By drawing on forms of *nagara* mobilism, popular trends and *shikôhin* cultures, Fujimoto suggests that an analysis of *keitai* culture demonstrates 'Japan's resistance to cultural globalization as well as Japan's leadership role in certain aspects of globalization' (ibid.: 92–93).

As Fujimoto persuasively argues, the rise of *keitai* cultures in Japan must be contextualised in terms of Japanese modernity and nationalism that precedes the introduction of Western modernity and links to a time in which Japan participated in the regional trade routes and attendant forms of transnationalism (Arrighi et al. 2003). Here we see that the 'accessory vs. necessity' debate that has plagued much of fashion – and particularly mobile phone customisation – highlights some significantly localised processes that are linked into broader formations of modernity and nationalism.

In the case of the introduction of mobile technologies in Japan, we can see that the rise of these technologies very much reflected major socio-cultural and economic shifts in post-industrial Japan post-World War II. The first wireless telephones in Japan date back to 1953 when the technology was deployed on ships. The role of the device as representative of a formalised group – through the *oyaji* – continued through the introduction of NTT's reportedly 'first car phone' in 1979,[2] then the 'shoulder phone' in 1985, and then the world's first handheld *keitai* in 1987 by NTT (Okada 2005: 42). For Okada, both the *keitai* and the pager began as representative of groups – particularly in the form of companies or organisations – whereby *keitai* and pagers were shared and symbolic of organisations rather than associated with individuals. The history of the *keitai* is one marked by the change from business or organisation to the individual.

As Ito notes (citing Matsuda 1999a), 'throughout the 1990s, during the adoption of mobile media, uses of the pager were extended into individual and personal purposes. This trend can be described as personalization' (2005b: 43). Okada notes that this shift parallels the history of landlines in Japan and identifies the 'dimensions of personalization also related to the individualization of television, radio, and other forms of mass media since the 1970s' (2005: 46). Here the collation between domestic technologies and the unilateral domestication between the user and the device are clearly integral to personalisation and individualisation (McVeigh 2003). This process occurs across the various platforms and genres of the *keitai* – namely visuality as in camera phones (Ito & Okabe 2003, 2005a), texting (Okada 2005) and aurality as in ring tones (Okada 2005).

For Okada, the shift in *keitai* cultures towards multimedia mobile devices is due, in part, to the active reappropriation and bricolage of the users (2005: 47). Acts of customisation, in which users perform modes of self-expression and individualism highlight (in keeping with McVeigh's notions of Japanese nationalism) the integral role individualism plays in national culture. In this light, one could argue that users forced telecommunication companies/industry into 3G convergences, not the reverse. In sum, this is indicative of a bubble-up, rather than a trickle-down, model of consumption, whereby users dominate the choice of adopted applications, rather than industry determining what 'killer applications' will survive. As Okada notes, personalisation was a key characteristic in users' adaptation of pagers and *keitai* both inside and outside the device. The remediated nature of the devices means that parallels formed, as Okada observes:

> . . . pager texting practices were the source of many of these patterns in *keitai* e-mail usage. The technical forerunner of *keitai* text-messaging functions was the numeric display on the pager, a service introduced in 1987 by NTT.
>
> (2005: 51)

The usage of *poke-kotoba* ('pocket words') to deploy numbers such as 0840 meaning *ohayo* (good morning)[3] saw a type of playful vernacular accompanying the user's appropriation of text in *keitai* emails. Along with textual and visual bricolage, Okada observes the importance of ring tones (*chaku-mero*) in 'the multimedia capabilities of *keitai*', a development that began in 1997 for the PHS service by Astel group (ibid.: 54). The multimedia elements of *keitai* culture occurred across various levels from texual, visual and aural features both outside and inside the device. As Okada notes, features such as ring tones and *keitai* cameras have been instrumental in extending modes of personalisation established by discourses around youth cultures utilising pagers.

In July 1999, Kyocera launched the first 'visual phone' in the form of the PHS VP-210 (Okada 2005: 56). The first *keitai* digital camera was launched by J-Phone, along with a service called *sha-mênu* (photo [*shashin*] mail). Okada argues that the development of the camera *keitai* was influenced by the introduction, in July 1995, of photo booths that made personalised stickers. Called the Print Club (*purikura*), this phenomenon cultivated a burgeoning subculture that developed around the archiving and collecting of the digital. As Okada notes, those who are avid users of mobile communications are also avid *purikura* collectors:

> Through *keitai* adoption, youth tried to express their individuality by putting *puri-kura* stickers on their handsets or adorning them with unique accessories and straps. The colour display promoted the practice of using favourite illustrations or photographs as *keitai* wallpaper. With the built-in camera, all these functions were incorporated in a single *keitai* terminal. In other words, the *keitai* camera has come to encompass the production of customized wallpaper and some of the functions of *puri-kura* stickers.
>
> (Okada 2005: 59)

As Keniji Kohiyama notes in 'A decade in the development of mobile communications in Japan (1993–2002)', the importance of personalisation as the central trend has dictated future trends by focussing on the personalisation of devices, connectivity and services (2005: 70). For Kohiyama, 'the personalization of mobile communication in Japan began with the pager' (ibid.: 71). One of the dominant forms of personalisation – in the form of user customisation – can be traced through the evolving socialising and individualising techniques of techno-cute from the pager to the *keitai*. Both were originally marketed at the *oyaji*, and then appropriated by the female user and inscribed with *kawaii* customisation, in the form of cute characters adorning and dressing up the device with 'kitten writing' and heavy usage of *emoji* [i.e. emoticons] (Kinsella 1995).

The early 1990s marked the beginnings of youth mobile messaging in the form of pagers (Ito & Okabe 2005d: 260), and it was a time in which *keitai* subscription surpassed pager subscriptions (Ito & Okabe 2005d: 213). Although 1997 saw the introduction by all carriers of SMS services, SMS could only be sent to users of the same carrier. In 1995, the first 100cc mobile phone was introduced, known as the PHS. April 1996 saw the first *keitai* text message service piloted by DDI with J-Phone (now Vodafone) introducing it in November 1997. By 1999 the '*keitai* IT revolution' (Matsuda 2005) was launched in the form of DoCoMo's i-mode, and the world was again enamoured by Japanese techno-soft cool, which looked a lot like the future for twenty-first century mobile media.

For Harmeet Sawhney, the industry and techno-nationalist rhetoric associated with the '*keitai* IT revolution' needs to be analysed. In 'Mobile communication: new technologies and old archetypes', Sawhney compares the i-mode systems with Minitel – the 'cynosure of the videotex era' (2004) or, put bluntly, a video format that failed dismally. For Sawhney, the i-mode ecology reflects the 'controlled-closed' system of French Minitel rather than the 'uncontrolled-open' architecture of the Internet. Sawhney thinks it wise to pause and reconsider the deluge of mobile technologies in terms of remediation; that is, the understanding that 'new' technologies are intertwined into a genealogy that draws on older technologies. As Sawhney notes 'i-mode has recently attracted much attention as a major breakthrough technology. It is invariably discussed in reference to the Internet since it is construed to be the first successful wireless Internet' (ibid.: 11).

From its launch in February 1999 i-mode went from five million subscribers in 2000 to 20 million in 2001 and 35 million in 2003 (NTT DoCoMo 2003). One of i-mode's main successes was its innovative technological and business ecology, whereby 'official' sites charge a few cents per transaction or via monthly subscriptions of about five dollars. This meant that content providers were paid, with DoCoMo keeping 9 per cent for every fee charge. As Sawhney notes, unlike the business-orientated European model of 3G, DoCoMo 'targeted mass youth market for entertainment services with spectacular success' (2004: 11). DoCoMo innovations included the 'always on' capability and charging 'bit-by-bit' (of downloaded information) rather than by time used. Its success in Japan was ensured by the lack of Japan's experience – and thus expectation – of the Internet. i-mode's closed and controlled network between users and content – defined as 'proprietary Internet' – provided a 'walled' version of the Internet. As Sawhney attests, 'one is led to wonder whether a "proprietary Internet" is an oxymoron' (ibid.: 12).

However, counterbalancing the controlled-closed system of i-mode's official sites is the subversive potential of the world of the unofficial. According to Ratliff, in 2002 i-mode had 2,000 official sites outweighed by 50,000 unofficial sites (Ratliff cited in Sawhney). The unofficial sites do not appear on the i-mode menu or go under DoCoMo's model for billing. It is in this world of unofficial sites that Sawhney sees the possibility for i-mode to be comparable to the Internet, asserting that perhaps it is 'in DoCoMo's interest to let go control over i-mode and let it blossom into a true wireless Internet' (2004: 13). But until that point, it seems DoCoMo's appropriation of the Japanese word *dokomo* (meaning 'everywhere') is a misnomer, because instead of possessing the open and ubiquitous architecture of the Internet (i.e. rhizomatic quality), it exemplifies corporate control over information and content services. As Sawhney warns:

> The success of i-mode has been celebrated as the birth of mobile Internet. It is widely seen as a transforming development that will greatly expand the reach of Internet and also make it more accessible to users in course of their day-to-day movements. In this euphoria, i-mode has been accepted, without much critical examination, as the latest avatar of Internet ... however (I) suggest that i-mode is actually an avatar of Minitel. The parallels between them are indeed striking.
>
> (ibid.: 14)

If DoCoMo were a person – and the book *DoCoMo: Japan's Wireless Tsunami* certainly tries to personify it by ascribing terms like 'passion' and 'love' – then *DoCoMo: Japan's Wireless Tsunami* would be a hagiography. In Chapter 1, entitled 'love', John Beck and Mitchell Wade explain the success of DoCoMo: 'With i-mode, DoCoMo made commerce on the mobile Internet compelling – so compelling that it is fast becoming universal throughout Japan' (2003: 1). The branding mechanism of DoCoMo is much like what it sells – it doesn't just sell technology, it sells content and lifestyle: in short, marking one of the first transitions globally into mobile media.

DoCoMo succeeded because, according to Beck and Wade, 'it tapped into the power of feelings' (ibid.: xiii). Here we see how companies thought it important to understand what Ito and Okabe (2005d) identify as the 'techno-social' elements. Mobile technologies are much more about sociality, intimacy and locality than they are about functional technology. Beck and Wade's discussion of i-mode in terms of emotion is key to the logic of Japanese customisation. The success of *kawaii* customisation has been its humanising of new technologies, as encompassed by McVeigh's techno-cute notion. In the case of DoCoMo, Beck and Wade argue, 'In the end, what sets DoCoMo apart is *passion*' (ibid.: xi).

Beck and Wade's book is an extended verbal advertisement for DoCoMo. It also highlights the pervasiveness of personalisation and customisation techniques as part of a post-industrial tool to deflect the decreasing role of humans in the process of production and manufacturing (Lupton 1994). Far from being a robot corporation, DoCoMo seems to be selling a type of neo-humanism. Just as users make 'warm' new technology, DoCoMo has tapped into the proclivity of users towards emotional customisation.

In other words, DoCoMo is trying to speak the diverse languages of the user, appropriating a bubble-up methodology to camouflage the trickle-down, controlled architecture of their version of *keitai* with Internet. However, in contrast to the global dissemination of Japanese customisation such as *Pokémon* and other *kawaii* products, i-mode's uptake was patchy – reflecting the diverse ways in which the relationship between mobile phones and the Internet is viewed in different contexts.[4] One way to make sense of the rise of *keitai* cultures is undoubtedly through the pivotal role played by *kawaii* cultures in the rise of personal technologies epitomised by mobile media.

Figure 4.3 Pedestrians with their *keitai* whilst waiting for the train. This picture vividly illustrates Fujimoto's *nagara* mobilism. Photo: Hjorth 2004.

Soft-ware and soft-where: the history of *kawaii* aesthetics and its migration into *keitai* cultures

> *Keitai* culture, together with high school girls' culture, *anime* culture, and character culture, has spread among the youth of Seoul, Taipei, Shanghai, and Bangkok. Western media are also predicting its penetration into the United States and Europe, invoking a version of 'yellow peril' theory or Orientalist prejudice.
>
> (Fujimoto 2005: 87)

The genealogies of *kawaii* customisation of technologies are, as conceptualised under McVeigh's rubric of 'techno-cute', disjunctive and contradictory (2000). At its core, techno-cute is about mixing two opposing cultures – one cold and unfamiliar, the other warm and friendly. Examples of the migration and adaptation of *kawaii keitai* culture have been noted in the region (Hjorth 2005a), in which fluctuating anxieties of Japanisation have been noted in places such as Taipei (Ko 2003). As noted in Fujimoto's argument, the genealogy of *kawaii*'s performative elements and ritualistic qualities overlaps with the changing role of mobile net telephony in the binding of social ritual in contemporary culture. But beyond the rhetoric about Japanese cultural capital and its formations in the region, what exactly is *kawaii* culture?

According to Kinsella's groundbreaking research, *kawaii* culture arose as a youth subculture in the 1970s as a means of self-expression and rearticulation, and as a reaction to the overarching traditions that were perceived as oppressive. Young adults preferred to stay childlike rather than join the ranks of the corrupt adults (Kinsella 1995). This phenomenon highlighted the way in which 'childhood' as a construct is conceived and practiced in locations such as Japan, with its premature adulthood, in contrast to the West (Ariés 1962; White 1993). Practices such as 'kitten writing' are examples of youths subverting Japanese concepts by intentionally misspelling words in acts of political neologism (Kinsella 1995). Kitten writing can be seen as earlier examples of *emoji* before it was institutionalised by industry as part of built-in *keitai* customisation. The *kawaii*, while stereotyped as a young female's preoccupation, and thus associated as female, was seen as traditionally asexual – that is, a gender without sex. Like the typical consumer, the *shôjo*, the *kawaii* was a female without sexual agency in a society where the *oyaji* was the national symbol post World War II.

As John Whittier Treat perspicuously notes, the *shôjo* signified a sexually neutral, consumption-focused female (1996). However, as *kawaii* culture married *keitai* scapes and was forged into virtual spaces, the *kawaii's* gender-without-sexual-identity took on new characteristics (Hjorth 2003a). Most notably, the *kôgyaru* deployed ironic appropriations of *kawaii* to infuse the gendered commodity with sexual connotations, thus transforming the *kawaii* into a gender *with* sexuality. This was predominantly enacted through *kawaii* customisation, whereby the cute was no longer deployed in an asexual manner.

Kawaii culture draws from the Japanese tradition for gift-giving, as well as providing a means to overcome the Japanese proclivity for shyness in social interactions (Kusahara 2001). The gift-giving genealogy is pertinent in *kawaii*'s translation into mobile telephony, highlighting and facilitating the gift-giving cartography of mobile telephonic social rituals and symbolic exchange (Taylor & Harper 2002, 2005). *Kawaii* culture's role in customisation articulates a type of social glue to the co-present online space of the *keitai*. It reminds users of the role of subjectivity in technological spaces; at its core, *kawaii* customisation domesticates the domestic technology. It transforms the technology into a socio-technology, bringing the role of the socio-cultural to the forefront of the technology.

Although both *kawaii* and mobile phone cultures are pervasive and ubiquitous, the studies addressing the overlaps are thin on the ground. In 'Individualization, individuality, interiority, and the Internet' (2003), McVeigh explores the usage of the Internet through mobile telephony in Japan. Conducting case studies based on interviews with Japanese university students, McVeigh focuses on what he argues is the dominating 'cyberstructure' in which the role of 'individualization' is central to this logic and practice. McVeigh's argument regarding 'individualization' both extends upon what Kogawa characterised as Japan's proclivity towards 'electronic individualism' (1984) and foreshadows Castells et al.'s argument that 'individualism, rather than mobility, becomes the defining trend of the mobile society' (2007: 251). McVeigh argues that mobile Internet technologies are indeed leading to a shrinking of social capital in which users only contact a small range of already existing friends – a phenomenon Habuchi Ichiyo has defined as 'telecocooning', whereby new technologies shrink, rather than extend, existing offline social capital (Habuchi 2005).

With Japan boasting the 'world's biggest Net-linking mobile phone market', the Internet and its attendant modes of spatiality and 'customisation' practices are indivisible from mobile telephony in that country (McVeigh 2003: 20–21). McVeigh asserts that the physical characteristics of mobile phones with Internet capacities – being small and 'handy' – have a lot to do with their overwhelming popularity. Three words that were most commonly used to describe mobile media practices were *ureshii* (happy), *tanoshii* (enjoyable), and *benri* (convenient). McVeigh's case study of university students suggests that mobile technologies are used to maintain existing relationships while jeopardising the possibility for emerging forms of social capital.

The growth in individualisation practices are discussed by Hirofumi Katsuno and Christine Yano in their study on the migrating genealogy of the *kaomoji* from early Internet usage, to pagers, and then to *keitai* (2002). Face marks are most commonly known in the form of 'smileys'; however it is interesting to note that while, in English, the smileys are vertical (i.e. :)) in Japanese they are horizontal (i.e. ^_^). With the proliferation of personalising modes of signification through mobile e-mail, photo e-mail and SNS, the growth in these forms of what Kinsella described as kitten-writing (1995) are becoming a leading form of techno-cute customisation of technological spaces. Katsuno and Yano argue that the *kaomoji* is not a form of post-humanism but rather neo-humanism; it domesticates and customises the technological spaces, reminding us that technologies are shaped by cultural and social processes. At the heart of *kawaii* customisation is its capacity to link the old and the new, the traditional and the contemporary, the individual and the collective in diverse ways. In this way, the *kawaii* undermines the Japan-as-robot rhetoric embodied within techno-orientalism; it is not about post-humanism but rather a clear demonstration of neo-humanist tropes.

One of the enduring features of Japanese customisation is its role in creating emotion and affect as a type of warmness in the coolness of technological spaces. For Tomoyuki Okada (2005) this is characterised by the importance of *yasashisa* ('kindness'); for Anne Allison it is *yasashii* ('gentleness') that makes Japanese character culture so easily consumed both within and outside Japan. In 'Portable monsters and commodity cuteness; *Pokémon* as Japan's new global power', Allison outlines the phenomenon of *Pokémon* both in Japan and US. In doing so, she not only helps revise the role of cuteness as a Japanese form of global commodity, but also explains how this allows us to see Japan's changing role on the global economy (2003). For Allison, cuteness isn't necessarily linked to the cliché of girlishness and the 'feminine'. It is – as she finds in interviews with both Japanese and US youth consumers of Japanese commodities such as *Pokémon* – deeply embedded in the idea of *yasashii* or gentleness. It is this concept of *yasashii* that Allison argues is 'precisely the word Japanese producers used to describe the marketing of *Pokémon* in Japan' (ibid.: 385).

The importance of customising and personalising the technology continued as the *kawaii* expanded into a broader form of aesthetics, provoking Allison (ibid.) to define the *kawaii* aesthetic as 'post-modern' (that is, culturally relative and contesting interpretations) by way of its underlying mode of *yasashisa* ('gentleness') in techno-cute practices. Koichi Iwabuchi sees phenomena such as *Pokémon* as selling a global image of Japanese technology as 'odourless' (ibid.). Iwabuchi's notion of odourless – as a type of flavour of Japanese cultural products at the level of global consumption – is both evocative and provocative, echoing as it does the essentialising 'Confucian capitalism' of the region at that time. However, in contrast to Allison's ethnographic focus, Iwabuchi's model focuses on the production and industrial side of the 'circuit' of Japanese 'odourless' culture, and neglects to gain a sense of the ways in which consumers – at grass-roots level – negotiate this odourless culture in everyday life.

Customisation helps to highlight the ironies and paradoxes associated with technologies, forging a new way to make sense of the shifting terrains. As Iwabuchi notes, the dissemination and indigenisation of Japanese popular culture globally 'articulates the universal appeal of Japanese cultural products and the disappearance of any perceptible "Japaneseness" ' (2003: 33). The 'odourlessness' of Japanese animation (*anime*), for instance, is underscored by the characteristics of *mukokuseki* ('statelessness') which Iwabuchi defines as 'the unembedded expression of race, ethnicity, and culture' that ensures that non-Japanese audiences are yearning for 'an animated, race-less and culture-less, virtual version of "Japan"' (ibid.: 33). As Iwabuchi observes in the case of the *Pokémon* global phenomenon:

> It can be argued that the yearning for another culture that is evoked through the consumption of cultural commodities is inevitably a monological illusion. This yearning tends to lack concern for and understanding of the socio-cultural complexity of that in which popular cultural artifacts are produced. This point is even more highlighted by the *mukokuseki* nature of Japanese animation and computer games.
>
> (ibid.: 34)

Although Iwabuchi's example of *Pokémon* addresses the paradoxes of Japanese global cultural production and its dissemination of a 'stateless' and thus imaginary 'Japan', for Allison his theory does little to comprehend the complexity of appropriation and domestication of *Pokémon* by both American and Japanese children. Allison's focus on micro case studies aligns her method for analysing Japanese global consumption with the work of Befu and his analysis of the global dissemination of Japanese people and commodities (2003). Drawing from focus groups and case studies, along with interview material with the designer of *Pokémon*, Tajiri Satoshi, Allison contextualises this growing commodity culture. Framing her arguments in terms of Japanisation and the associated Japan-as-global-power rhetoric, Allison makes a claim for *Pokémon's* fame as part of general global economy shifts whereby a 'decentering of cultural (entertainment) trends once hegemonised by Euroamerica (and particularly the United States)' becomes prevalent (2003: 394–395).

This de-centring is also identified by Fujimoto in this section's opening quote. Fujimoto observes an element of occidental fetishisation of *keitai* cultures that are underpinned by 'fear' that the west could be 'Japanised' by such paraphernalia (2005: 87). Taking *Pokémon* as a symbol of 'cuteness' (*kawairashisa*), Allison analyses how this translates for consumers – both Japanese and also American. Pointing to the history of *kawairashisa* in Japan as a site for the 'imaginary', from its link to traditional Japanese culture to its commodification in the 1970s with the likes of Hello Kitty, Allison's argument that 'in the millennial play product[s] Japan is selling – and using to sell itself – on the popular marketplace of global (kids') culture' links back to this genealogy (2003: 382). According to Allison:

> But with kid hits like *Pokémon*, Japan is becoming recognized for not only its high-tech consumer goods, but also what might be called postmodern play aesthetics. Japan's achievements here signal another important change in its brokering of the globalised landscape of culture/economics. Unlike the policy it has adopted in postwar times of

culturally neutering the goods it sends overseas in order to assure their
marketability, Japan has marketed *Pokémon* as clearly 'Japanese.'
National origins are imprinted rather than effaced here, indicating a shift
away from what Iwabuchi Koichi has called Japan's policy of 'de-
odorizing' the cultural aroma of its exports.

(ibid.: 383)

Here Allison's discussion of the global dissemination of Japanese products
seems to reflect the 'glocal' techniques utilised by the likes of Sony Walkman.
However, her reading of Iwabuchi somewhat underestimates the complexity
encompassed by the odourlessness analogy. This is partly to do with her
deployment of the 'post-modern', a contested notion at the best of times.
Here Takashi Murakami's 'superflat' (2000) can prove illuminating. As
Murakami acknowledges, the 'post-modern' in the West has a different
ontology to the 'post-modern' in Japan, which he argues can be linked to
traditional *nihon-ga* (Japanese narrative painting) of the seventeenth and
eighteenth centuries: an argument for Japanese modernity that parallels
Fujimoto's discussion of *keitai* customisation and *nagara* mobilism as an
extension of *shikôkin*.

For Iwabuchi, the dissemination of Japanese popular culture globally, specifically in relation to *Pokémon*, ensures an erasure of Japanese-ness that is, in
itself, a type of Japanese-ness; a process whereby the *mukokuseki* encapsulates a type of 'un-embedded' culture that is undeniably linked to a virtual
imaginary of Japan. In other words, Iwabuchi's notion of odourlessness is, in
itself, a form of odour. The virtual imaginary is inextricably linked to the
actual, or put in terms of McVeigh's techno-cute, the Western projections of
the techno-orientalist/post-humanist views of Japan have been subsumed
into the Japanese global imaginary, underpinned by neo-humanism (as symbolised by the recombinant role of *kawaii*).

As Allison notes, although many American children consuming *Pokémon*
knew it was Japanese, this had little to do with their satisfaction with the
product. However, their introduction to Japan via *Pokémon* has now led
many interviewees to be interested in Japan and in other 'cool' Japanese
goods. I would argue this is another clear illustration of what Iwabuchi calls
the *mukokuseki* odourlessness of many cute Japanese products; an odour that
is distinctive and yet so 'flexible' and 'gentle' that it can be translated into
different cultural, social and technological contexts. It is this co-present (of
the state and yet stateless, virtual and yet actual) capacity, built into *kawaii*'s
multivalent customisation of mobile net telephony, that makes it a type of
odourless odour. As Allison identifies, 'it is this polymorphous, open-ended,
everyday nature of *Pokémon* that many of its Japanese producers or com-
mentators refer to under the umbrella of "cuteness"' (2003: 384).

Although *kawaii* customisation is still occurring outside the *keitai* in the
form of cute characters hanging off the device, it is the arising forms of
kitten writing in Internet communities such as the Japanese SNS mixi that

illustrate extending grammars of personalisation. As aforementioned, the role of mobile media in Japan as part of the rise of online communities is pivotal and distinctively reflective of Japanese 'electronic individualism' (Kogawa 1984). Thus, in interviewing respondents about their mobile media practices, I also talked to them about their relationship to online communities and SNSs such as 2ch (channel 2 or *ni-channeru*, the second BBS forum in Japan)[5] and mixi.

Unlike 2ch, which consists of anonymous posting and is used by many millions of Japanese to discuss public issues and to source information, mixi is more frequently accessed and maintained with a strong sense of community (rather than information gathering like 2ch). Both sites are accessed predominantly via the *keitai* and usually during the long commuting periods that are part of everyday life in Japan. The power of 2ch as a reference point for millions of Japanese everyday was vividly highlighted by the *Train Man* (*Densha Otoko*)[6] incident of 2004/5 – a true story that later became a series of *manga*[7], a television drama series and a feature length film.

The story begins when an *otaku* was on a train and witnessed a pretty young woman, whom he calls 'Hermés', being harassed by a drunken man. The *otaku* intervened and ended up fighting with the drunk, while Hermés escaped. The *otaku* and Hermés later becaome friends, and the *otaku* soon fell in love with Hermés. As the *otaku* had little experience of being in love, he sought out advice on 2ch – an action that would result in one of the largest

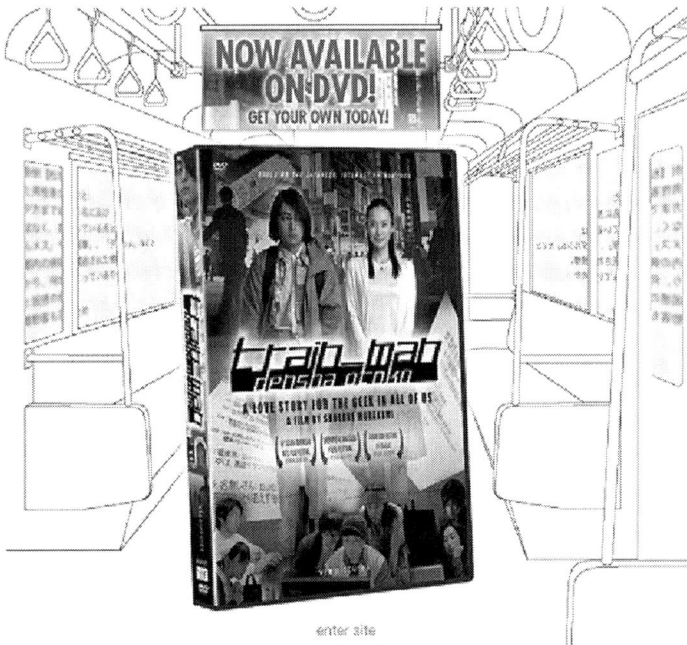

Figure 4.4 DVD of *Train Man* (*Densha Otoko*).

and longest posting threads in 2ch's history. Suddenly, millions of Japanese were giving the *otaku* advice, his life and relationship with Hermés becoming of national significance. Within just over a year later, a popular television drama series (September 2005 and September 2006), a movie (June 2005) and *manga* (October 2006) had been released under the titled *Densha Otoko*.

It was this story that finally transformed and rehabilitated the reputation of the *otaku*. The symbolic home of the *otaku*, Akihabara (electric city), quickly became gentrified, as the *otaku* became one of Japan's unsung heroes. This changing perception of the *otaku*, in comparison to the *kôgyaru*, also highlighted the ways in which gendered performativity in Japan was being reconfigured around new mobile technologies. While the *otaku* represented a man highly obsessed with media – and thus, in the case of *Densha Otoko* a new type of Japanese hero – the *kôgyaru* represented the negative female version. Through these two gendered types we can see the key role technology plays in the construction of their identity in Japanese culture.

Unlike the Western SNS MySpace, where anyone can sign up, mixi requires an introduction from someone who is already a member. This introduction model helps to establish a level of trust between new and established members. Moreover, many users enjoy mixi because it allows friends to connect synchronically and is thus less intrusive than phone calls and email. As Yuji Mori observes in his 2005 study of mixi users, many experience great satisfaction at being able to share stories and information. Mori noted that there were two types of mixi communities – large communities of over 20 members and smaller, more intimate groups. In the quantitative study many respondents noted that they preferred the small, under-20 member communities, where they could focus more on ongoing sharing and dialogue.

By contrast, the over-20 member communities tended to feature more information distribution and exhibitionism. Mixi's use of icons to denote community (and social, cultural and economic capital) associations, along with ASCII and *kaomoji*, were significant forms of customisation for respondents. I will discuss this in further detail in the case study. However, firstly I will contextualise the particular role of the Internet – synonymous with the mobile Internet – in Japan.

Kaomoji and beyond: emerging emotologies in mobile Internet usage

The *keitai* is a locus that integrates Japanese subcultures. Young people use the latest J-Pop (Japanese pop music) for ring tones and Japanimation (Japanese *anime* and *manga* characters) like *Pokémon, Doraemon*, and *Sen/ Chihiro* from the movie *Spirited Away* as motifs for wallpaper and *keitai* straps. A *keitai* is for them more than just a tool – it is something that are highly motivated to animate and to customize as a dreamcatcher, a good luck charm, an alter ego, or a pet. I might even call it an idol or a fetish and regard it as an animistic handy object (*tokko*) stretching a spiritual barricade (*kekkai*) around the body.

(Fujimoto 2005: 87)

Figure 4.5 A *keitai* adorned with a plethora of Hello Kitties. Each Kitty represented a different location in Japan and the respondent spoke about each one, fondly identifying who she was with when she acquired it. In this way, *keitai* customisation extends on objects such as the charm bracelet, in which the owner is reminded of times and places shared by looking at the mementoes. Photo: Hjorth 2004.

As Fujimoto observes, it is impossible to explore the role of the *keitai* without detailing its customising culture. As the mobile technologies became more sophisticated with the rise in mobile Internet, the *kawaii* culture so prevalent outside the *keitai* device begins to be deployed increasingly within the virtual spaces. The usage of *kaomoji, gyaru-moji* and *heta-moji*[8] becomes even more prevalent when mobile Internet is used for SNS sites such as mixi. Before discussing my sample study of respondents I will outline the role of Japanese Internet and, specifically, mixi. While the Internet forum 2ch is the dominant forum used in Japan, mixi is the dominant SNS. Many of my respondents checked 2ch occasionally but identified more strongly with mixi as a site that they accessed, and contributed to, regularly. This was partly because it enabled them to construct an ongoing online presence.

The growth in new forms of *kaomoji* and *emoji* as reflective of communities in Internet *keitai* spaces is clearly represented by the various mutations of Shift_JIS art. As a subset of ASCII, Shift_JIS – abbreviated to SJIS or AA (ASCII art) – utilises the Japanese keyboard to make art works that are deployed to communicate feelings and emotions between users. Unlike ASCII, which is restricted to 95 characters, SJIS is not limited in characters and has taken on various forms of representation that reflect subcultures as well as being a popular mode of communicating on 2ch. The usage of SJIS, extending modes of kitten writing, are used to reflect the particular character-istics of a community by representing specific group's interests and emotional dialect and connection as well as a broader modes of virtual expressions for contemporary Japanese users of 2ch.

Figure 4.6 An example of SJIS art that features regularly in 2ch.

Source: Wikipedia Japan.

As mentioned above, the global attention that Japan gained for its mobile innovation and convergent media (Matsuda 2005) was in part due to its initially slow uptake of the Internet. This meant that users were not familiar with accessing the Internet from a PC, and were therefore able to adapt easily to convergent *keitai* Internet usage. In 1994 the Prime Minister's office established a section for policy management group as IT, which in 1998, under the rule of Prime Minister Mori, was given the title of IT Strategic Headquarters (ITSHQ). In 2000, ITSHQ drafted a law in the IT fundamental act and in 2001 it drafted an action plan entitled 'e-Japan Strategy'.

According to an MIC (Ministry of Information Affairs and Communication) survey in 2006, the highest rate of Internet subscriptions – amongst the 116 million population – was via the *keitai*, with over 80 per cent of the population subscribing. Whereas Internet access was difficult and expensive in the early to mid-1990s, this latency allowed Japanese users to jump a step in the convergence chain. Having never reached high penetration rates for Internet access via PCs, Japan was able to leapfrog to high levels of Internet access via the *keitai*. The Internet was first introduced in Japan in 1984, with the JUNET connecting Keio University, the Tokyo Institute of Technology and the University of Tokyo. By 1997 the telecommunications giant NTT initiated i-mode walled garden versions of the Internet nationwide. By September 2001, SoftBank (Yahoo!) had opened Japan's ADSL service. However, the popular use of the Internet really came with the introduction of mobile Internet services, which allowed users to access the Internet whether they had a PC or not. To this day, the *keitai* remains the most popular way of accessing the Internet.

The role of the Internet in Japan's grappling with twenty-first century post-modernity was an important issue throughout the development of MIC. In 2001, its policies were defined by 'the accelerating IT revolution: a broadband-driven IT renaissance'. In 2003 MIC focused on 'building a new, Japan-inspired IT society' in order to position Japan globally as a key player. The MIC White Paper noted:

> Japan is seeing rapid progress in developing telecommunications infra-structure, which is indicated by the fact that broadband services in Japan have become the most inexpensive and fastest in the world. From now on, it is more important for Japan to realize a society that leverages optical communications, mobile communications, information appli-ances, etc., over which Japan has a sizable edge on an international scale.
>
> (White Paper 2003)

By December 2004, MIC announced its policy aim of achieving a ' "ubiqui-tous network society" (u-Japan) in which "anything and anyone" can easily access networks and freely transmit information "from anywhere at any time" by 2010' (MIC White Paper 2005). By 2006, MIC had implemented policies aimed at stimulating and forging Japan's socio-economic systems with IT. The wording of these policies were indicative of the importance of robust techno-nationalist rhetoric in trying to ensure Japan's techno-soft cultural capital in the region as a centre for twenty-first century postmodernity – a label that was being strongly contested by Korea's lauded techno-nationalism (West 2006) and 'broadband centre' (OECD 2006).

According to Ken'ichi Ishii (2004), there is a distinct difference between the usage of the mobile Internet and PC Internet in Japan. The key distinc-tion is that while the high users of PC Internet tend to spend less time with family and friends, high users of mobile Internet are the opposite, tending towards very active interpersonal skills and socialising capacities (ibid.: 56). Moreover, mobile phone users tend to be more vivid and honest about their offline identity online and there is a more seamless relationship between online and offline identity and relationships. According to Daisuke Tsuji and Shunji Mikami (2001), mobile email usage between university students enhances sociability. The fact that more than half of mixi users predominantly access it via mobile Internet, as opposed to PC Internet, suggests that these people prioritise sociability and are adept at maintaining intimate ties.

Keitai @ convergence: case study of sample respondents

My case study of sample respondents was conducted in July 2004 and again in December 2005. In the 2004 interviews my sample group consisted of 20 university students, staff and administrators, half male and half female respondents aged between 18 and 29, at Chiba University. I will discuss these findings and then move onto the 2005 sample study group. One of the outstanding observations from the interviews was that the respondents did not match the stereotypical young savvy user. Rather, almost half had phones that were between one to two years old and with basic functions. Some phones did not have accessory functions such as camera. I asked one male respondent, aged 22, whether he felt the need to update his phone to keep up. He stated:

> I see my phone as just that – a phone. I use it to call my mother and family. As I've moved away from home to study at university, I see my phone as helping to keep me connected with my family. I miss them. The longer I have been here (at university), and more friends I have made, the more useful the phone has become.

In the sample group, it was the female respondents that tended towards having newer models of phones and using them in more multimedia ways. One female respondent, aged 20, noted:

> My parents bought me this phone for my birthday. I wanted a phone with a good camera because I don't have a camera and I thought it would be useful. When I first got the camera phone I took lots of pictures, just because of the novelty. After a while, I stopped taking so many pictures. Now I just take photos when I'm with friends and we are having fun so we can remember it. I like to mobile email my friends. But I don't like the Internet on the phone; I prefer it on the computer.

In terms of customising their phones, many respondents had *keitai* straps with the female respondents tending towards *kawaii* characters, while the males conventionally had the strap supplied by the phone company. One male respondent aged 19, however, had a quite feminised phone with a cute character hanging from the strap and a wallpaper image of *kawaii* characters. When I asked him why he had chosen those forms of customisation he replied:

> I want a girlfriend. I think if a girl sees my phone she will think I am caring and gentle. My phone is often on display on the table when I'm sitting down and so I think that if I'm talking to a girl she will think I'm cute.

Here we see that through the image of the *kawaii*, this male respondent is transgressing norms in order to make their *keitai* a possible vehicle for connecting them with a female in the hope of potential intimacy. Another male respondent, aged 25, had a girlfriend. His camera phone images, which he stored on the phone, were a collection of images of her. He stated that the cute phone strap was a present from her, and that she had the same on her phone. He stated:

> I spend most of my spare time with my girlfriend. And when we are not together I think of her. I like having images of her on my phone because in class I can look at them if I want to see her. Sometimes when we are together we will go through our images on each other's phones. Her camera phone is better and she tends to take more pictures so then she sends them onto me. Sometimes when we are apart she sends me an image of herself. It just makes me want to see her in person even more!

One female respondent, aged 27, liked her *keitai* as a device for convenience. She predominantly used it to access train timetables, movie screenings and general informational services. Beyond its convenience, she seemed cautious about what she viewed as growing dependence on technologies. As she noted:

> My *keitai* is useful. It helps save time when organizing things. It helps to fill in time when I'm waiting. But I do worry about being too dependent on such technologies. Sometimes I leave the phone in my room, just to test how much I need it. I have friends who see the phone as very important. I don't think the phone is important, it is just convenient. Family and friends are important, not the *keitai*.

The function of the *keitai* as a time 'filler' has been noted in numerous studies on the 'micro-coordination' (Ling 2004) However, in the case of Japan, such activities as emailing for social confirmation is significant in maintaining forms of etiquette both in the workplace and personal life (Ito 2002). As noted in Matsuda's study of *keitai* usage between mothers and their children, the *keitai* is often used to re-confirm pre-existing rituals and practices. For example, a child will email their mother to notify her that they are coming home even though they are arriving at the usual time. It is significant that this respondent identified the role of the *keitai* as maintaining existing social relations. Her ambivalence about not transferring the importance of relationships onto the actual *keitai* device is a very pertinent one, especially given the increasing role of the mobile media in maintaining full-time intimacy.

For some, the *keitai* affords them both the means and interest to develop themselves creatively. Another female respondent, aged 20, noted how having a camera phone had led to her developing an interest in photography. She stated:

> I never had a camera. I occasionally bought disposable cameras for special occasions so that I could help photograph it. But I never felt comfortable taking pictures. I wasn't very good and would be embarrassed showing my images. But over the last year since I purchased

the camera phone I take more pictures and see my skills improving. I can always delete bad photos and try again. I am now thinking of doing a photography course, something I would have been scared to do before.

Here we are provided with a snapshot of how the multimedia capabilities of the *keitai* are affording users accessibility to modes of creativity and production they would have not explored otherwise. This sample study also reflected findings by Ishii (2004) – discussed earlier in the chapter – on how certain usages of technologies reflected the user's ability to socialise. The few users who used mobile Internet were much heavier users of the *keitai* than the respondents, who mainly the *keitai* used as an extension of the *denwa* (landline phone). These respondents had over 100 people listed in their address book, as opposed to the group average of about 50, and actively kept in contact with at least five of the contacts more than once a day. They also seemed much more open about disclosing their feelings towards the *keitai*, and their emotions in general.

One of the striking characteristics of this sample group was the way in which customisation had begun to move into internalised forms of expression rather than external ones in the form of cute character phone straps. Perhaps in response to the growing convergence and multimedia possibilities of the phone, respondents tended towards customisation modes such as camera phone images and stored emails deploying *emoji*. But rather than reiterating McVeigh's observations about individualisation and interiorisation, the respondents' relationships to the images and messages were very much about giving high priority to interpersonal relationships. As the male respondent mentioned earlier noted, the phone became a treasure chest of mementoes to be continuously updated, shared and relived with his girlfriend.

In this sample study, it was hard to find evidence to support blanket statements about emerging forms of individualism and sociality, apart from noting the shift towards internalised modes of customisation. As an extension of users' identity and social network, the phone was obviously being used to express modes of subjectivity. For a variety of reasons, we can see a persistence of gendered (most notably, feminised) customisation being deployed by both male and female respondents.

Only two of the 20 respondents in the sample study had mixi sites, mixi having only launched its mobile service in September 2005. At this stage, mixi's membership numbers had reached two million (escalating to eight million in 2007). The female user had been using mixi for about six months, while the male respondent had set up his own page only one month prior to the interviews. The female respondent found mixi useful in helping to get reconnected with old school friends and people she had lost contact with. But she also noted that she had met a couple of new people, who were friends of friends, via the mixi site. The male respondent felt it too early to be conclusive about his observations. However, he noted:

> It's fun, but I think I need to spend more time accessing and adding to my page. Some of my friend's pages are really interesting, so I feel inspired to do the same. But it's hard to find time in between study.

For both male and female respondents, the need to be 'true' to their identity in the actual was pivotal in their engagement with the Internet and *keitai* emailing. The use of *kaomoji, emoji* and montaging between generic *kawaii* characters, and the deployment of respondents' own pictures (i.e. taken via the camera phone) were important in customising/personalising the technology and the co-present spaces. Customisation, particularly in the form of camera phone pictures combined with *kawaii* frames, was not just a form of self-presentation and regulation but also a way of familiarising and personalising the de-personalised space of the technologies for friends. This form of techno-cute was exaggerated in female respondents' customisation, but it was also evident in the male respondents' customisation. The 'post-modern' and 'playful' characteristics of *kawaii* culture assigned by Allison (2003) were appreciated more by female respondents, whereas male respondents tended to be more literal in their understanding and practice of representing identity.

For many of the respondents, the growth in multimedia possibilities afforded by the *keitai* was allowing them new ways to interact with their friends. From customising camera phone images, to uploading them on their mixi website, to printing them out as stickers (*purikura*) that could be stuck on their best friends' phones, many found that communicating via the *keitai* ancillary customisation modes (web or hard copy) provided them and their friends with many meaningful experiences. The *kawaii* aesthetic, so fundamental to early *keitai* and pager customisation, rather than disappearing, seemed to be becoming increasingly mainstream.

In the second sample study conducted in December 2005, there was a marked shift towards the multimedia capacity of the *keitai*. This time the respondent group were not limited to being graduates of Chiba University, but included graduates from a variety of differently ranked universities. This difference undoubtedly opened up the diversity of cultural capital of the respondents who were on various educational tracks. Of the 20 respondents surveyed, aged between 20 and 36, half male and half female, many seemed to make use of such applications as the camera phone and mobile Internet. Over half of the respondents had mixi sites, which they accessed and updated on a daily basis via both PC and their *keitai*. When discussing the pleasures of using mixi, many commented on its usefulness for maintaining connection with existing friends. The respondents did note, however, that the function of connecting members – via the selectable icons that were indicative of the members' social and cultural capital – allowed them to meet people with similar interests and reconnect with long-lost friends (eg from primary school). As one female respondent, aged 27, noted:

I predominantly use mixi mobile because it means that I can be on a train and connecting with friends who aren't physically there. It is also more direct than email and I can do it anywhere without disturbing others around me. I find mixi is really useful. I can regularly connect with friends and chat. I can also pursue my interests with other members and we can share information. For example, I'm very interested in gardening, and through mixi I have met others who share the interest and can tell me things I didn't know.

In this respondent's comments we see that mixi affords her a type of 'communities of co-presence' (Ito 2002) in which the virtual is used to reinforce the importance of actual presence and place. Other respondents used the SNS to further their interests and, in many cases, meet new people online with similar interests, who would sometimes become offline friends. Mixi's particular form of social networking could be viewed as a precursor to Western SNSs, such as Facebook, which attempt to create 'communities of feelings' via various forms of social, capital and economic capital.

For one male respondent aged 30, connecting with people via mixi rather than mobile emailing had become an important part of day-to-day life. He observed that he liked the fact that mixi was direct (synchronous) but not intrusive, and that it emphasised community. He stated:

I work a lot and this means that I do not see friends as much as I would like. Using mixi means that I can still keep updated and connected with

Figure 4.7 An example of a mixi page. Courtesy of Yuji Mori.

them so that I do not feel so distant from their lives. I feel good when I use mixi. I like it better than mobile emailing, as I am not just connecting with one person or a group indirectly but connecting at the same time to other friends online. I mainly access it at lunch or in the evening, as do my friends.

Members of mixi did not email using their mobile as often, suggesting that they preferred the sense of community of mixi to the synchronicity and 'networked individualism' of some mobile applications. An increase in camera phone usage was also notable, in particular with the male respondents. This was partly due to the fact that camera phones were becoming increasingly common in Japan and that many respondents, when updating their phones, ended up having a camera phone whether they desired it or not.

As outlined in Part I, the growth in mobile media applications such as camera phones needs to be understood in terms of the remediated quality of mobile media customisation. One of the distinctive characteristics noted in this Japanese case study was that there was a split between those who 'shared' camera phone images, and those who kept them for personal consumption. One male respondent, aged 35, viewed the camera phone as 'like having a camera with you everywhere'. He mainly took photos for his own personal consumption. His photographic style was highly professional in its composition and ambience.

The two images above show how the respondent used the camera phone to document times when he was physically alone. These images were taken for his own personal consumption. His camera phone images, carried around with him everywhere, operated as a portal between his two homes – London and Japan. As a graduate student who studied in London, he found himself missing the day-to-day life of Japan. Hence, the image of Japanese-style utensils reminded him of everyday life in Japan. The second image, his wife (who remained in Japan during his studies) on webcam, was an attempt to capture the co-presence to savour afterwards. This respondent mainly

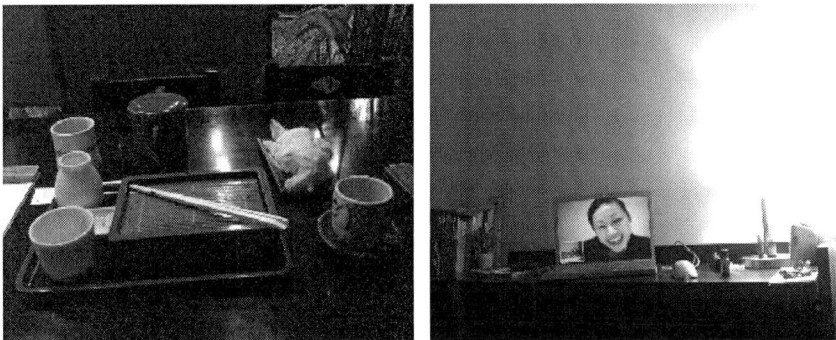

Figure 4.8 Images from left: the end of a solitary dinner; webcam wife.

stored his images on the camera phone so that he could look at them at any time.

By contrast, a 27-year-old male respondent used his camera phone as a way to share experiences with friends. Unlike the poetic gestures of co-presence in the previous respondents' images, these images are about staged and intertextual interventions. In this way, this respondent used the camera phone like paparazzi, deploying confrontational tactics. He saw camera phone image making as a way to extend the performativity of the everyday, often taking pictures of the 'unreal' or 'uncharacteristic'. In this way, he defied Koskinen's (2007) argument that camera phone image-making is about taking the banal; rather, he tried to capture the extraordinary (often staged) and sublime of the everyday, which he then shared with friends sending via *keitai* picture email.

The respondent explained that he liked to share these images by *keitai* email as well as storing the images on his *keitai* (as wallpapers) to show to his friends in face-to-face scenarios. He and his friends shared an interest in taking 'funny' images and sending them as 'silly jokes'. The first image is

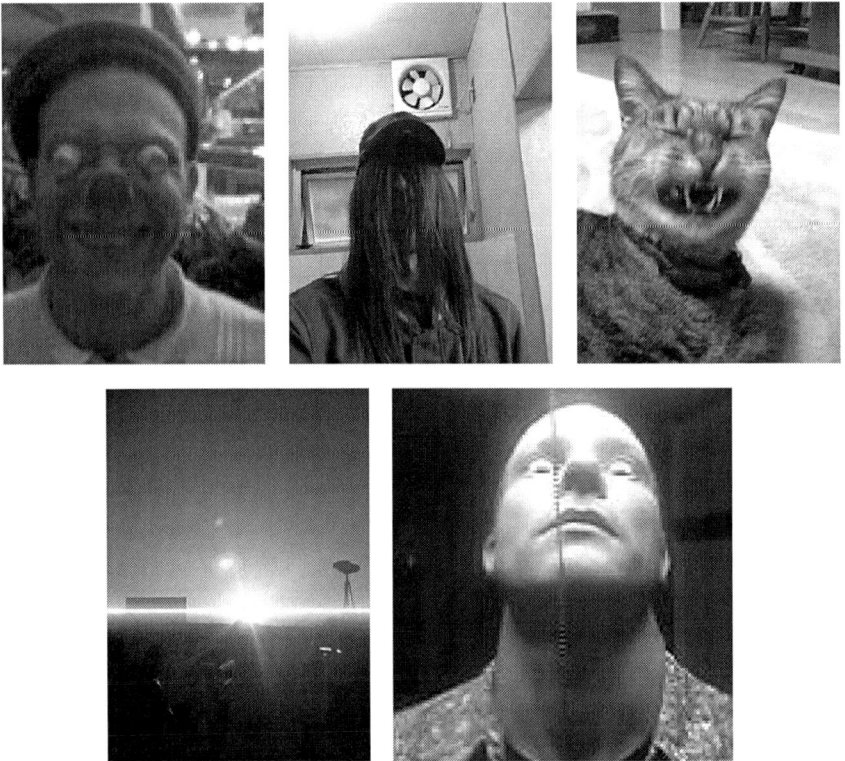

Figure 4.9 Clockwise from top left: friend's funny face, self-portrait as *The Ring* monster, scary cat, friend as 'exorcist', sunset at Aichi World Expo, 2005.

indicative of the respondent's interest in making humorous performative interventions in everyday social situations, in which he and his friends often dressed up and pull strange faces. In the second image, the respondent has dressed up to imitate the monster in *The Ring* (*Ringu*, 1998, directed by Hideo Nakata).

In the third image, taken of the respondent's family cat, the respondent had wanted to take a *kawaii* picture of the cat, but the result was exactly the opposite. In the fourth image, one of his foreign friends attempts to be scary: the respondent sent the image to friends under the heading of 'exorcist'. The respondent was also playing with the notion of the 'evil Westerner'. In the final image, with quite a different tempo and pitch from the other images, the respondent was at the Aichi World Expo 2005 when he was surprised by a beautiful sunset. He felt that part of the beauty stemmed from the fact that the image produced was very different from the actual light he had witnessed.

For all of the respondents in the second sample study, the customisation inside the phone had taken priority. This is not to say that the *keitai* were not customised outside with *kawaii* phone straps. Rather, as the *keitai* became more and more multimedia capable and the newer, larger models made viewing images and the Internet more viable, respondents started to engage more fully with various applications in different ways. Moreover, the *kawaii* customisation once denoted by *kawaii* character hardware attached to the phone had become more complex and migrated into the electronic and virtual UCC spaces inside the phone. *Emoji* and ASCII style customisation were on the increase as respondents attempted to 'domesticate' the domestic technologies. Moreover, both male and female respondents were deploying these forms of UCC *kawaii* self-expression, with little differentiation between the gendered usage.

Cybercute: conclusion

Kawaii customisation clearly demonstrates Bourdieu's reflections of the process of cultural knowledge and tastes as encapsulated in the concept of 'cultural capital', and its negotiation and contestation at the level of the embodied practices that he defines as 'habitus' (Bourdieu 1984 [1979]). Cultural capital is informed by one's education, background, and family, whereas habitus is the negotiation and contestation of cultural capital at the level of the embodied practices. Practices such as customisation operate simultaneously at the level of the local and global, individual and society, exhibiting modes of both interiority and exteriority. Put simply, the practice of customisation clearly demonstrates alignments and contradictions between hegemonies of cultural capital and contingencies of habitus as demonstrated by the respondents. *Kawaii* customisation has witnessed some shifts, most specifically in such as spaces as *keitai* Internet. These shifts could be seen as indicative of the global phenomenon of UCC and Web 2.0 'produsers' however, the tactics of customisation are imbued by the local – highlighted by the particular role played by *kawaii* customisation in this case study.

There are various ways in which the *kawaii* is changing, as a barometer of the way in which customisation is being redefined in an age of UCC. Firstly, the *kawaii*, explicitly tied to character culture, was changing in accordance with shifts in new technologies and global tendencies. This change paralleled the shifting status of the prime consumers of characters: women. Both female and male respondents are both actively using *emoji* and ASCII in order to customise their co-present communication, whether telephonic or virtual. The rise of hyperfeminine customisation can be clearly viewed in the domination of JISC in 2ch.

Secondly, the notion of *kawaii* occupies a particular relationship to adulthood and sexual identity rather than the frequently desexualised category of childhood as often assumed by adults. Finally, the *kawaii* appropriation within mobile telephonic and associated Internet spaces is generating neohumanist (rather than post-human) models for rethinking mobility, mediation and convergence. This can be witnessed inside Japan with the burgeoning of the virtual community mixi, which deploys *kawaii*-like writing, SJIS and emerging *keitai* practices such as camera phone images to customise and socialise the technological spaces.

As the *keitai* became increasingly multimedia, many respondents noted spending less time customising outside the device and more time customising content inside the *keitai*. With the Internet being accessed predominantly through the *keitai*, the relationship between Internet and *keitai* UCC blurs. In particular, while observers watching the rise of the *keitai* in 2000 had bemoaned the demise of older media such as *manga*, we can see that *keitai* UCC has re-ignited people's interest in remediated media such as *manga* while at the same time shifting the gendered associations, with an increasing number of women reading and writing *manga* (particularly adapted from *keitai shôsetsu* or 'portable novel'). Often enduring stories will move from *keitai* UCC to films and *manga*. The rise of the *keitai shôsetsu* – a short story written for the *keitai* that, in some cases, is translated into other forms of media such as films and *manga* – is a great example of the power of UCC in Japan. Although *keitai shôsetsu* were initially written by professionals, by the mid 2000s everyday users had begun to be inspired to write and disseminate their own *keitai shôsetsu*. In late 1999, around the launch of i-mode, maho-island (*maho* meaning 'magic') began its UCC-related service with four channels for *keitai* UCC – novels, poems, photos and music. By 2007, nearly four million different *keitai shôsetsu* had become hard copy. With one million *keitai shôsetsu* being produced in 2007 and 1.9 billion page views per month, maho-island has become an exemplary case of the popularity of UCC. But what does this *keitai* UCC mean for female users?

Although UCC may provide the everyday user with a voice and also models for intimacy, interactivity and dialogue between authors and readers, one of the big problems is that it exploits the user's creative and social labour often without remuneration. However, it seems Japan is providing a different picture for the future of UCC in which media content distribution companies such as maho-island take an active role in encouraging and fostering talent and media literacy programs. For example, maho-island has an annual award for UCC – a winner of the best *keitai shôsetsu* round can win one million yen plus a publishing contract, a runner-up getting 500,000 yen and a publishing contract. In 2007, maho-island featured nearly 3,000 *keitai shôsetsu*, over 3,000 *keitai* poems, 1,600 photos and 300 songs in its four UCC channels.[9] While this paints a picture of an active UCC scene in which *keitai* cultures are nurturing new talent, how does this reflect female empowerment through such technologies?

It seems pertinent to turn to the *keitai shôsetsu* – a phenomenon marked and shaped by the dominance of female authors and audience. One of the key features of *keitai shôsetsu* is that they tend to follow a diary-like, confessional autobiographical model – very much reflecting Berlant's observation of the growing 'publicness of intimacy' (1998) being forged through the highly personal and intimate media, the *keitai*. These novels inspire users to become writers as well as being part of a broader trend towards the professionalisation of UCC particularly through its adaptation and translation into other media such as films. One example is the famous adaptation of the most popular and famous *keitai shôsetsu, Koizara*, into a movie. Written by female *keitai* novelist 'Mika', the story became so popular that it was soon adapted by female director, Natsuko Imai, into a popular film. This is one example of the hyperfeminine characteristics of the *keitai* that make it so suitable for female users to become audiences and creators of these new, but remediated, forms of creativity.

Figure 4.10 The website of the UCC organisation maho-island, featuring the advertisement for *Koizara*, the movie (directed by Natsuki Imai, 2006).

Through *keitai* UCC, we can see many examples of female users finding inroads into creative activities. Thanks to UCC-orientated organisations such as maho-island, these users can be empowered on various levels – sharing and collaborating on stories as well as potentially making a career, and gaining professional recognition in the form of book publishing or film contracts. Far from *keitai* cultures eroding the significance of older, remediated media such as *manga* and film, they are providing new material for, and interest in, adapting stories by everyday users. In the case study, female respondents were active in customising both the exterior and interior of the device. The tendency of customisation to be cute – whether in the form of actual *kawaii* deployment, or through *emoji* – was evident. The fact that 'kitten writing' was now part of the industries' gender scripting (i.e. the *keitai* came with increasingly varieties of *emoji*) highlights how the UCC feminised practices have become institutionalised. In this sample study through SNS models such as mixi, there would seem to be a shift away from the traditional *kawaii* customisation modes. It was apparent in this case study that many respondents, when they did share customised mobile media such as camera phone images, did so with a small group of friends predominantly through *keitai* email.

While the diverse modes of *kawaii* customisation manifest in all forms of mobile media, both external and internal, they demonstrate that the overtly feminised mode of customisation is no longer the sole prerogative of female consumers. Returning to my question at the beginning of the chapter, it seems that the increase in feminised modes of customisation is being utilised as much, if not more, by male respondents. The 'cute' is no longer just about female performativity. In the rise of mobile media, both male and female respondents are using hyperfeminine modes of customisation. Unquestionably, female users are feeling more empowered by various forms of customisation from camera phone images to SNS such as mixi, and are partaking in UCC spheres of creativity such as *keitai shôsetsu*. However, in their 'produser' deployment of mobile media, there are still questions about how much this will translate to more employment of women in the creative industries. The example provided by maho-island, and especially the female-written and directed adaptation of *Koizara*, suggests that the hyperfeminine qualities of *keitai* cultures are affording female users not only forms of expression and identification but also the potential to make inroads into the creative industries in Japan.

Many female respondents spoke about certain pleasures associated with the creative dimensions of mobile media, although many noted the double-edged sword of mobile media customisation as a time trap. On the one hand, mobile media allows users to 'fill in time' with rhetorical forms of expression and co-presence that strengthen existing relations. On the other hand, mobile media customisation can overtake other forms of everyday practices. Undoubtedly, the shift in gendered relations in Japan is reflected in mobile media practices – but this empowerment and visibility of women, and the rise of hyperfeminine modes of customisation, also highlights a central paradox for contemporary women in Japan. In one way, mobile media provides them with more freedom and opportunities for creativity on various levels. Alternatively, this freedom and creativity can operate to reinforce the increasingly flexible and casual work practices, whereby women are expected to occupy many roles simultaneously – mother, professional and prime carer.

What will inevitably unfold in the following chapters is the ways in which different locations reflect values and attitudes through their adaptations of specific mobile technologies and attendant modes of gendered customisation. The adaptation and migration of *kawaii* customisation techniques can be found in various locations, and as I will show in the following chapters, this often involves feminised customisation that is not necessarily a version of 'Japan'. As the region moves unevenly towards convergent media spaces and increasingly divergent forms of co-presence, we can see many examples of customisation. For example, in the next chapter we will see that Korea's dominant SNS – Cyworld mini-hompy – is adorned with cute customisation ranging from avatars to cyber-gifts. Characters such as Pucca seem to re-enact the techno-cute aesthetics of Hello Kitty. However, such a comparison overlooks the complexity of the region's politics. Unlike *kawaii* customisation in Japan, which rehearses electronic individualism, cute culture in Korea's mini-hompy reflects emerging patterns of intimate communities in Korea and a reworking of the symbolic and literal function of the family within Korean nationalism (Cho 2000).

While Japan has dominated the region historically in terms of innovative customisation, we are now witnessing a plethora of divergent forms of customisation, especially as people and localities try to negotiate the co-presence of mobile Internet spaces. *Kawaii* customisation needs to be seen as part of a broader phenomenon of imaging communities that are domesticating and localising the new technological spaces. These localised cartographies of personalisation are reflecting shifts in gendered intimacy and labour. For example, the dominant shift towards customising modes such as camera phone practices is being led by females. While this might be seen as female consumers reiterating mass media's objectification of women, the actual ways in which these images are contextualised and shared go beyond the well-cited negative usage by the *kôgyaru*. Rather, it is inciting female users to become the photographers, not just the photographic objects. Moreover, men are adopting this feminisation of technological spaces as a mode for humanising the technology.

This chapter has sought to address the role of the Japanese mobile media user, and specifically the female user, not as a celebration of the sovereign consumer but rather to articulate some of the micro and bubble-up effects between respondents and mobile technologies. As one of the most heralded mobile centres with a legacy of producing technologies for the consumption of the region, and with its rise of female users concurrent with the creative deployment of *kawaii* customisation of *keitai*, the Japanese superpower model provides an interesting case study for the 'future' of gendered mobile media. We can see how localised forms of customisation – particularly through the *kawaii* – reflect certain types of nationalism as outlined by McVeigh in the introduction of this chapter. By exploring the customisation of mobile media in the region, we will gain insight into the burgeoning imaging communities that see the gender imbalance around consumption and creation of technologies, and the stereotype of passive Asian femininity, being challenged and transgressed.

5 Engaging rings

The *haendupon* and intimate communities in Seoul

The contrast between modern Korea and Japan, the first site of mobile media, is not without historical irony. As examples of techno-nationalist agendas and global centres for mobile media innovation, the two countries represent very different contexts for Confucius revivalism, *vis-à-vis* capitalism, in the region. The historical conflict between Japan and Korea is now performed metaphorically through the soft capital of mobile media, with both locations vying for global dominance of the symbol of twenty-first century post-modernity, the mobile phone. Once dominating the region with its particular flavour of cultural capital in the form of J-pop, Japan's popularity has been eclipsed by the rise of K-pop in locations such as Hong Kong and Taiwan. Lauded as the country with the greatest broadband access by the OECD (2006) and a key example of twenty-first century modernity, it is easy to forget that Korea's global rise has been relatively recent.

The fact that Korea was under Japanese rule from 1910 to 1945, and that during this period Korean culture and language were largely suppressed, is not lost on the contemporary Korea that boasts the world's ninth-largest economy in terms of GDP. Two events of the twentieth century have left an indelible mark on Koreans in general: the colonial experience that destroyed a 500-year dynasty, the *Chosôn* (1392–1910), and the subsequent division of the country into North and South. The resulting Korean War (1950–1953) was one of the bloodiest conflicts in the country's history, and played a major role in shaping the national identity of post-modern South Korea.

Since the Korean War, South Korea's government and major companies (*chaebôl*) have worked hard to regain a sense of national pride and global strength. Both were weakened in the Asian economic crash of 1997, in which the South Korean Reserve Bank had to be rescued by the IMF (International Monetary Fund). Part of Korea's rapid rise to become the third-largest economy in the region has been due to the robust governmental and industry focus on new technologies and on embracing twenty-first-century post-modernity. The IT policies of Korea, in the form of techno-nationalism, have been noted as the best in the world (West 2006), and the IT industry in Korea represents over 15 per cent of GDP (MIC Korea 2006). On average, Koreans are online at least 13 hours per week (NIDA 2006) with email, online communities and information. The most dominant usage of the Internet is participating in online communities from Cyworld mini-hompy, followed by online massive multiplayer gaming (called MMORGs).

Since the 1997 economic crash, both Korea and Japan have worked actively to invest in the pioneering of mobile technologies that are now been exported worldwide. This investment in various forms of governmental and industrial policies and infrastructure such as broadband has seen both countries gain attention globally as representing possible futures for mobile media. The aforementioned shift from J-pop to K-pop exemplifies the ongoing re-orientations of the region, which are now increasingly centred on techno-soft capital.

In this chapter I explore the particularity of Korean camera phone practices and its relationship to online presence, and how these practices inform localised modes of gendered co-presence, intimacy and community that, in turn, reflect and contest traditional notions of nationhood. In order to understand the specific modes of cartographies of personalisation, we must understand how these challenge hypermasculine conflations of new technology, family and the nation state. As I argue, behind Korea's global image of innovative mobile technologies and exemplary twenty-first century broadband capabilities are the changing face of key mobile media 'produsers', women.

In contrast to other locations discussed, the connection between the *haen-dupon* and online/offline communities is essential in making sense of the mobile media phenomenon particular to Korea. Unlike Japan, where mobile Internet dominates, Koreans access the Internet principally via the desktop, and especially in the context of social rooms such as *PC bang* (PC room). Virtual communities such as Cyworld mini-hompy ('Cy' meaning 'between' worlds or 'relationship' world) – with over 18 million members of Korea's 48 million population – are very much embedded in everyday life, and reflect a strong sense of community that, in turn, highlights Korean techno-nationalism. Koreans are by far the most voracious users of the mobile phone through a variety of applications (Mitomo et al. 2005). They use voice calling much more often than Japanese and Australians, and use SMS/ emailing more often than the Hong Kong people and Australians. Korea also has the highest camera phone usage. These differences suggest that through mobile media we can gain a sense of constructed and performed contemporary 'Korean-ness'.

The symbolic dimensions of mobile technologies as part of Korea's self-projection as a twenty-first-century centre for post-modernity cannot be understated. As Jung-eun Hong (2007) notes in 'Mobile phone machine and its discourses analysis', the naturalisation of Korea as a 'global leader' for mobile technologies has been all pervasive. Since the 1997 IMF financial rescue, Korea has constructed well-orchestrated production on various levels from the techno-national to the personal, to locate the mobile phone – and specifically Samsung's Anycall – as symbolic of Korea's twenty-first-century post-modernity.

The construction of Korea as a powerhouse of post-modernity and global-isation *vis-à-vis* mobile technologies has resulted in the following 'discursive strategies': the *Hyoja* (meaning a son devoted to filial duties/industry) as the symbol of the mobile phone industry, the 'normalizative idealization of Samsung Anycall as a national mobile phone', and the 'justification of neo-liberal deregulation of mobile phone industry' (Hong 2007: n.p.). The symbolic dimensions at the level of the national are also reflected through the gender performativity in which both males and females utilise technologies to represent their femininity and masculinity (Na 2001). However, through gen-dered customisation of mobile media, we can see both a reinforcing of gender stereotyping – as well as providing new modes for transforming gendered performativity – contributing to the symbolic role-playing of national identity as family.

As aforementioned renowned feminist and anthropologist Cho notes in 'You are entrapped in an imaginary well', Korea in 1990s experienced a period of pivotal 'high-growth economic development' that became exemplary of a 'economic miracle'. Korea's example of successful turbo-capitalism was viewed as a type of 'Confucian capitalism' that contested the West's 'Protestant capitalism'. However, the 1997 financial crisis – resulting in the IMF financial rescue – transformed the tune of optimism into a state of national disgrace. According to Cho, the undulating rising and falling of Korea over the short period of the 1980s to 2000 had major implications on the feminist movement in Korea. Most notably, stereotypes about gendered labour were revisited, whereby men became aligned with production and the public sphere while women were seen as responsible for the family and con-sumption. As Cho observes:

> What about the women's movement? As the anti-dictatorship social movements of the 1980s began to decline, feminists who were unable to establish their own voices within the major social movement circles felt that their time had come. They were excited to organize the 'revolution-ary energy' that remained from the 1980s *minjung* (people's) movement. However, consumer capitalism, which arrived too rapidly as a result of 'compressed' growth, has drawn a new generation of women into fashion shows, department stores and fancy cafes. The feminist movement ended up losing the opportunity to advance into the next stage. Much energy was wasted because of such out-of-synch timing, and as a result, the gross inequities and unbalanced nature of society has become more severe.
>
> (2000: 54)

With the momentary lapse of the Korean feminist movement came the phe-nomenal growth of young female professionals – with high disposable income perfect for mobile phone consumption – characterised by the 'missy syndrome' (Y. Lee 2000). The missy syndrome began in Japan in the 1980s and marked a new epoch in constructing strong but feminine 'modern' female types. By the early 1990s, the phenomenon had spread to places such as Korea – adopted by both housewives and professional women under the image of the new 'modern' woman. Integral to this phenomenon was the role of technologies, particularly domestic technologies such as cars, televisions and radios (followed by the mobile phone) (Na 2001). Advertisers became more aggressive with their campaigning for the female consumer, particularly by shifting the female typecasts and design to reflect this new type of femininity.

The backdrop to the new turbo-capitalist consumerism – in which Korean youth were beckoned into hypermodernity in which conservative gender stereotypes around labour applied – was undoubtedly the 'hegemonic powers' (Cho 2000: 57) in which the notion of *kukmin* (a member of the nation) and *kajok* (family) was conflated. The re-deployment of the 'united we survive, divided we die' slogan – which became 'public discourse during the nationalist resistance movement against Japanese rule' (ibid.: 59) – saw relevance again within the nationalist trope of the *kukmin kajok*. The conflation of Confucian capitalism with *kukmin kajok* also ensured the elimination of individual citizens within a civil society. As Cho avows:

> In the official discourse, the nation, the state and the people are one and the same. South Korean society was extremely successful in manufacturing a 'majority' consisting of middleclass, middle-aged male members of society. Until recently, the mass media frequently used expressions such as *kukminjok chongso* ('popular sentiment' or 'national sentiment') which functioned to block the emergence of alternative opinions and imagination, and these incantatory phrases contributed to producing the uniform subjectivity of the *kukmin*. The phrase *wihwagam chosong* ('promoting discord') has also played an important role in suppressing the emergence of any new voice.
>
> (ibid.: 59)

Cho argues the post-IMF crisis led to mass media campaigns focusing upon fathers as the symbol of the family in the public sphere. In the economic slump it was women – not men – that were the first to be laid off their jobs (ibid.: 62). Men became synonymous with *kukmin*, while women were relegated to the 'marginalized' *kajok*. In an earlier work, 'Marriage stories in a male-centred republic' (1996, cited in Cho 2000), Cho notes that the impossibility of separating Korea's rapid modernisation from the naturalisation of conventional gender roles within the *kukmin*. According to Cho, major gendered 'differences exist between men and women's perceptions of the family' (1996 cited in 2000: 62) whereby men identify family as the paternal lineage, while women's family includes both paternal and maternal formations. In her detailed analysis of the gendered subjectivities under the reconstitution of *kukmin*, Cho compellingly argues that it were the ideologies surrounding the *kukmin* – such as gender inequality – that led, in part, to the 1997 crisis in the first place. As Cho conjectures:

> Why do people not realize that if the *chaebol*-centred system is a source of problems, then the male-centred privilege system must also be harmful to the economy? Why is it still possible to indulge in the reckless notion that structural reform is possible without reform of existing gender relations that have divided men and women into 'the leaders of society' and 'the home'?
>
> (Cho 2000: 64)

Unquestionably, in Korea's twenty-first-century turbo-capitalism the *kukmin kajok* has become embedded within the country's rise as one of the global leaders in mobile technologies. Samsung becomes synonymous with *kukmin kajok*, which, in turn, is constructed around a male-centred culture. This is highlighted by Hong's (2007) aforementioned discussion of the *Hyoja* as the symbol of the Samsung Anycall embedding within notions of *kukmin kajok*. As S. D. Kim notes in his two defining chapters on Korean mobile practices, the success of mobile technologies within Korea was ensured by their ability to reinforce and re-enact existing cultural practices: in short, amplify notions of *kukmin kajok*.

In 'Korea: personal meanings' (2002), Kim discusses how practices such as cronyism not only continued but thrived through the rise of mobile communication. In 'The shaping of new politics in the era of mobile and cyber communication' (2003), Kim analyses the role of mobile technologies as a form of resistance to traditional hierarchies, exemplified by the case of the President Roh (2002) elections that featured 'people power' through mobile technologies. Through the Roh case study, Korea's 'Confucian capitalism' becomes unstable as notions of the civil society and democracy come to the forefront; here we are provided with a challenge to the hegemonic power of notions of *kukmin kajok* in which, as Cho had argued, quashed any notion of civil society. Echoing Kim's argument, Kyongwon Yoon (2003, 2006) challenges the techno-orientalism prevailing around projections of Korean youth as cyber-kids by discussing the ways in which mobile technologies both rehearse and resist existing familial and hierarchical relations within contemporary Korean everyday life.

Thus mobile phones – as both symbolically and materially representative – function to expose, reinforce and possibly subvert gendered norms around the family–nation–technology paradigm. Consequently, the exploration of mobile media practices can provide insight into residual and emerging forms of the *kukmin kajok* in the performing of familial intimacies. For example, how do these practices reflect salient or emerging forms of gendered intimacies and the relationship to the family–nation–technology paradigm?

As in other locations, gender inflections in Korea are defined through the feminising mode of customisation. As noted with some of my respondents, we can witness explicit instances of this with youth when they form new heterosexual relationships. Male partners *surrender* their phone to their female partners for customisation. Both inside and outside the phone, the female partner adorns their male partner's phones with images and icons; the phone becomes the repository for the relationship, with the female as dominate gatekeeper of what gets remembered and forgotten, saved and shared. Here we see the female partner – potential wife – taking control of the symbolic *kukmin kajok*, the *haendupon*.

As the *haendupon* is always present and visible, the reterritorialised – feminised – male *haendupon* demonstrates not only to others that he is 'engaged', but, also, operates as a constant reminder of the co-present partner. On the one hand, the customisation by the female could be seen to entail her control of the relationship, but, on another level, could this be viewed as a further exploitation of women's emotional labour and symbolic of the woman as prime consumer? Regardless, this is an example of how the *haendupon* is an integral part of emerging patterns of intimacy with Korean heterosexual romance, and thus a repository for analysing gendered power relations that, in turn, could reflect new constructions of the *kukmin kajok* within twenty-first-century post-modernity.

Alternatively, the surrendering of the phone can be read as symbolic of the ways in which new gendered familial patterns are being transferred onto one of the icons of intimacy and post-modernity, the *haendupon*. Most certainly, as an index for Korean nationalism around technology and the family, the relinquishing of control on behalf of the male of his *haendupon* to this female partner to adorn in whatever way she likes could be read as indicative of changing gender, technology and power paradigms in Korea. As I have argued elsewhere in the book, customising of the mobile phone (both outside and inside the device) is an activity that not only humanises (as in the case of cute customisation) but also feminises – *vis-à-vis* emotional labour – the technology, emotionally imbuing the object and transforming it into a socio-technology.

In each different location – drawing from localised forms of gender performativity – mobile media customisation reflects and rehearses emerging modes of postmodernity. This process, returning to L.H.M. Ling's (1999) argument regarding post-industrialism's hypermasculinity that operates dialectically with hyperfemininity, is explicitly apparent in the poignant role and practice of mobile media in Korea. The mobile phone, through various forms of localised customisation, becomes the battlefield in which a sense of identity and community are contested; and Korea is no exception. As we shall see, the ways in which phones are feminised are as diverse as the customising practices. In Korea, unlike other locations in the region, we see a particular form of customisation that strives to reinforce traditional notions of Korean national identity, and that sees the mobile less as an extension of individual identity than as a repository for social and family networks and possibly, a motif for changing gender formations within the configuration of the *kukmin*.

In this chapter I shall outline the socio-cultural context of Korea and the importance of technologies in emerging forms of techno-nationalism in the country's global projection of post-modernity. As it is impossible to discuss *haendupon* practices in Korea without discussing the interconnected space of virtual communities such as Cyworld (in which much of *haendupon* material is shared and stored), this chapter will also outline the role of this community in re-enacting offline identity and sociality. I will then move on to a case study with university students to explore the micro-politics of customisation – and in particular camera phone practices – and what this means to individual users.

Figure 5.1 An example of DMB advertising for the TU phone. Note the use of a traditional Korean colour scheme (*saekdong*) as the colour test screen background. Photo: Hjorth.

The space between connection and contact: the South Korean technoscape

In Korea, Internet and mobile telephonic spaces are helping to progress Korean forms of democracy (Kim 2003: 325). For Korean sociologists S.D. Kim (2003) and Cho (2004), the rise of a specific type of democracy in Korea has been supported in part by new technologies such as mobile phones. In particular, in Seoul one can find two types of youth sociality predicated around two convergent technological spaces: firstly that of the *haendupon* and secondly that of the Internet through virtual communities (such as Cyworld's mini-hompys) and online multiplayer games and their attendant social spaces (i.e. *PC bang*). This usage of technological spaces is *not* about substituting the virtual for the actual but rather about supplementing actual relationships. The relevance of the technology is linked intrinsically to maintaining face-to-face social capital. As Yoon (2006) observes, the rise of *haendupon* technology in Korea after 1997 was linked to the rise of youth cultures and their often subversive use that saw them labeled 'Confucian cyberkids'. Parallels can be made between the 'youth problems' associated with the rise of mobile technologies in both Korea (Yoon 2006) and Japan (Matsuda 2005), and the reorientation by government and industry to rectify the negative press.

As a broadband 'centre' with the world's highest penetration rates and fastest speeds, Korea represents a prime example of innovative and emerging convergent mobile technologies. While Digital Multimedia Broadcasting (DMB) – commonly known as mobile TV – was orchestrated as a partnership between Japan and Korea, it was in Korea that the technology (after an initial lag in uptake) was successfully adopted, unlike Japan where the implementation was stalled. Interestingly, although the advertisements were targeted at the designated ideal consumer – a male between 19 and 30 years old – it was the female users who predominantly adopted and re-appropriated the technology (S.D. Kim & Hjorth 2005). Now, Korean companies such as PandoraTV and afreeca allow users to have their own broadcast channel that they and others can view on their mobile and PC. This growth in UCC content and distribution systems also marks a period in which Korean users are shifting from being defined by activities of 'scooping' or *per-na-ru-gi* (i.e. copying or transferring of other people's content) to new patterns of actively creating their own content (Yoon cited in Yoo 2008). The coordination between mobile and PC usage of the Internet to access and update virtual community sites has seen the emergence of new forms of UCC in Korea (Yang & Park 2005).

In 2007, Korea launched its 4G (fourth generation – multimedia features such as broadband) mobile media service, while other locations in the region such as Melbourne and Hong Kong were still predominantly using 2.5G. From September 2005, the streets of Seoul were decorated with posters advertising a DMB mobile phone content competition by TU mobile. Utilising traditional Korean rainbow-like colours of the *saekdong* (that are also, coincidentally, similar to a TV test pattern), these ads were clearly converging Korean national identity – *kukmin kajok* – with its prescient state as 'global DMB centre'; postmodernity was performed through the role of the traditional – both via images of such iconography as the *saekdong* to images of families. The *haendupon* as the vehicle for representing, maintaining and rehearsing the quintessential role of the 'family' in Korean-ness was unmistakable. The use of the *kukmin kajok* within the discourse of mobile technologies clearly serves to reconnect the traditional with post-modernity and, in turn, resurrect a revision of Confucius capitalism.

The role of technology is bound up in the way in which Korea exports its mobile technology products (such as Samsung and LG) globally, and also the way in which the local market of 48 million consumes local technologies and service providers. Paralleling the industry and governmental regulations that nurtured Japanese local industries and ensured innovation on a global scale, the consumption of mobile technologies in Korea is tightly bound to explicit and implicit forms of nationalism. The hardware and software components are made in Korea, serviced by Korean telecommunication companies, and with conservative estimates of 78 per cent penetration rates, the Korean mobile phone success story has taken global centre stage.

It is incredible to consider that Korea, just ten years after the IMF financial rescue, has secured itself a leading position of soft capital, both globally and specifically in the region. By scripting itself as a centre for convergent digital technologies, Korea is rewriting its history through the global currency of ICT production. As Chua notes in his discussion of the trans-Asian flows of products and cultural capital in the region, the boom in 'consuming Korea' is part of a growing phenomenon he defines as 'communities of consumers' (2006). Revising Anderson's notion of 'imagined community' to describe the associations and perceptions surrounding identification with a nation state, Chua's geo-political imaginary has discarded any allusion to the concept of the citizen in constructions of nation state. As Chua highlights, today's citizens are undoubtedly defined by the choices they make as consumers; a notion that is prevalent in much of the 'produser' rhetoric around texting, and the way in which television has been the first to exploit this via SMS voting on reality TV programs. Moreover, in the case of Web 2.0 and the rise of social software, 'imaging communities' might be a more apt rubric for incorporating the 'prosumer' rhetoric in constructing cultural differences and distinctions.

In March 2006, Samsung released a ten mega-pixel camera phone in Korea that would revolutionise the 'digital divide' surrounding quality, and thus content, between camera phones and stand-alone digital cameras. One of the dominant differences in the relationship between the two was linked to the depiction of, and association with, official and unofficial occasions. The camera phone was always there, 'on hand' (both literally and metaphorically), to capture the trivialities of everyday. By contrast, the high-resolution stand-alone camera was brought along purposely to events deemed 'special'.

A second difference related to the *context* for sharing. The camera phone had 'sharing' built into its logic with quick functions such as MMS, Bluetooth and the facility for uploading to a blog almost instantly. By contrast, the digital camera had to be taken to an often-stationary computer and then uploaded. The launch of the ten mega-pixel camera phones represented the *connection* of these worlds. No longer would the camera phone images just be trivial and 'fun'; they had the potential to be printed in high resolution, blurring the world between amateur and professional digital photography. The introduction, in 2007, of a new breed of quasi-professional camera phones that have professional lenses and capabilities such as LG viewty (which allows users to create and edit movies and upload to UCC sites such YouTube) along with workshops and competitions for UCC, high-lighted the visible push towards Korea as a global showcase for leaders in UCC.

As a burgeoning centre for innovative technologies and with a conspicuous usage of technologies in the everyday, Korea's capital Seoul could be viewed as a showcase of techno-nationalism. The projection of 'dynamic Korea' (the tourism slogan used from 2005 to 2007) is one that has infused notions of technological innovation with the rise of power associated with Korea as a nation. In September 2006, approximately 6.4 million *PC bang* operated on the dominant Korean Internet portal, Daum. Almost 80 per cent of Koreans use these *PC bang* as a social space, participating both online and offline for an average of over six hours per week (NIDA 2006). With over 20,000 *PC bang* gracing the second levels of most commercial buildings, and with over one-third of Korea's population spending hours per day online in the virtual community of Cyworld mini-hompy, one could be mistaken for believing that online identity and relationships were surpassing offline sociality. However, although Koreans do, in general, have much trust in technological spaces such as the Internet as a site for reliable information and democratic com-munication, the online is still no substitute for offline sociality.

As Yoon's ethnographic study of young people's use of mobile phones noted, the mobile phone helps to reinforce physical contact and exchange (2003). In Hjorth's and Heewon Kim's (2005) ethnographic study on youth using Cyworld's mini-hompy community, it was found that virtual connect-ing was always about the *need* and desire to be *connected* on various levels and *never* about *substituting* for face-to-face contact. Thus the overlap between virtual and actual was inevitably about offline relations and connections. Such functions as 'search people' allow users to reconnect with old friends they have lost contact with. In Florence Chee's persuasive ethnography on *PC bang* and the politics of online multiplayer games, she argues that these spaces are *social spaces* that are viewed as 'third spaces' between home and work (2005). For Korean youth, of most whom still live at home before getting married, these *third spaces* operate as private spaces to connect with other people. As Jun-Sok Huhh (2008) observes, the *PC bang* ensured the success of online games in Korea by nurturing both the culture and the business side of the industry; thus the online game is seen as synonymous with the *PC bang*.

Parallels can be made between, on the one hand, the rise of the webcam and 'reality' aesthetics associated with the hand-held camera in television and film, and on the other, the rise of the camera phone and sharing internet communities such as Cyworld, MySpace and YouTube. As a convergent communicative media premised on the logic of gift-giving, the various ways in which camera phone images can be 'stored', 'shared' and 'saved' are, as Ito and Okabe (2005b) note, relevant to how the images are read and con-textualised. With the low resolution giving further 'authenticity' and 'realism' to the 'voice of the people' aesthetics, the camera phone provides a glimpse into the user's personal world – a genre and technique that is also being quickly outdated with the rise of the high-resolution and superior lens. Examples of new advanced technologies can be found in the aforementioned LG Viewty that are concurrent with the rise in reality TV programs and in the UCC.

For D.H. Lee, in her ethnographic work into the ways in which camera phones are used in Korea, these practices reinforce while at the same time provide contingencies for female empowerment and new ways of seeing and presenting (2005). As Lee notes, by 2004 mobile phone penetration rates were around 75 per cent, with 36.1 million people owning one or more handsets. The role of the phone and mobile media in Korean everyday life is all-pervasive, with users upgrading their phones every ten months on average. In 2004, 73 per cent of the *haendupon* sold were equipped with built-in digital cameras, and by the beginning of 2006 it was virtually impossible to buy a *haendupon* without the integrated camera. As Lee and Sohn note (2004), the changing representational codes and accessibility for image making and dis-tribution are affording opportunities to groups that were previously excluded from the domain, most particularly women. In their study, Lee and Sohn found that women were more 'active in adopting new multi-media functions of the mobile phone' and that 'their willingness to adopt such functions is significantly stronger than men's' (ibid.: 1).

With Korea's status as the most broadbanded country in the world, the relationship between online and offline is seamless (although offline

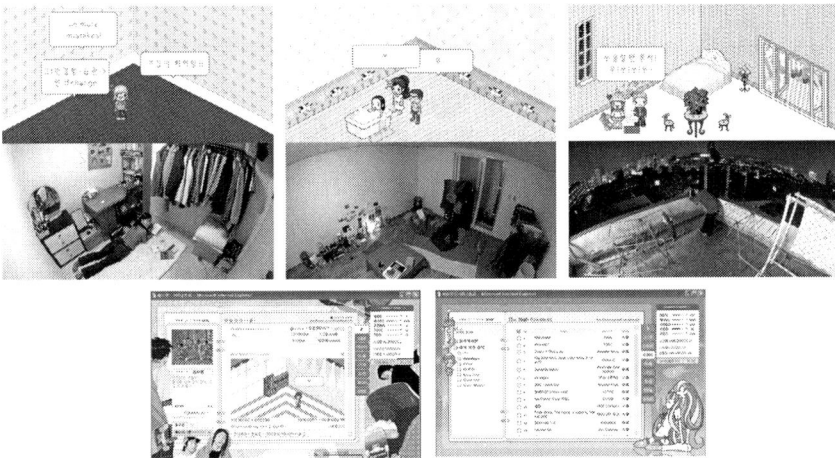

Figure 5.2 (a–e) Examples of Cyworld mini-hompy (featuring the mini-room) – imaginary and actual. The three top images are by artist Emil Goh, in which he compared people's mini-hompy with their offline living spaces. Figure 5.2d is an example of male performativity and personalisation in Cyworld mini-hompy, whilst Figure 5.2e is indicative of a female user's Cyworld mini-hompy.

communication is still more highly valued). Online virtual communities such as Cyworld mini-hompy are used by both young and old, and play a significant part in most people's everyday lives. In order to gain some insight into the dominant mode of *haendupon* practices (particularly camera phone use), this chapter will provide some background into the rise of Cyworld communities providing context for the final section on the case study of camera phone users.

Home page: locating Cyworld in Korea and beyond

> Money pours in when the Cyworld population goes on a decorating, gift-buying or music-downloading spree to adorn their 'room'. The more attractive and interesting the room, the more visitors it gets. And in Cyworld, popularity equates to fame and success. The site even measures sexiness and friendliness, which it gauges by the number of gifts a person gives or receives.
>
> (Cameron 2005: n.p.)

> Cyworld's social networks are already known in East Asia, with an especially strong base of 18 million users in its home country of South Korea. Riding that success, Cyworld is expanding operations worldwide. Among its targets is the United States this August. While MySpace is currently atop the social networking heap, Cyworld comes with seven years experience and backing from SK Communications, a subsidiary of South Korea's largest wireless operator SK Telecom.
>
> (Jacobs 2006: n.p.)

In August 2006 Cyworld launched its web community, mini-hompy, in the US. Having already launched in Asian locations such as China, Japan, and Taiwan, Cyworld's launch into the US marked a move towards a global web community. However, what is significant is its ongoing role in Korean daily life. Over 90 per cent of 20–29 year olds have mini-hompys, with 92 per cent of mini-hompy users updating at least once daily (Kanellos 2006). In 2006, Cyworld recorded annual profit figures of 100 million US dollars per year, three times the revenue of Western equivalents such as MySpace (ibid.).

As one of the first Korean telecommunication companies, Daum provided the first free emailing in 1997. Daum established the first Internet cafés in 1999, and they now boast over 50 million users. The rise of online communities through Internet cafés from 1999 to 2002 was epitomised by *pyein* (geeks) using the online community site Cyworld (dubbed *cypyein* as in cy-geeks). Cyworld was started by four KAIST (Korean Advanced Institute of Science and Technology) graduates in 1999, using modest means. It was acquired by telecommunications giant SK in 2003 for US$8.5 million. In 2003, 11 million Korean households (75 per cent of all homes) subscribed to broadband, which helped ensure Cyworld's success in reaching in the general public. This year Cyworld is expected to generate over US$140 million in sales (Jacobs 2006).

Cyworld's success has been, in part, attributed to dominant lifestyle trends in Korea such as the ubiquity of high-rise blocks of flats, and the easy accessibility of broadband coverage. The importance of Cyworld is demonstrated by the fact that over one-third of Korea's 48 million people regularly use their own and visit friends' mini-hompy. In Cyworld friends are called *ilchon*, a concept once used to denote one degree of distance from family members in a traditional Korean kinship (i.e. one's mother is one *chon*). Cyworld has re-branded its cyber-rooms with the notion of *ilchon* and non-*chon* to denote 'friends' and 'non-friends'. *Ilchon* can gain more access to their fellow *ilchon* information and be invited to visit their cyber-room. *Non-chon* can only gain cursory access.

As one of the first Internet companies in Korea, the purchase of Cyworld by Korea's super-giant, SK communications only furthered its popularity. Featuring a photo gallery, music section, message board, guest book and personal bulletin board, what differentiates a mini-hompy from a blog is its 'mini room', in which the users' avatars can be housed. This mini room often reflects the users' own lounge room (or ideal one) and users can invite friends – in the form of an avatar – to visit. In this way, the mini-hompy tries to replicate users' offline lives offering a seamless and instantaneous connection, thanks to high broadband speeds. However, this seamless co-presence between online and offline does not reflect a blurring of hierarchies; rather, users indefatigably state the importance of offline contact over online connection (Hjorth & Kim 2005). Moreover, as noted in Chapter 4 in the discussion of the Japanese mixi site, although visual differences between the two SNS are striking – mini-hompy sites are adorned by the cute, while the mixi site is relatively austere – there is an obvious emphasis on community rather than on the 'networked individualism' that Barry Wellman (2002) notes in Western equivalents such as MySpace.

The emphasis on being faithful to one's offline identity has ensured the success of Cyworld in Korea (Hjorth & Kim 2005). Moreover, the gift-giving culture of Cyworld, where friends can buy each other music or cheap cyber-gifts (such as furniture) for their mini-room with the Cyworld currency *dotori* (meaning acorns), reflects already existing customs of generosity and hospitality that are key to traditional Korean identity (S.D. Kim 2002). Mini-hompy avatars (all of which are 'cute', although they can be customised according to the user's mood) provide a vehicle for negotiating co-presence between online and offline spaces. Thus, far from eroding the importance of place and actual contact, they help facilitate traditional forms of intimacy. What is evident from the outset is the underlying force of traditional forms of sociality, which sees *PC bang* not just as a place to play online multiplayer games but often also as a place for connecting socially (Chee 2005). The usage of technology to reinforce forms of intimacy and contact is so prevalent that even 'co-present' (virtual and actual) spaces such as Cyworld's mini-hompy reflect a strong correlation between online and offline identity and relationships.

Avatars, as Cyworld's mini-hompy attests, express the investment – both financially and emotionally – of users in co-present communities, and their willingness to maintain them. Although the cute may elude to polysemic interpretations, this multiplicity should not be mistaken as a type of apoliticality. Rather, with the rise of global social networking spaces such as Cyworld, Gaia and Habbo Hotel, the cute is linked to types of sociality that in turn are linked to specific forms of locality. Behind the 'cuteness' is a struggle to humanise social technological spaces, to reassert intimacy regardless of technological interference. Although the use of the cute in the west has been associated with child's play (White 1993), in the Asia-Pacific the cute is part of a neo-human mode of customising and socialising new media spaces (Allison 2003; Hjorth 2003a).

Interrelated with its role in humanising and personalising technological spaces, cute customisation, has, as I discussed in Chapter 4, a history infused with gift-giving. It is not by accident that both cute customisation and mobile technology re-enact gift-giving traditions (Taylor & Harper 2002). This is particularly evident in Cyworld, where social networks and the norms of reciprocity and trustworthiness that arise from them play a central role. Gift-giving in Cyworld can operate on various levels – the gift of visiting someone's mini-hompy; the gift of leaving a message in the visitor's book and then the return gift of answering it; the gift of asking someone to be your 'cyberrelative' or cybuddy; the gift of sharing a photograph with someone; the gifts of *dotori* or buying cybergifts for one's home page. All these processes can be seen as contributing to individual social capital (Hjorth & Kim 2005), and collectively may lead to growing social capital among Cyworld users.

Nevertheless, much of the gift-giving associated with Cyworld occurs in ways that cannot be accounted for by the Cyworld register (an official register on each mini-hompy site) of the user's popularity, charm and generosity. The 'unregistered' gifts that are perhaps the most significant are the gestures of presence and co-presence that can be found in *haendupon* practices, such as uploading camera phone images of friends for them to share and relive. Having contextualised the role of gift-giving in the overlaying between online and offline *haendupon* practices (and particularly camera phone images), I will now turn to the micro-politics of the everyday user in a case study of 34 Korean University students, conducted from September to December 2005.

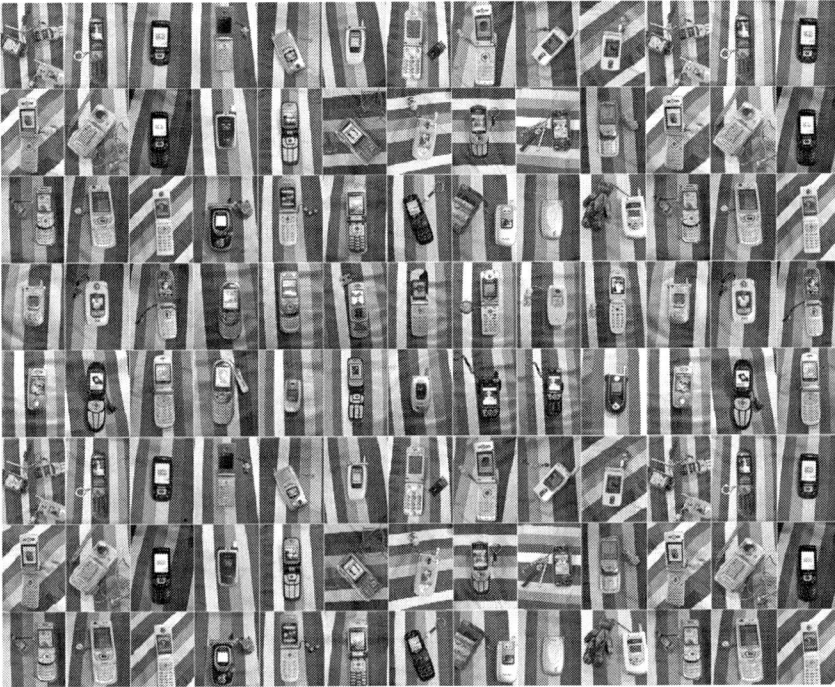

Figure 5.3 Some examples of students' *haendupon.*

Snapshots: a case study of Korean university student camera phone practices

From September to December 2005 I worked with 34 university students at Hallym University, asking them to analyse and comment on their own mobile phone practices. The student group, coming from various disciplines such as English, Media and Communication, Sociology, IT and Film/TV, consisted of roughly half female and half male participants aged from 18 to 29 years. Most were born in Chuncheon, apart from two Filipino exchange students. In a series of workshops where students worked individually and collaboratively to compare and contrast, I asked students to reflect on their textual, aural and visual uses of the *haendupon*. Beginning with lectures on the ways in which theorists and practitioners have explored mobile media and ICTs, the seminar moved on to workshops where students were asked to conduct presentations based on three small research projects.

In the first project students worked individually to analyse their own camera phone images considering notions and motivations such as Ito's and Okabe's three 'Ss' – sharing, storing and saving. The second project consisted of students working in pairs with each partner documenting their own camera phone images for a week and then giving their data to their partner for further analysis. In the third project, students documented their *haendupon* practices for a week and then gave the data to their partner for further analysis. This project also included students conducting interviews with partners after analysing the data, to draw more personal meaning from the material.

This sample study of Korean university students is simply a 'snapshot'. It is not meant to be representative of Korean young people in general. Rather, it is an attempt to sample actual users rather than perpetuate media stereotypes about the sort of young techno-savvy Korean users who Yoon has dubbed 'neo-Confucian cyberkids' (2006). It should be noted that over the four months, students developed strong trust and friendship bonds that led to increasingly open discussion and sharing. Students also participated, both in collaboration and individually, in collating and analysing the data.

I encouraged the students to think creatively about how a phone is personalised – including 'naming' their *haendupon* – and how their practices conformed or deviated from conventional or well-documented genres. It should be noted that the gender differences in usage were linked to age differences, as most of the male students were older than their female counterparts because they had completed at least two years' military service before university. The students were asked to consider various factors in their camera phone practices, namely the 'what' (subject/theme), where and why the pictures were taken, and how the images were consumed, shared and stored. Students were also asked to consider the two simultaneous levels on which MMS functions, namely the individual and social levels. On an individual level, respondents used camera phone images mainly as mementos. On a social level, they used images to share with others both as a performance of their identity and to maintain intimate ties.

In the workshop, all but two students had camera phones. For over 80 per cent of them, their current camera phone was their first. For most of those recipients, the camera facility was not a consideration in purchasing their phone; many just bought the camera phone because it was hard to buy a phone without a camera. The gender of the recipient did inflect the types of images that were taken, but only marginally. For many of these users, the sharing, storing and saving of camera phone images was primarily via peer-to-peer *haendupon* and secondly via mini-hompy sites, with sites such as Flickr barely registering. Most students explained this by noting their dissatisfaction with the low resolution of the camera phone, which rendered it suitable only for capturing 'trivial' everyday events and moments.

For most of the respondents, camera phone use genres fell into the following categories: everyday, special occasions/places, friendship/family, self-portraiture, and favourites. Although the categories were often generic in their deployment of well-known analogue photographic genres, the three 'Ss' were an important distinguishing factor in how and why respondents took their camera phone images. Many respondents described their camera phone usage as 'personal', thus rendering and contextualising their images not only in terms of earlier epochs of personal photography but also in terms of a loose style of 'documentary'. By this I do not mean the omnipresent 'god-like' documentaries that claim a type of authenticity and objectivity, but rather the genre that has been hybridised by reality TV, and that has now become a common way of digital storytelling. It should be noted that the low-quality DIY-type aesthetics of current camera phone imagery are similar to that of the web-cam, and hence have a similar 'voice of the people' (seemingly unedited, immediate) authority. Respondents were clearly aware of the subjective elements in their mobile digital storytelling techniques, acknowledging the 'newsworthy' and 'ambient intimate co-presence' aspects of their images in the way they stored and shared them. I will now turn to the projects that the students conducted in pairs.

In the first case study, consisting of two males in their early twenties, the participants noted three main categories of camera phone images also – 'routine', 'special occasions' and 'family and friends'. In the routine category, both respondents had taken images of themselves performing their day-to-day activities such as playing an instrument or documenting specific sites of everyday practice (such as the university café). The students did not share these images, and only one of the images was stored in a site other than their mobile (in this case, the mini-hompy). Many of these images would have been deleted, had they not saved them for the workshop. Motivations for taking photos varied from wanting to monumentalise a significant moment to techniques of time-filling (or 'killing time') when waiting. The 'special occasions' images, depicting subjects such as the respondents standing next to a favourite car/motorbike at a special motor expo or images of people, were more likely to uploaded onto their mini-hompys for friends to view. The 'family and friends' images were more actively archived and stored in sites such as mini-hompys for re-presentation and sharing.

One of the distinguishing features of the photographic style of the male respondents was that the images, regardless of content, were taken from a distance of at least 1.5 metres with the subject firmly in centre frame. This suggests that either the camera phone picture is taken on a timer or there is another person present who has taken the photograph for the respondent. Here we see one of the most enduring documentary modes being utilised: the omnipresent, god-like position. It was the mode in which the images were taken, rather than the subject matter, that differentiated between the genders, as was evident from the second case study pair of two late teens/early twenties female respondents. The genre of *sel-ca* (self-portraiture) was predominantly unfavoured by the male respondents who viewed it as 'narcissistic' and feminine; one male respondent noted, 'that's what girls do'. However, many examples of *sel-ca* could be found equally in male and female respondents' *haendupon*. The significant feature of *sel-ca* was the way in which it was shared; predominantly sent by female respondents to best friends or their partners to further give a sense of co-presence.

In the second case study, the two female respondents had no categories for their images and stated that they stored most of them either on the phone (sometimes as a screen saver) or on a mini-hompy page. One of the respondents had images of herself, her friends (with customised drawing on paper that she had documented), a detail of a piece of pizza and a detail of drinking at a bar. Most of her shots were taken from less than one metre, suggesting that the respondent was also the photographer (in most instances, this was the case) and that, in some cases, she was alone (for example, one photo was taken while her friend was in the restroom). Some of her images were blurry and obviously taken hastily, with emphasis on expressing a mood or feeling in a *cinema vérité* manner akin to photo-booth performativity or quasi (self) paparazzi style. This mode demonstrates one of the ways in which the camera phone can be used as a tool for creating and re-presenting oneself through expressive digital storytelling techniques.

The other participant's images featured the same style of *vérité* phoneur imagery, taking detailed images of food that was obviously a shared meal. Here food – deeply symbolic in the Korean sense of family and sharing – is used to convey a sense of friendship, conversation and connection. On the individual level, these moments are about meeting with friends, the importance of actual physical contact being symbolised by the way in which pictures of food denotes the other senses like smell, taste and touch. On a social and cultural level, eating and exchange rituals around food are an integral part of Korean tradition.

The third case study, consisting of three male respondents, illustrated how the camera phone user can play the role of both director and subject, akin to Jenny-cam aesthetics (the phenomenon of a 15-year-old girl who made her life accessible 24–7 on the Internet). In this context, the low resolution of the camera phone images gives authenticity, implying a type of self-confessional and deeply personal genre that fulfills voyeuristic tendencies to witness the 'voice' of the everyday. In a world saturated with glossy and hyper-real images, this type of DIY aesthetics has a growing audience of neo-realists. Combining the *vérité* phoneur aesthetics of the second case study with the more open frames and distances of the first, these respondents saw the camera phone largely as medium for creating postcards, that is, for making 'I was here' statements.

Two of the respondents used their phones to 'document' themselves in expressive and playful ways, for example by using camera angles that echoed the 'fly-on-the-wall' documentary style. The third respondent used the camera frame in a more intimate way to emphasise expressive and humorous gestures. Many of his close-up shots were intentionally ugly, with the respondent playing the 'mook' (an MTV [music television videos] category for young rebellious men). These three respondents did not have 'steady' girlfriends, so most of their photos revolved around themselves and their friends.

There was an obvious contrast, in both subject matter and mode of framing, between respondents with and without long-term partners. This was the most distinguishing factor in the usage of camera phone images, more so than gender/age differences. Whilst the single respondents tended to play to imaginary (or, if the images were uploaded and shared, real) audiences of their friends, respondents with partners almost exclusively took and shared photos with their partner. One noticeable difference in camera phone usage between genders was the height at which the camera was held. Over 85 per cent of the images taken by male respondents were taken with the camera held at either eye height or below. By contrast, female respondents took almost all pictures of themselves and friends with the camera held above eye height. This positioning suggests equality, as well as an intimacy, between the photographer and photographic subject.

In the fourth case study, one of the two participants had a partner. Most of the photos that he had taken consisted of him making 'cute' faces for his girlfriend (either taken by her, or taken by him with the thought of sending them to her), pictures of himself and his girlfriend, or pictures featuring just his girlfriend. In one of his photos, which he had categorised under 'favourite', his girlfriend's eye took up the whole frame. In another, categorised under 'myself', his girlfriend sits in the foreground taking the picture of herself and the respondent in the distance. While the girlfriend looks into the camera, he looks lovingly at her.

This male respondent clearly demonstrated his 'engaged' status by customising his phone. He used the omnipresent image of the girlfriend's eye (his phone hence also becoming her "I") as a screen saver for a couple of months. Interestingly, almost all of the images stored on the phone were either *sel-ca* sent from his girlfriend or pictures his girlfriend had taken on his *haendupon*. Even the few pictures he had taken, she had orchestrated him to take. Here, the role of the female as official gatekeeper of the memories of the relationship is not only regulated on her *haendupon* but also her male partner's *haendupon*.

His girlfriend had also 'feminised' the phone by including a cute phone strap. Here we see how the male's phone can be taken over by the girlfriend as a metaphor for his commitment to her. When asked what the function of the screen saver image was he replied that it not only helped to remind him of her but also reminded him to call her. It is interesting to note that the phone and its customisations also function symbolically as an extension of the hand,

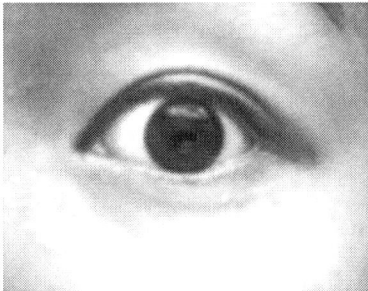

Figure 5.4 An example of an image from a 'coupled' male student's phone.

representing an engagement ring. As the phone is often placed on tables and is on constant show, it signals to intimates and strangers that the owner is 'engaged'.

In case studies five and six, consisting of female pairs, the overt use of emotion to describe and categorise camera phone practices was evident. Deploying overtly 'feminine' customisation from pink colours to cute characters, these two case studies seemed to be 'performing' and playing with gender stereotypes. This can be partly explained by the aforementioned age difference between the genders. In case study five, the two female respondents used the categories 'loveliness, 'joy', 'sadness', 'surprise' and 'tension' to classify their camera phone image genres. With their phones overtly customised in a feminised style – from the phone's cute appearance lacquered in a light pink colour with 'cute' characters attached, to the 'pretty' ring tones and camera phone images, these two respondents were knowingly 'performing' gendered identity through the phone.

Both respondents preferred the immediacy of voice calling to SMS, and took photos mainly when they were bored. Many of their photos were of themselves or school-related activities, and most were taken from above to further emphasise the aesthetics of the 'cute' and smallness so prevalent in girl culture *sel-ca* in the Asia-Pacific. In particular, the holding of the camera from above also was seen to make people look more attractive, making eyes look big and bodies small. These photos were often sent to the students' boyfriends or uploaded to Cyworld. This case study was curious in that the respondents seemed to play into stereotypical images of 'girl' camera phone practices. However, they did so with a hint of irony, toying with modes of gender performativity as if it were a mask. One respondent noted, for example, that much of 'girl' culture customisation was unnecessary and could be easily lived without.

In case study six, which consisted of two female respondents, the images were more focused on the social rather than the individual. Unlike the self-documentary style of case study five, half of the case study six images are of school-related activities and the other half are of 'cute' moments. Although the school-related activities were often taken from a distance or included a lot of background in the shots, the cute images consisted mainly of detailed shots of poetic gestures of the everyday – from a curtain on a bus window to two full glasses and a sleeping dog. While the subject matter itself is more inquisitive and exploratory rather than conventionally cute, respondents often discussed motivations for taking the picture 'because it is cute'. Here, we gain a sense that what constitutes the cute is a deeply personal and ambivalent category. On the one hand, there is the manufactured cute as in cute cartoon characters or cute tactics (such as photographing from above to make the eyes look big and the rest of the subject small). On the other hand, the ubiquity of 'cute' as a rationale or motivation dominated female respondent's discussion of *haendupon* customisation. Undoubtedly, the pervasiveness of cuteness dominated the respondent's images, even casting a shadow over her photos with her partner. Over half of her images were shared with her partner and stored in both the *haendupon* and mini-hompy.

Case study seven consisted of a male and female student pair who entitled their presentation 'Jack and Jill'. The two respondents' *haendupon* practices – from SMS to camera phone images – were vastly different. While 'Jill's' usage of voice calls was over double that of Jack's (Jill made 160 calls per week compared to Jack's 50), they used roughly the same amount of texting and camera phone images. In the camera phone images they playfully started to cross over, with Jack featuring in Jill's and Jill in Jack's. Jack's images consisted predominantly of himself playing the 'fool' while Jill's included many friends with very few of herself (she appeared mainly in Jack's photos). While both camera phone practices took on a DIY aesthetic, Jack's content was much more about playing the role of 'detective' whereby much of his images were viewed as 'clues' to his life and identity; whereas Jill's were more conventional snapshots of friends. It was interesting to note that in their reflexive play of camera phone practices, it was the female respondent who took the role of director, while the male respondent played the actor in what seemed to be Jill's theatre. Here we see the way in which female respondents will often rise to the role of the gatekeeper of mobile media, undoubtedly influencing the types of images that are taken, shared and saved. The role of emotional labour is significant in that it marks both an continuation in the exploitation of 'female' labour as well as the growing importance of emotional labour in the cultures and currencies of mobile media. This example also provides an alternative model of the *kukmin kajok*.

In case study eight, also consisting of a male and female pair, culture/ ethnicity becomes a dominant factor in *haendupon* and camera phone practices. The male respondent, a Filipino exchange student, took around seven camera phone pictures a day compared to his female Korean counterpart who only took approximately one image per day. As an exchange student, his practices differed greatly from his partner's in that he preferred texts (not just because texting was a dominant practice in the Philippines, but also because it helped with the language barriers and comprehension). He also took many more scenic – including landscapes, meals and people – photographs in which his gaze shifted often between that of tourist, traveller and anthropologist. For example, he photographed the first meal he was able to order in Korean. His photographs were well-considered and well-composed, in contrast to the 'snapshot' feel of many other respondents' images.

Here the camera phone becomes privy to the acquisition of a new culture and the slow adaptation to its everyday practices. The camera phone becomes an aid for negotiating the two worlds – accompanying him on his journey through a new culture and everyday life, and providing documentation to be sent back to family and friends in the Philippines. Hence the camera phone becomes a photo journal and takes on an ethnographic feel. This respondent never sent his images via MMS, but instead uploaded them to multiply.com and imageshank.us. By contrast, his female partner uploaded most of her images to Cyworld to share images with friends.

The choice here of different sharing and distribution networks is quite telling in terms of the contextualisation of the images. Both respondents saw the need for and stressed the importance of mobile phones in maintaining and facilitating social relations and everyday activities. The female respondent noted that 'almost all of my important private things are in my cell phone . . . Even though I am not interested in making my cell phone pretty, I think of it as precious.' The male respondent concurred by replying, 'It's very, very important for me because it's a necessity rather than a luxury. Being detached or apart from it is like being detached from the world.'

In case study nine, again featuring an exchange student from the Philippines but this time a female, the two respondents engaged in a playful discussion of the *haendupon* as an extension of one's identity by giving names to their phones. They also analysed what type of phoneur they were in terms of Plant's (2002) definitions, which outlined different birds as analogies for mobile phone behaviour. The female Korean respondent categorised herself as the Plant 'swift talker', owing to her constant moving, multi-tasking and intensive use of both voice calling and texting (in four days she had received and sent 58 SMSs, 73 voice calls and three photos). This respondent preferred video camera phone footage to still images, and like the rest of the case study groups, preferred to upload images onto Cyworld (particularly images of her and her boyfriend). She never sent MMS. The only messages she saved were 'sweet messages' from her boyfriend (her phone had a storage limit of 50 SMSs). Her still images and video footage consisted mainly of her life with friends and especially with her boyfriend. The female Filipino exchange student identified herself as Plant's 'solitary owl/calm dove', as she preferred 'company and counsel . . . non-mediated face-to-face communication', and took and made calls discreetly, making sure she was 'quiet and modest when using the mobile phone'. Unlike the male Filipino exchange student, her camera phone practices were less about landscapes and artful photo-journalising and more about capturing the significance of moments or events when she was in contact with others.

This relinquishing of control and responsibility of the *haendupon* by the male to their female partner could be read in two ways – either as reinforcing the role of certain technologies being gendered, or as indicative of a re-scripting of gendered associations. In the first case you could argue that the handing over of control to the female is due to the gender stereotypes of women's innate sense of fashion, in which the *haendupon* is viewed as the ultimate accessory. An alternative view acknowledges the dynamic and shifting role of domestic technologies in the shaping of, and reconfiguring, gender divisions. For example, through the rise of mobile media, more female respondents are extending their interest in becoming 'produsers' – rather than just consumers – of new media. Korea is a prime example (D.H. Lee 2005).

For all respondents, the most exposing forms of personalisation were the

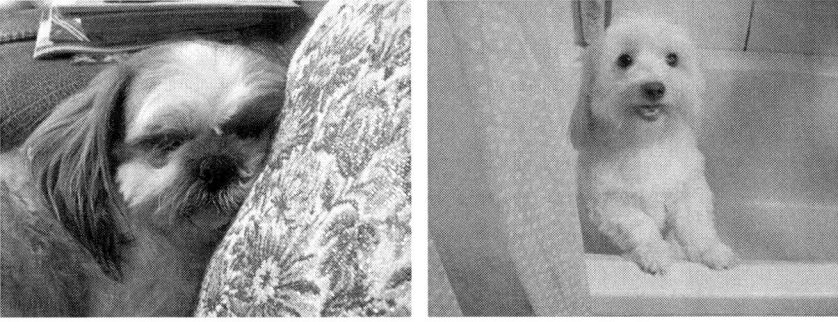

Figure 5.5a Typical 'everyday' and 'special occasion' images as denoted by 'cute' (photos by female participants).

colourings; that is, the musical tune that replaces the dialing sound that callers hear when ringing the person who owns the phone. This sound contextualised the identity and feelings of the user for the incoming caller, and hence there was great emphasis on getting the *colouring* right. A wrong *colouring* could send out a wrong impression to the incoming caller that could possibly have devastating results on the conversation and relationship. The *colouring* flavoured potential conversations, comtextualising the receiver, and creating ambience for the caller. Respondents talked about how important it was to choose the right *colouring* so that it could put the incoming caller in the 'right' mood. All of them had customised the *colouring* function because they believed this made the greatest impression on others about the user's tastes and values.

In their camera phone images, respondents frequently used objects as analogies for special moments and for gestures of friendship and sharing. Although there were examples of the typical 'group of friends' shot, many of the participants preferred taking pictures of surrounding objects to create more connotations – for example, the ritual of sharing food with a best friend; the first two beers ordered on the night of celebration with friends; the sharing of coffee between two best friends. Here we are reminded of the work of Korean

Figure 5.5b Some of the typical 'everyday' and 'special occasions/friendship' images (photos by female participants). Here, the symbolic role of food and its relationship to sharing with friends/family is demonstrated.

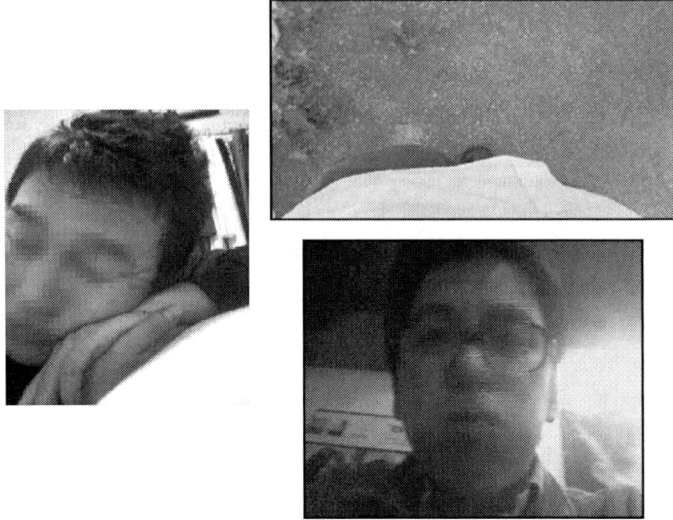

Figure 5.6 Modes of realism or 'reelism' are most obvious in webcam or user-as-media-producer techniques.

director Hong Song Soo, and particularly his film *Oh Soojung* (English title *Bride stripped by the bachelors, even,* after the Marcel Duchamp artwork), in which the documenting and re-enactment of 'everyday' rituals around food symbolise a notion of *kukmin kajok*.

The pictures above on the left and right illustrate two types of media mimicry. The image on the left illustrates a form of 'paparazzi': the respondent's girlfriend took it while they were sleeping, to blackmail them afterwards. The photo on the right is an example of 'self-paparazzi'. Often portraiture and self-portraiture mimicked the conventions of mass media such as paparazzi, but occasionally they slipped into metaphoric gestures on the sublime nature of the everyday, as can be seen in the centre image. The image on the left was taken by the student's girlfriend and then stored on her phone, while male participants took the other two shots.

The fact that women have control and regulation of their male counterpart's phones means that they become central figures in scripting which memories and experiences get recorded and which ones do not. Even though camera phone practices are considered *unofficial*, it is these unofficial images that will become part of the documentation of a UCC epoch, thus rendering them *official*.[1] This visual storytelling not only impacts on a family's memories and thus identity; it has the power to become indicative of particular national ideologies. In this way, women are not only taking a proactive role in contemporary everyday media cultures, but are also acting as gate keepers for their partners by determining the contents of the new family album through the *haendupon*. This process highlights that the *unofficial* discourse I call *imaging communities* that inevitably feeds into the future *official* 'imagined communities' (Anderson 1983) of Korea and, in turn, into emerging forms of national reorientation both in the region and globally. As witnessed in the mobile media practices of couples, the *haendupon* is no longer the site of *Hyoja*, but rather, it is the female partner that 'reterritorialises' the device inside and outside. These forms of unofficial imaging communities that, by the very publicness of the *haendupon*, become part of the public discourse, most certainly demonstrate that the male-centred correlation with the *kukmin kajok* no longer fits so conformably as it once did.

Female respondents did not participate in the same visual depictions as male users, they also were also more likely to describe and categorise their *haendupon* practices – and particularly their camera phone genres – in terms of emotions. One of the key factors that distinguished between male and female users and between singles and couples was the context in which they stored and shared images. For the study group as a whole, the dominant mode for sharing, storing and saving was via the users' mini-hompy pages. Half the respondents took on average about three images per week, most of them shared rather than kept for personal consumption. The more social the respondent, the more they were an active camera phone user – both taking images and sharing. This is indicative of the fact that Korea is a very social culture. Although the two dominant modes of camera phone use were self-expression and the 'documentation' of special events/occasions, these genres were always about a negotiation of the social and remaining faithful to one's identity both on and offline.

Camera phone images are being used to highlight significant gestures and moments in everyday life. As the participants of the sample study noted, the documentation of everyday images helps to give significance to what might otherwise be deemed 'mundane' and trivial; it gives significance to the every-day. In this way, camera phone images are articulating the very paradox of the everyday – navigating both the trivial and the sublime that constitutes the everyday. Camera phone practices are part of general customisation modes; one's phone becomes a battlefield where the technological is always relocated in the social and intimate. As mobile media grows into a multimedia device with multiple capabilities, the competing customisation practices between the industry generated (i.e. gender scripting) and UCC become more dialectical.

The role of co-ordination with other virtual spaces, most particularly mini-hompy, also dominated discussions. Many of the respondents spent on average one to two hours per day updating and accessing theirs and others' mini-hompys. It was seen as an important activity, particularly for those respondents who wanted to keep in contact with high-school friends or family that they had left to come to university. One of the conspicuous issues that arose in my discussions with the students – an issue that is central to the rise of UCC – was the amount of (free) time and effort users spent in shaping content. This raises the question about who profits – in various senses of the word (for example, emotionally or financially) – from UCC. The very vehicle of freedom for self-expression and representation can also be a further entrapment of the user's emotional and creative labour.

Many respondents suggested that there was an involuntary element at times, in the sense that they felt compelled to keep up, keep connected and keep updated: to be part of the emerging *kukmin kajok*. This negative side of UCC customisation was most strongly felt in the Korean case study. This can be explained, in part, by fact that online communities such as mini-hompy have been running the longest in Korea.[2] While the respondents felt the need to keep in contact with friends and family as one of their highest priorities in life, sometimes the upkeep was unrelenting in an age premising the significance of the 'instant' and 'immediate' as symbolised by Web 2.0.

Specifically, some of the respondents expressed great concern that some of their professors had started a mini-hompy, and thus might be able to access limited sections of the students' sites. Some felt that it made them less likely to disclose parts of their lives to the general community (*non chon*). Here we see that the power relationships experienced offline are replicated online. Moreover, many respondents noted the various 'costs' – both economic and immaterial – in maintaining one's mini-hompy and online reputation.

Much time can be spent – either through a PC or *haendupon* – updating one's mini-hompy so that it is current. Faced with the risk of becoming yesterday's people, users can feel compelled to perpetually re-present the present. Gift-giving and participation (often under duress) has a trade-off. Much time can be spent maintaining connections through the various modes of gift-giving, so that users can feel enslaved to community in order to maintain a presence. However, it is important to recognise that gift-giving has always involved, and been a product of, social hierarchies (Mauss 1954); the existing forms of social structure, power and obligation are clearly performed within mobile media practices (Taylor & Harper 2002, 2003, Berg, Taylor & Harper 2005).

In the 2005 study, respondents noted that they would often immediately upload their camera phone images of an event to their mini-hompy for friends to view and share. However, this often meant that they deferred the experience of being present by taking the role of social recorder; some respondents mentioned that they felt that the burden of social responsibility in capturing the experience accurately meant that they could not enjoy the moment as they needed to be detached. They noted that this resulted in them only later fully experiencing and enjoying the moment by viewing the images afterwards and thinking about the pleasure they were giving to others ('emotional labour'). This phenomenon of temporal disjuncture is what I have referred to elsewhere as fast-forwarding to the present – through the trope of 'fast-forwarding present' (Hjorth 2006a) – in terms of Japan in Chapter 4. Here we see that the presents of gift-giving co-presence can jeopardise the experience of presence in the present. This paradox of convergent media, whereby users find themselves labouring (free) more around 'everyday' moments in an attempt to maintain intimacy, was evident in many of the respondents' comments. One female respondent, aged 26, noted:

> At one stage I felt like all I was doing was taking pictures, editing and uploading them. This was initially fun because my friends enjoyed seeing them – especially when they were in them! However, I found the novelty wore off and then my friends came to expect me to be the photographer, which meant that I couldn't enjoy the event as much because I had to concentrate on getting 'good' pictures of everyone. One time I pretended the camera wasn't sending pictures properly so someone else would take the responsibility.

When asked how much time they spent daily taking, customising and uploading camera phone images, many respondents thought they spent at least one hour per day. The compulsion to continuously update, maintain and check their own and other mini-hompys proved time-consuming, with many respondents claiming to spend at least three hours per day. This reflected stories in the media of *Cyholics* who have confessed that they became addicted to mini-hompy and ended up living predominantly online, in the world of Cyworld, rather than using the online to reflect offline relationships and contact.

In the case of Cyworld, the *haendupon* is often deployed in many parts of the process in accessing and updating mini-hompy pages, despite the fact that many use the PC in a location where wireless broadband is readily available. Far from being liberated by a device that allows them to update and check their mini-hompys while on the run, many users expressed frustration about the constant pressure to be present in Cyworld, complaining about the 'need' to take the *haendupon* everywhere, even to the boiling hot sauna of the *jimjil-bang*. For one male respondent, mini-hompy is about identity that is not bound to individualism but rather to a sense of community. He states:

> A lot of girls use the mini-hompy in a more obsessive manner. I think it can be a competition sometimes. I know a lot of girls who spend a lot of time on their mini-hompys. Some mini-hompys are amazing. In Korea, accessing the Internet all the time anywhere is part of everyday life. The Internet is about community.

As the above respondent notes, the Internet is very much 'part of everyday life' and thus the offline/online distinction can seem artificial. One female respondent stated:

> Many younger people like to spend a lot of time decorating and updating their mini-hompys. It is like a fashion. But then, isn't youth about great demonstrations of expression? If it weren't the mini-hompy, it would be clothes or music.

While highly engaged mini-hompy users may soak up their funds (in the form of *dotori*) by perpetual customisation, one could argue that they are not being exploited or duped, but rather are using their mini-hompy as a creative and performative space. For another female respondent, the cost of such performativity never outweighed the importance of face-to-face. She suggests:

> I think the upkeep of a mini-hompy can become obsessive, but I think that is more to do with the actual person and who they are in the real than the mini-hompy changing a person's character. There is often a story in the paper about a person who dies in a *PC bang* after playing a game too long . . . but often it is the same story repeated. I used to be more interested in my mini-hompy but now I just see it as one form in many to communicate to my friends and family. I'm not interested in spending too much time online. Neither are my friends. It is just what you do when you can't see each other.

For another female respondent, the mini-hompy was an important way to keep disparate groups of friends cohesive. She stated:

> I like to keep in contact with school friends and university friends. My mini-hompy is convenient in that way because I can put my recent images there and always keep friends involved. It has been a way for me to keep in contact even though we are all going different career paths.

When asked about how the cute graphics of Cyworld functioned, she replied, 'I like the characters and design. It's playful. It makes the communication more fun and less formal.' The playful possibilities of customisation, both with mini-hompy and through mobile media such as camera phone practices, were an important part of why respondents continued to utilise customisation despite the fact that it could monopolise much time and effort. For the respondents, being social and bonding with a community was central to life, whether it was online or offline. Respondents saw online customisation as important in maintaining offline relations. Whether via the *haendupon* or mini-hompy, respondents identified that maintaining a strong sense of community, rather than individualism, was integral to Korean everyday life.

Although stereotypes abound about female preoccupation with cute customisation, the reality is somewhat different. Both male and female respondents engaged with the cute politics of mini-hompy on varying levels. Cute is a key mode of engaging into gendered intimate relations for contemporary Korean youth in the case study. The rise of cute graphic games such as *Kart Rider* (an online racing game) is indicative of females starting to engage in previously male-dominated areas such as gaming, not just as players/consumers but as producers of content. Customisation was seen as an important way to bring emotions to the technological spaces so that the 'imaging community' of a networked society could reflect the dynamism of day-to-day cultural practice. The imbuing of these spaces with emotionally driven customisation was not indicative of individualisation, rather it was about reflecting the importance of emotion to Korean national culture, which is predicated around the idea that it is one symbolic family. That is, the *kukmin kajok*. However, the once-hypermasculine notion of the *kukmin kajok* is being challenged by the hyperfeminine cartographies of personalisation.

What is also noticeable is that the customisation of the phone is very much linked to the social relationships of the user and particularly to their intimate relationships. Many respondents featured their partner as a screen saver on the phone, and the hanging accessories were often presents given by their partners; although both of these practices were orchestrated by female respondents for both their own and their partners' *haendupon*. Female partners often advised their male partners what predominant accessories should be brought, what images should be on the phone and what ring tones and *colourings* should be used. Here customisation takes the form of 'presents of presence'. Thus camera phone practices – as with mobile phone practices in general – are remediations of older media and social rituals and play an undeniable role in maintaining and reinforcing the importance of family as part of Korean national identity. The agenda, and gender, of the *haendupon kukmin kajok* conflation is undoubtedly being challenged.

Figure 5.7 An example of a 'feminine' phone owned by one of the female students.

Freeze frame: reflections upon *haendupon, kukmin kajok* and gender

As noted earlier, gender differences in the university context are often fused with age issues with most of the male students being at least two years older than their female counterparts. While some distinctions could be noted, such as female users preferring images of cute objects or meals, there was little difference overall. Almost all respondents shared their images predominantly via the community context of the mini-hompy or face-to-face. The striking difference was between single and coupled users. As stated above, once in a couple, often the male user's phone would become the object for and of the female partner. The male user's *haendupon* would be used to store images and texts from the partner, and females would often buy matching customisation objects to adorn both their own and their boyfriend's *haendupon*. Female partners often took *sel-ca* and sent them to their male partner's phone; whereas the images of the male partner were often taken by the female on her phone. In both scenarios, the female plays the key role as photographer and gatekeeper for the images and memories of the relationship.

In a country where the notion of family is synonymous with Korean nationalism (*kukmin kajok*) and the phone is symbolic of South Korean postmodernity (in striking contrast to North Korea), the female is playing an integral role. The relinquishing of the male respondent's *haendupon* to his partner, to be feminised, both signals to others that the male is engaged and constructs the *haendupon* as a repository for the relationship memories and constant presence. While similar parallels could be found in other locations, the explicit nature of this was most evident in the female's adornment of her partner's *haendupon*. This phenomenon signals a challenge to the *kukmin kajok* as it has previously been orchestrated around the *Hyoja haendupon*.

As Gye (2007) notes in the case of Nokia, current camera phone advertisements replicate the ads of the Kodak camera in urging people to capture the moment and thus render it worthy of official collective memories. This scenario is exemplified in the case of the launch of video-calling in Seoul in 2007, in which companies such as LG clearly target couples – maintaining co-presence – as the dominant users. The politics of realism and its impact upon divisions between official and unofficial, private and public are overtly inflected by gender. Just think about how the Kodak camera – a symbol for the rise of vernacular photography – gave the everyday bourgeoisie a way in which to further mythologise, and thus naturalise, their gendered performativity (especially in terms of performing familial roles like father or mother).

Gye's discussion of camera phone imagery drawing on the familial conventions initiated by the Kodak camera is particularly apt in the case of the single and coupled modes of photographic performativity observed among the sample study of Korean university students. For participants who were single, their friends became part of the metaphoric family, often predominantly emphasising one type of gender performativity (i.e. 'masculinity' in the case of male participants, 'femininity' in the case of female participants). For coupled participants, the phone was pivotal in signaling engagement and collecting mementoes of the 'new' family. Here we see how the mobile phone is adapting already existing cultural priorities and rituals.

As can be noted by the differences between single and coupled students, the phone was both literally and symbolically used to denote, capture and operate as a perpetual presence for loved ones. For couples, females took control

Figure 5.8 An example of advertising in which the multimedia capabilities of the *haendupon* are customised by cute character culture. The symbol of *Hyoja*, Samsung Anycall, in this image has been reterritorialised by the female 'produser'.

over the male's phone and customised it inside and outside to signal that the male was engaged. For singles, the phone operated as a site for gender performativity amongst female friends or male friends. The divide between male and female was overt, with female users filling their *haendupon* and mini-hompy with images of themselves performing with female friends and often performing 'femininity' in an exaggerated manner. Although males, too, performed 'masculinity', they tended to take fewer photos and share less than their female counterparts.

I want to return to an earlier proposition about whether gender inflects the mode of realism of camera phone practices. Female users tended towards taking more images especially around the ritual of eating food with friends, and tended to take more paparazzi images of friends that they uploaded to their mini-hompy pages and shared. Moreover, customisation patterns were markedly different between singles and couples. Male partners often surrendered their phone to their female partner, who would adorn the inside and outside of the *haendupon* with reminders and signifiers of their relationship.

In this way, the *haendupon* became symbolic of the engagement ring, being a constant presence for users of their partner as well as signaling in public that the user was 'taken'. The *haendupon* became the keeper for the relationship, storing images and messages as the relationship grew. For single users, the *haendupon* became a shrine for friends and moments shared. In general, females were much more proactive in documenting shared moments and then

Figure 5.9 An example of hand-made customisation. Photo: Hjorth.

distributing them via mini-hompys or MMS. In addition, female users also took more images of objects and animals that they liked, which were often either for personal consumption or for sharing in face-to-face situations. The sharing of camera phone images and participation in communities such as a mini-hompy reterritorialises place and maintains the importance of social networks that demonstrate burgeoning roots and routes of imaging communities.

Through the *haendupon* we can see emerging patterns of intimacy that, in turn, reflect new gendered formations within the *kukmin kajok* framework. Gendered labour and intimacies are being performed and archived through the *haendupon*, making it a poignant symbolic and material practice for emerging mobile and immobile imaging communities, which, in turn, reconstitute the *kukmin kajok* within twenty-first-century postmodernity. Undoubtedly, the emerging forms of mobile media imaging communities – with their overtly hyperfeminised modes of personalisation – are reconfiguring Korea's 'imagined community' in the twenty-first century.

6 Nostalgic mobility

Memory and the mobile phone in Hong Kong

Competition is so intense that voice call prices are amongst the cheapest in the world. On 2002 ITU data, monthly usage of 300 minutes costs over US$80 in Japan but less that US$20 in Hong Kong. It is probably much lower today due to continuing vicious price wars. It is actually cheaper to make a voice call than send an SMS in Hong Kong. As a result, Hong Kongers have never developed a 'data habit' and 2G operators are focused on the struggle to survive rather than investing in future services with uncertain prospects.

(Waters 2004: n.p.)

While Tokyo and Seoul exemplify the Asia-Pacific's mobile revolution, other locations in the region such as Hong Kong and Melbourne present a different picture. Although these two latter cities have little in common apart from a shared history of British rule, the uptake of mobile technologies has been similar, particularly in terms of the relatively slow uptake of 3G technologies. Both locations have moderately small and fragmented markets serviced by numerous global telecommunications companies and service providers (unlike Korea and Japan, where the service provider sector is largely monopolised). Although Hong Kong and Melbourne have similar mobile phone markets – even having the same advertisement campaigns for brands such as Motorola and Nokia – the existing socio-cultural and linguistic differences between Hong Kong and Melbourne result in two very differing models of mobility.

In the case of Hong Kong, mobile media – from customising the outside of the phone with cute characters to the particular modes of camera phone practices – provide multiple avenues for nostalgia and reterritorialisation. In the reconstruction of Hong Kong post-1997 handover, the role of nostalgia has often been discussed as a site for displacement and resistance to territoriality while, at the same time, operating to create a sense of cultural identity in the face of 'weak nationalism' (Ma 2000). Much of the literature on Hong Kong and the politics of nostalgia from Eric Kit-wai Ma (2005) and Rey Chow (2000, 2007) has focused upon cultural practices such as film to evoke and rehearse nostalgia as a trope for Hong Kong post-modernity. As I will demonstrate in this chapter, the role of the nostalgia deployed by mobile media practices can provide more acuity – particularly from a micro-imaging communities perspective – into the current reterritorialisations of Hong Kong's imagined community within the twenty-first century.

For Gilles Deleuze and Félix Guattari, contemporary globalisation has caused 'bounded' territories to undergo processes of deterritorialisation followed by reterritorialisation (1986). In the case of Hong Kong, this tripartite process is complicated, with spectres of previous geo-imaginary identifications haunting. The return back to the 're-boundedness' symbolised by the 1997 hand back to China is not without nostalgia (Ma 2005). In this oscillation, complicated by spectres of prosperity and memories of economic mobility, mobile media practices such as camera phone play an important role. As Koskinen (2007) has noted, camera phone practices operate to perpetually reterritorialise a sense of place upon an individual and personal level that, in turn, reflect new forms of identity, expression and community. This can be seen in the ways in which imaging communities continuously reterritorialise, providing roots and routes. These imaging communities produce affective reterriatorialisation for users, a process of reterritorialisation that differs from the Deleuze and Guattari's model.

For Ma, nostalgia is central to Hong Kong's identity (2005). This nostalgia is performed upon both an *individual* and *collective* level, and through both *unofficial* and *official* media. There is a tendency to believe that there were times in which things were better, most notably before the return to China and the simultaneous financial crisis of the region in 1997. This nostalgia is re-enacted through the procurements of commodities, particularly mobile phones, and the ways in which they are used. For Chow (2007), the deployment of nostalgia by Hong Kong's New Wave directors such as Wong Kar-wai, can be viewed as part of a general trope of the 'recurrent sentimental' in Chinese cinema from fifth-generational filmmakers like Chen Kaige and Zhang Yimou in the mainland, to Ang Lee in Taiwan. As Chow argues, 'the persistence of a predominantly affective mode' – a mode she calls 'sentimental' – is a key aspect within contemporary Chinese cinema that seeks to challenge Western precepts of modernity (ibid.: 14). For Chow, it is the Chinese notion of *wenqing zhuyi*, which translates to 'warm sentiment-ism', that is one of the enduring affective techniques in Chinese cinema (ibid.: 17).

As Chow notes, in Wong's *Happy Together*, nostalgia 'is no longer an emotion attached to a concretely experienced, chronological past; rather, it is attached to a fantasized state of oneness, to a time of absolute coupling and indifferentiation that may nonetheless appear in the guise of an intense, indeed delirious memory' (ibid.: 51–52). According to Chow, the deployment of nostalgia differs greatly from the fifth-generation Chinese directors of the People's Republic who use it as a 'form of cultural self-reflection, by way of stories about traditional China', to Taiwanese and Hong Kong directors who evoke nostalgia through 'a fascination with legendary eras with clear moral divisions' exemplified by martial art (kung fu) movies (ibid.: 52). In particular,

Wong's deployment of nostalgia seeks to reterritorialise people and places, what Chow describes as 'a flawless union among people, a condition of togetherness in multiple senses of the term. This is a condition that can never be fully attained but is therefore always longed for' (ibid.: 53).

Unquestionably the affective politics of nostalgia Chow attributes to Wong's films are instrumental in the cartographies of personalisation of Hong Kong mobile media. From customising the outside of the device with transnational commodities such as Japanese *kawaii* culture that rehearse *wenqing zhuyi* to the collecting of camera phone images and texts, mobile media practices re-enact the notion of nostalgia so particular to Hong Kong everyday life; this, in turn, reflects the particular characteristics of Hong Kong post-modernity. These practices occur upon various levels – symbolic and material, individual and collective – as an affective reterritorialisation. Moreover, these cartographies of personalisation are marked by a sense of a longing, togetherness and reterritorialisation that cannot, as Chow observes, be fully attained. Through the various imaging communities of Hong Kong's mobile media we can see a deployment of the *wenqing zhuyi* to understand and chart Hong Kong's 'hyphenated' modernity.

In Hong Kong, mobile media practices extend the modes of nostalgia as a site for resistance to the reterritorialisation forces of the mainland. Through the re-enacting of nostalgia within mobile media cartographies of personalisation, Hong Kong mobile media 'produsers' are able to map new imaging communities that are dissimilar to mainland practices. The role of mobile media for Hong Kong users is that it creates various imaging communities that are, in turn, reterritorialising Hong Kong's imagined community in twenty-first century post-modernity. Mobile media imaging communities allow users to re-enact the intimate and the everyday in order to make sense of broader geo-political reterritorialisation, as well as operating as a site for resistance against mainland ideologies.

I argue that these practices are about constructing multiple imaging communities that overlay various forms of nostalgia as users try to grapple with Hong Kong's emergence with the twenty-first century 'Global South'. From Hong Kong's economic boom of the 1990s as a British colony, followed by the return to China in 1997 and the subsequent financial crisis, to being part of China's role as the new 'Global South' or twenty-first-century post-modernity, the mobile phone has been consistently present and increasingly part of the ways in which Hong Kong re-orientates itself away from the mainland. Thus the mobile phone clearly harbours much symbolic and material implication in Hong Kong's reterritorialisation within twenty-first-century post-modernity.

A vivid illustration of the symbolic role of the mobile phone as an icon for post-modernity (and thus also as a harbinger for problems associated with such 'progress') can be seen in the now infamous *bus uncle* video (http://ie.youtube.com/watch?v=2AzD5H7vWek). In this video, a young man demonstrated his dissatisfaction with the older man talking loudly on the phone by patting the older man on the shoulder; an action that results in the older man taking exception to this and beginning to berate the younger man. Here, the etiquette around mobile phone usage is central to the discordance between generations; thus mobile phone etiquette is seen as a central motif in the grappling of the Hong Kong of the past (symbolised by the old man) and the future (represented by the young man). The older man's diatribe at the young man begins about the mobile phone and then quickly shifts towards his general anger towards contemporary Hong Kong life. Ironically, the whole incident was recorded by another passenger on their mobile phone.

The *bus uncle* video became famous as an allegory for tensions around post-modernity; it was widely circulated both within Hong Kong and globally in late 2005, and became symbolic of the new phenomenon of UCC and everyday photo journalism through the mobile phone. There have even been discussions about a feature-length movie version of the *bus uncle* video. The phone, in this scenario, was indicative of a schism between the generations, between different epochs. The phone thus becomes the repository for a clash between two different Hong Kongs. It is the symbol of hyphenation par excellence. Thus to explore the mobile phone in Hong Kong is to explore the clash between eras and cultures around the perceived 'imagined community' that is Hong Kong. The phone operates not only as a repository of the particular version of Hong Kong's capitalism today, but also as a loaded symbol of social mobility and as a vessel for the nostalgia (Ma 2005) that constitutes an integral part of Hong Kong's 'imagined community' today.

In this chapter I begin by foregrounding contemporary Hong Kong

Figure 6.1 Two examples of mobile phone advertising in Hong Kong in 2004. The advertisement on the left is for a phone by the US company Motorola, which attempted to market its products as Japanese using language plays on 'moto' and imitating Japanese models. Motorola dropped its J-pop references in 2006 in the wake of the burgeoning Korean wave. The image on the right is of a billboard in Causeway Bay (a main shopping area) advertising Samsung (Korea), featuring a Western female model. Identical campaigns for both models were run in Australia. Photo: Hjorth.

post-modernity. In the section, ironically called 'Made in China', this chapter engages with some of the challenges and displacements Hong Kong experiences in the light of China's role as the twenty-first century new 'Global South' (Dirlik 2007). I reflect upon China's dominating role in the mobile industries as both a global manufacturer and site for immense consumption, and how Hong Kong is configured in this dynamic in which we can see 'one country, two systems' (Yue 2005: 155) clearly evidenced in the divergent mobile industries. I then explore the role of nostalgia as part of mobile memories within Hong Kong geo-spatialisation. In 'Upward mobility', I explore the role of mobility – as evocative of both *The New Rich* and as the symbol of twenty-first century mobile media in Hong Kong. In this section I also survey the literature on Hong Kong mobile media – a relatively overlooked area in light of the mainland's burgeoning phenomenon. This is followed by a case study of users in the section 'Yearning for *wenqing zhuyi*'. I conclude with 'Imaging @ communities' in which I reflect upon Chow's trope of the *wenqing zhuyi* in the context of emerging mobile media 'produsers' in Hong Kong. As I argue, Hong Kong's particular cartographies of personalisation can be keenly witnessed in the deployment of *wenqing zhuyi* upon various symbolic and material levels of mobile media to provide context and a sense of place within Hong Kong's hyphenated modernity.

Made in China: Hong Kong reterritorialised

In Hong Kong the streets are awash with a cacophony of one-sided conversations, many a person sporting earplug devices that see them performing a version of interiorisation by walking along and apparently talking to no one. In part due to initial pricings that saw voice-calls being cheaper than SMS, Hong Kong appears seemingly untouched by the SMS phenomenon. Unlike places such as Japan where speaking on phones on public transport is a social *faux pas*, public transport and urban spaces in Hong Kong are a cacophony of everyone talking; but not talking to each other. Ambient co-presence is the name of the city soundscape. Just as the sounds of the city reflect the contemporary *phoneur* (Luke 2006) phenomenon, so too do the images of Hong Kong contemporary life. Many shops sell a plethora of post-production mobile phone customisation, a mix of authentic and pirated cute paraphernalia. In a city consumed by fashion, the mobile phone operates like an extension of the body with users continuously updating their phone with the latest accessories to coordinate with their attire.

Through cartographies of personalisation, Hong Kong mobile media users not only re-inscribe and reterritorialise place through mobile intimate gestures from SMS and camera phone pictures to customising the outside of the phone as an extension of one's lifestyle and identity, but the socio-technology also allows them a space in which to negotiate new formations of the urban in an epoch of Global information cities. Indeed, recently China has gained much attention in scholarly, as well as journalist, circles for its economic and technological growth that signals the new 'Global South'. However, Hong Kong sits uncomfortably within that model, demonstrating the role of mobile industries and regulation to reflect differences between the mainland and Hong Kong. Thus, on the one hand, it is problematic to draw on literature regarding mainland China in the context of Hong Kong. On the other hand, by drawing from the discussions occurring around mobile communication in China we can glean how Hong Kong's 'satellite modernity' is placed within this framework.

The Hong Kong we know today was founded in 1841 after the Opium Wars caused by the British East India Company's growing exchange of opium for Chinese commodities and goods such as silver. After the Chinese authorities' ban on opium, smuggling became rife. With the Chinese becoming increasingly fearful of British military threats, Hong Kong was ceded to the British under the Convention of Chuen Pi in January 1841. Under British rule, Hong Kong prospered and many companies transferred from Guangzhou to Hong Kong, further empowering the colony as the key Asian port. Conflict between China and Britain escalated, resulting in the Second Opium War that led to Britain acquiring the New Territories. Hong Kong continued to grow and became a portal between Chinese and European production.

However, the Asian financial crisis in 1997/1998 was a turning point. The handover to China left many businesses feeling uncomfortable about the future, not just of Hong Kong, but, given the city's pivotal role both in the Asia-Pacific, also globally, as it changed Hong Kong's role as portal between east and west. Since 1997, Hong Kong's GDP growth has averaged 3.6 per cent; it took Hong Kong eight years to reclaim, in 2005, its 1997 GDP. At that time Hong Kong's GDP per employee was US $52,000 – compared to Singapore's US $54,000, Japan's US $66,500 and Korea's US $34,000 – with its key capital exports being driven by the private sector (OECD 2006). Hong Kong's economy is in sharp contrast to Korea's, with its emphasis on manufacturing industries that dominate over China's and India's in terms of quality and branding. The expatriate population living in Hong Kong – who were a major force in the economy – have been on the decline as companies have begun to relocate employees to other mainland Chinese cities. For example, since 1997, the number of Japanese nationals has fallen by 39 per cent, British by 22 per cent, Canadian by 20 per cent, and American and Australian by 16 per cent. Hong Kong today is still a location grappling with its British colonial history and its current positioning as part of China and China's role globally.

Hong Kong, as a 'satellite' (Ma 2000) location whose identity is 'hyphenated' (Abbas 1997) between its status as a former British colony and current identification with mainland China, has undergone much reorientation in both a regional and global context. Once the portal between East and West, Hong Kong's history as a 'satellite modernity' (Ma 2000) is most aptly symbolised by its consumption of mobile technologies. As Yan and Pitt (1999) observed, writing just after Hong Kong's return to mainland China, the differences between Hong Kong and mainland China could be clearly seen in their approaches to telecommunications deregulation: Hong Kong followed the British telecommunication model (with such technologies as BlackBerry and service providers like Hutchison and Vodafone), whereas mainland China followed the Chinese model (Chinese service providers and handset manufactures). Consequently, the mobile industry in Hong Kong, as in Australia, is a smorgasbord of global brands all vying for domination.

One only needs to glance at the multiple and competing brands and companies in the Hong Kong telecommunications market to be made aware of the ways in which Hong Kong's hyphenated identity oscillates between its British and Chinese affiliations. Unlike the Korean and Japanese markets, in which the hardware, software and service providers are predominantly home grown, Hong Kong – like Australia – is serviced by a multitude of global products. Common brands include Nokia (Finland), Motorola (US), Samsung (Korea), LG (Korea), 3 (UK's Hutchison), SonyEricsson (Japan and Sweden), Vodafone (UK) and BlackBerry (UK). Some of the advertising campaigns by 3, Samsung, Vodafone and Nokia in Hong Kong were replicated in Australia. The significance of the mobile phone in Hong Kong is highlighted by the fact that in 2004 the territory recorded penetration rates of 117 per cent, and that by 2006 over one million subscribers had 3G handsets (OFTA 2006).

Hong Kong and Australia were both less willing than Tokyo and Seoul to embrace the shift of the mobile phone into multimedia device – that is, the shift from 2 to 2.5G (Hjorth 2005b, 2006c; Goggin 2006a). In Hong Kong, this shift was largely orchestrated by two UK companies – BlackBerry and 3 (Hutchison). Australia was also set to have BlackBerry, but when negotiations failed between Blackberry and Australia's main service provider, Telstra, Telstra adopted i-mode. The result was a marketing disaster as i-mode failed in the Australian market due to poor uptake (this will be discussed in more detail in Chapter 7). Another parallel between Hong Kong and Australia was the relatively slow uptake of mobile phone features such as camera phones in both locations. In Hong Kong, respondents preferred voice calling to SMS; in Melbourne, it was the other way round.

Mainly because of the low price of voice calls and the initial lack of SMS interoperability between service providers, Hong Kong has been slow to participate in the global SMS phenomenon (Goggin 2006a; Lin 2005). By contrast, with high voice-call and cheap SMS prices, Melbourne has been swept by the SMS and now the MMS craze. So why has Hong Kong, given back to China in 1997, not followed suit in what Weigui Fang (2005) calls 'China's culture of the thumb'? Perhaps this cultural inertia, discussed by Yan and Pitt (1999) was not just a product of service costs and industry and service provider regulations, but is also symbolic of Hong Kong's liminal socio-cultural practices. Thus, in order to understand the role of the mobile phone in Hong Kong it is important to address the hyphenated feelings of the people and the role that nostalgia plays in media images, identities and thus identification.

> Wireless working-class ICTs are low-end wireless solutions serving the 'information have-less', an emerging social group that plays a definitive role in the lower strata of the evolving network society. The informational city is a particular mode of social and economic restructuring through the utilization of ICTs, resulting in the rise of 'the space of flows' as a predominant way of urban organization.
>
> (Qiu 2008: forthcoming)

The role of wireless ICTs within the construction of twenty-first-century post-modernity has largely been a product of 'unique characteristics that are informed by its own social, historical, and institutional pedigree' (2008: ibid). While Qiu's work refers to ICTs in mainland China, undoubtedly this influences the types of practices and techniques of reterritorialisation adopted in Hong Kong as part of the construction of the new 'Global South', in which the mobile phone plays central roles both upon *symbolic* and *material* dimensions. Just as Chow identifies two distinct deployments of nostalgia in filmic trope – one by the mainland, the other by satellite modernities such as Hong Kong or Taiwan – we can also witness different affective imaging communities within mobile media. Just as Hong Kong cinema of 1997 signaled China's 'one country, two systems', ten years on one could argue that the mobile phone operates as the lynchpin between the 'two systems' and two geo-spatial memories.

According to Yue's discussion of Wong's *2046* as an analogy for contemporary Hong Kong's 'migration-in-transition', 'the formulation of the ethics of the self, as a cultural practice of the border, is pertinent to Hong Kong and its "one country, two systems" transition' (2005: 155). This practice of the border is evident in the ways in which Hong Kong differentiates itself from mainland – most particularly through the emotologies of *wenqing zhuyi*. Through the saving of camera phone images and SMS predominantly for personal consumption or sharing via face-to-face rendezvous, the cartographies of personalisation of mobile media are a process of perpetual reterritorialisation. According to Yuezhi Zhao (2007), China in the twenty-first century consists of at least four worlds in one, these worlds differing across political, economic, socio-cultural, ideological scapes that are played out within mobile media practices.

> At the turn of the century, wireless communication finally began to spread into the lower social strata, a decisive development that brought sea change to the telecom industry. Before long, in October 2003, the number of mobile subscribers surpassed that of landline. As of November 2007, China boasts the world's largest national mobile subscriber population of 531 million, while the US has 250 million subscribers. Considering that the upper- and middle-classes are smaller in China, it is clear that the information have-less accounts for a main chunk of the Chinese mobile market.
>
> (Qiu 2008: forthcoming)

For Qiu, Chinese domestic services catered by Little Smart (*xiaolingtong*) offer the demographies he calls 'China's information have-less', 'new ways of connectivity' whereby this may not only be their first mobile handset (*shouji*) but also their first phone and personal ICT device. As Qiu succinctly surmises, 'the growth of the Chinese informational city has been closely related to the development of wireless working-class ICTs' (ibid.). He argues persuasively that this occurs across various levels that, in turn, resonate within the practices of the urban space and concomitant techno-social processes. The rise of wireless working-class ICTs both affords and reflects these new formations of the information city; so much so that Qiu argues that contrary to 'the informational city is a dual city' (ibid.), in the case of China, the information city 'seems to be quite different, given the dissimilar position of this particular society in the macro historical streams of industrialization, urbanization, and globalization' (ibid.).

For Qiu, the rise of new cheaper mobile technologies such as Little Smart have seen working-class demographies shift from nodes in the manufacturing process to core creative consumers, and that this dynamism should be read as emblematic of new socio-economies in China that are embodied by a mobile communication revolution. In the case of Hong Kong, it is undoubtedly the migrant workers that are making the most of these new cheaper mobile technologies. Over ten years on from *The New Rich* (Robison & Goodman 1996) with its symbol of the man on the mobile phone, we are now confronted by various socio-economic mobile media creative classes. However, the localised practices of imaging communities reflect particular middle-class cartographies at play – for example, while the practice of camera phone imagery has continued to grow in Hong Kong, its prevalence in the mainland is restricted to higher–income users (Gai 2007). All awhile SMS continues to be popular in both Hong Kong and the mainland, and with it, new forms of creative enterprise are occurring.

> Sending an SMS is the most vital function of mobile phones in China, which is also the reason why it is the primary 'pillar' of commercial services of the telecommunications industry. Otherwise the term 'thumb economy' wouldn't have been coined in China.
>
> (Fang 2006: n.p.)

As Fang observes in 'China's culture of the thumb', 'over half of all SMS transferred worldwide is sent by Chinese users'; making it a phenomenon difficult to ignore. He notes that the dominance of SMS in Chinese everyday life is leading to new forms of cultures and SMS creativity. The debate around 'SMS literature' – as either a continuance of the tradition of reading in an 'audio-visual era' or simplification of the poetics of great literature – is no more apparent than in China. With the abundance and significance of SMS, new forms of profession have been born – from the SMS writer to cell phone soap drama producers. Fang notes, as 'more and more Chinese mobile

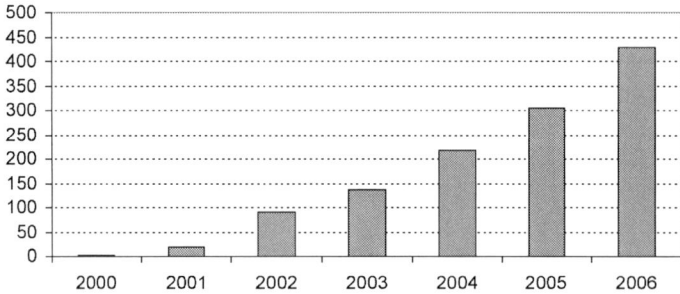

Figure 6.2 Annual SMS traffic volume in China, 2000–2006 (billion messages).

Source: MII Annual Statistical Reports on the Telecommunications Industry (cited in Qiu 2008).

phones and internet users seem eager to download and pass on SMS texts, the demand for "creative" SMS writers is surging' (2006: n.p.). From birthday poems to Lunar year greetings, SMS has become the new Chinese cracker – a short message wrapped up in wisdom or, at least, a bite (byte) of sweetness. The year 2005 also saw the first Chinese TV series of five episodes of five minutes made for mobile phones entitled *The Promise*.

> In present-day China, the Internet is often referred to as the fourth among the communication media and SMS as the fifth. SMS is of course just a service, but with further development of mobile technology it is not entirely unthinkable that the cell phone will evolve into a genuine fifth media platform. Today, China's mobile phone users say that SMS is part of their daily needs and that sending a short message is as natural as opening an umbrella in rainy weather.
>
> (Fang 2005: n.p.)

The rise of the equivalent of the hallmark industry in SMS highlights just one example of the ways in which mobile media not only help to create outlets for creative expression but, also, financial remuneration. Parallels could be made between the SMS writers in China and the *keitai shôsetsu* writers in Japan. Moreover, the relatively recent rise of SMS in China (due to various factors such as price and service provider incompatibilities) could be seen to parallel the significant role such mobile media practices make in the constructions of intimacy and the postal metaphor in Melbourne. However, in the case of Hong Kong, we see that SMS, concurrent with practices such as camera phone imagery, operate to help re-inscribe a sense of history with that of geography. The role of mobile media as a tool for reterritorialisation – as identified by Koskinen (2007) – cannot be underestimated in the case of Hong Kong's hyphenated identity.

As Qiu observes, such media as SMS have seen the phenomenon of 'SMS writers' (*duanxin xieshou*) who regularly produce large quantities of messages to be sold or spammed to mobile subscribers. The rise in SMS from its beginnings in 2000 in China has seen its annual traffic volume has gone up from 1.4 billion messages in 2000 to 429.7 billion in 2006 (Qiu 2008). With this phenomenal rise, new forms of labour markets are being created – some exploitative, while others, like the best *duanxin xieshou* or gold farmers of online multiplayer games such as *World of Warcraft (WoW)*, can provide the 'have-less' with an annual income in one week (Chan 2006: n.p.). Through mobile media, some are able to enter new worlds of virtuality and creative industries.

> China's Internet audience has, for the most part, given sites like Facebook and MySpace the cold shoulder. Even local Chinese sites like Xianonei or 51.com have failed to establish big national followings. What may seem on the surface to be a stubborn backwardness on the part of the Chinese, however, could also be interpreted as a viable alternative to Western-style social networking . . . In many ways the big difference in China can be summed up in three words: instant mobile messaging. The low proportion of home PCs has made the mobile phone the preferred Internet-access device. And Chinese clearly prefer instant messaging – chatty, real-time communication that takes place via PC or cell phone – as opposed to ordinary e-mail, in which you never know when your correspondent might respond.
>
> (Liu & Zoninsein 2007: 40)

As Melinda Liu and Manuela Zoninsein note in 'These Surfers do it their own way' (ibid.), the preferred uptake of IM, as opposed to SNS, in China is the result of various factors including its cultural and techno-national aspects. This techno-nationalism is partly orchestrated through government policies and regulation around communication technologies in which virtual spaces such as the Internet are often a site for censorship and subversion as has been noted in the case of Google censorship and the blogging debates that ensued. But it is the cultural factors, such as the desire for quick and instant contact; that makes IM so popular.

In contrast to the US, Korea or Japan, only 13 per cent of Chinese own a PC. Demographics are significant too, with 'the average age of China's 172 million Internet users [being] 35, seven years younger that their 211 million Americans counterparts' (ibid.: 40). While the age distribution of users influences what they want from the Internet, cultural differences are most apparent – Chinese users generally associating it with entertainment, whereas their Americans counterparts use it much more for information gathering. Moreover, almost half of the approximately 39 million Chinese bloggers have broadband access, as opposed to only 14 per cent of US bloggers. This virtual co-presence of cyberspace is but an overlay of the mobile co-presence enacted by mobile media users in which the imaging communities' practices of SMS, IM and camera phone images are deployed to locate users in a particular geo-political place. This is particularly the case with Hong Kong and its satellite modernity that oscillates between various geographies and histories of Hong Kong's past, present and future.

Mobile memories: the geo-political imagery of Hong Kong

Arguably, Hong Kong is going through a romantic revisionism that sees nostalgia as big business. From before the handover in 1997 until now, the role of nostalgia as technics of mobility specific to Hong Kong has remained unabated. For Andrew Murphie (2007), technics of mobility structure and organise contemporary life in which work, life and fused into one. As Murphie identifies, through mobile technologies we can gain a sense of the broader forms of labour and intimacy. Mobile technologies provide insight into 'new technics of mobility' – that is, arising social, collective and subjective processes of living in the contemporary world (ibid: n.p). Technics of mobility refer to the flows of capital, labour (of which intimacy is one form), gender and the associated multiple forms of mobility and immobility. These technics are augmented and expressed through the unofficial imaging communities (such as mobile media practices); In the case of Hong Kong, imaging communities reterritorialise place through a sense of nostalgia.

Filmmakers such as Wong and Stanley Kwon have been exemplary in deploying these affective economies within their imagery and content (Chow 2007). Through the rubric of nostalgia, industry and government in Hong Kong have been able to reterritorialise a sense of geography and historicity to offset Hong Kong's 'weak nationalism' (Ma 2005). This nostalgia industry is eloquently discussed in Ma's fascinating analysis of a TV commercial by Hong Kong and Shanghai Bank, in which Ma discerns Hong Kong's re-inscription of nostalgic historicity (Ma 2005: 139). In the commercial, faux documentary-style aesthetics (including black and white photography) are deployed to construct a sense of nostalgia of a bygone era, in which an image of a solitary fisherman waiting to catch fish evokes a time when relationships between production and consumption were not in the thrall of globalisation. In a time of national uncertainty, harking back to a time (real or imaginary) when things were simple and good is an important ideological tool in solidifying the imagined community of the nation state. As Ma advances, the image of Hong Kong depicted in the TV commercial re-advertises a Hong Kong identity through an ideological filter of nostalgia. The picture presented by this incomplete piece of propaganda is one in which a lone fisherman is up against nature – a situation that Ma identifies as exemplifying 'decontextualised' (2005: 142) and 'atomized' (ibid.: 149) individualism.

> In the discourse of postmodernism, a nostalgic aesthetics is said to have contributed to the disappearance of historicity by its hybridization of styles of different historical periods and its destabilizing use of parody

and creative intertextual references. However, this argument should be applied to the Hong Kong case with extra care. In the years before and after the handover, the upsurge of nostalgic media in Hong Kong has been characterized by a strong commitment to modernity.

(Ma 2005: 139–140)

The politics of nostalgia, and its re-inscription through aesthetic and visual economies, is omnipresent in user customisation of mobile phones in Hong Kong. As most users are not as techno-savvy as those in Tokyo or Seoul, the customisation of predominantly 2.5G phones in Hong Kong (unlike the multimedia spectacular of 3G phones in Tokyo) is saturated with nostalgic intra-Asian products. The mobile phone thus becomes the battlefield for consuming global flows: from the inside to the outside, post-production and pre-production customisation, it becomes the playground for the owner's various forms of social, economic and cultural capital to be performed. For example, a Japanese *Pokémon* hangs from a US Disney Mickey Mouse, whilst a ringtone keeps the tune of a Korean TV drama. This is but one example of the politics of mobile phone customisation as consumption practices.

In particular, the cultural capital attached to these commodities operates as a repository for reminiscence on two levels. Firstly, on a micro level it operates as an individual's return to a bygone time when things were fun, akin

Figure 6.3 Some examples of young people's customisation of mobile phones. The trend for wearing the mobile phone like a lanyard highlights its role as an extension of the user's identity and as a title for others to read. Notice how both phones are customised with cute character attire that transforms them into a teddy bear or stuffed animal. Photo: Hjorth.

to imaginings of childhood. Secondly, on a macro level the objects purport to represent types of nostalgic cultural capital; for example, Japanese *kawaii* culture began to be consumed in Hong Kong in the 1970s, a period of economic boom that continued until the downturn (and handover to China) in 1997. Thus the consumption of such cultures revisits a time when Hong Kong was experiencing great financial power and stability. As Ma convincingly argues, the use of 'nostalgia media' is a technique for creating a 'collective consciousness', particularly in difficult times, forming a type of 'imagined community' filter through which a nation can look back and yearn for a bygone golden age. Moreover, the personalising techniques of such customisation as *kawaii* clearly re-enact the techno-cute whereby 'cute' makes warm the coldness of new technologies.

The ability of the mobile phone to articulate and play an instrumental part in Hong Kong's cultural and economic repositioning operates at both a micro (individual) and macro (collective) level. The construction of nostalgia does not necessitate participation in 'retro' design. It can manifest in the ways in which one imagines the future. For example, the genre of science fiction is less about an actual future than about fears from the past. In many of the advertising images disseminated in Hong Kong, the mobile phone symbolises a type of fun and freedom – ensured by attendant economic stability – that can be seen as reminiscent of those times of prosperity pre-1997. The consuming of Japanese mobile phone *kawaii* culture can be seen as an attempt to reconnect to commodities that were consumed by previous generations, when Japanese cultural capital dominated the region and the region's economy expressed a sense that 'things were good'.

Arguably, one of the reasons why Korean popular culture has become big business in locations such as Hong Kong and Taiwan is the way in which it evokes nostalgia. In the case of the Korean dramas such as *Winter Sonata*, the plots and characterisation are *über* melodramatic and highly sentimental. Characters are always searching for 'pure love' and the familial archetypes are highly conventional. Through these highly emotively charged narratives, saturated in nostalgia in which Korean Confucianism is romaticised, Hong Kong consumers can indulge their fantasies and collective fictions (via Internet fan communities) about another time (Erni 2006). As Ma argues, Hong Kong's identity is about trying to grapple between once being part of the British colony and a 'satellite modernity' for high capitalism and globalisation, to re-orientating itself within mainland China's nascent and divergent form of capitalism. This re-scripting, according to Ma, oscillates around a couple of key questions, namely, 'How does Hong Kong negotiate its identity by repositioning its past and future connection with the globalized economy? Does Hong Kong's globalized position trigger or dilute the desire to consume a local past?' (Ma 2005: 138).

The mobile phone then, as symbolic of global and local flows and of new forms of individualism, and yet salient to already existing forms of sociality, could be the repository for Hong Kong's ignited nostalgia and re-historicising.

The mobile phone, with its history as a device once signifying the socially and economically mobile, has a double reading in the light of Hong Kong's re-enchantment with a time of economic and cultural mobility from the 1970s onwards – a mobility that was derailed by the handover and the financial crisis in 1997. As a metaphor for high capitalism, and yet used to maintain forms of communities and perform techno-nationalist agendas, the mobile phone is also a vehicle for Hong Kong users to perform their own nostalgia practices whilst reflecting national ideologies. Although the imaging communities of individual mobile media practices might be about engaging with the personal, these practices also inflect the socio-cultural that, in turn, speaks to particular national imaginaries about what it means to live in contemporary Hong Kong.

For example, one female respondent had Japanese cute characters hanging off her phone. She and her friends liked Japanese *kawaii*; it reminded them of being children while also tapping into her and her social network's interest in the cultural capital of Japan. This is a key example of the power of *wenqing zhuyi* with the *kawaii* making 'warm' and 'sentimental' mobile technology: in sum, indicative of Hong Kong's cartographies of personalisation. The consumption of Japan in post-1997 Hong Kong has undergone some changes, particularly as a result of the region's backlash against Japan that was most recently played out through the rise of the Korean wave in China in 2001 (Cho 2005). As Cho notes, such phenomena as the Korean wave function to reflect processes of globalisation as well as 'the formation of "Asian" subjectivities across national boundaries' (ibid.: 149). The consumption and alignment with particular products demonstrated respondents' relationship to various forms of the intimate – between friends and family, but also allegiances between various national and transnational 'communities of consumers' (Chua 2006). These are also allegiances to different forms of regional imaging communities, and how they reflect transnational and national imagined communities. Often respondents were aware of the multiple imagined communities associated with consumption practices.

The aforementioned female respondent noted being hypersensitive to the possibility that others could misinterpret her bricolage techniques as childish. This misinterpretation was particularly prevalent in the case of work situations in which she could be judged by others for her consumption practices. When asked to elaborate, she said that an older fellow employee might not understand her interest in Japanese *kawaii* culture and may just think she is childish. With this in mind, she would remove her customisation in work situations, in a fine instance of engaging with Goffman's (1969) analogy of front-stage and back-stage (the front-stage can be viewed as the public performance, the back-stage the private). For the respondent, the *wenqing zhuyi* and pleasure around *kawaii* customisation was a back-stage activity between intimates, despite the front-stage position often granted to the customisation (except in work scenarios). Here the symbol of mobile phone functions as a site for *wenqing zhuyi* and mobility between stages, spaces and time.

The nostalgia mobility – that is, a nostalgia for a bygone time and the need to create collective fiction for national clarity – as Ma argues, is a new type of historicity that romanticises both a period of time in Hong Kong and its relationship to the region. However, this nostalgic mobility is also closely tied to the ways in which the 'circuit of culture' around the mobile phone is practised in Hong Kong. For example, the telecommunication industry has been governed by Eurocentricism, with British companies such as Hutchison and Vodafone dominating, and Eurocentric phone brands such as Nokia and BlackBerry ruling the market until recently. It is through the mobile cultures of production, industry, advertising and consumption that one sees the legacy of British colonialism and a reinscription of its import-ance through conscious and subconscious mobile patterns. A Nokia phone adorned with *Pokémon* seems to encapsulate the nostalgic yearning for a type of economic mobility now gone: a clash between Hong Kong's relationship in intra-Asian flows and as an ex-British colony.

The various brands and industry-created customisation highlight users' identification with certain histories and the products' countries of origin. For example, when BlackBerry was first launched, it did not have the capacity to send Chinese characters via email. This meant that users had to prefer English to Chinese, signifying identification with Hong Kong as a British colony, rather than as part of mainland China. The fact that BlackBerry is also a

Figure 6.4 An example of nostalgic customisation in the form of Hello Kitty mobile phone customisation. Hello Kitty became very popular in Hong Kong in the late 1970s and 1980s, a time when Hong Kong experienced great upward mobility. Photo: Hjorth.

British company that is popular in the UK means that by consuming Black-Berry in Hong Kong, issuers are partaking in a type of historicity. This preference for European brands in a market that is dominated by European (particularly British) service providers makes Hong Kong's cross-cultural identity and identification seem closer to that of Melbourne than of Seoul or Tokyo.

But do hyphenated practices such as having a Nokia phone adorned with *Pokémon* represent the new modes of Asian consumption as outlined by Chua (2000), whereby consumption is no longer equated with Westernisation and moral bankruptcy? What meanings can we draw from this metaphor of nostalgic mobility? How does the metaphor conceptualise notions such as individuality and sociality? As a signifier of post-modernity and as both a symbol and a literal manifestation of mobility, the mobile phone has the ability to further instill myths around the politics of nostalgia and Hong Kong's time of upward mobility. The dressing up of one's phone and choice of brand become an active process in what Ma sees as a celebration of upward mobility and high capitalism – factors that have worked to under-mine national discourses. It seems appropriate at this point to review some of the relatively new research exploring mobile phone usage in Hong Kong.

Upward mobility: mobility in Hong Kong

In Lo Shiu-hing's 'Hong Kong: Post-colonialism and political conflict' chap-ter in Goodman and Robison's *The New Rich in Asia*, the author describes the anxieties and consternation that prevailed in Hong Kong one year before the 1997 handover and the concurrent economic crisis.

> In short, the rise of the new rich, especially the middle-class segment, has had the immediate political impact of stimulating elite participation and conflicts. Moreover, the local bourgeoisie has sought to compete with middle-class liberals by forging an alliance with both middle-class nationalists and the mainland Chinese bourgeoisie in Hong Kong.
>
> (Lo 1996: 172)

Lo predicted that the impending political reform in the SAR and the resulting decline in democracy (with its co-dependent link to Western capitalism) would have an immense impact on the livelihoods of the Hong Kong 'new rich'. Lo talks about the *guanxi*, a Chinese phrase for personal relationships, as being a commodity that was highly desired by the capitalists of Hong Kong. As Lo notes, the local bourgeoisie and the new rich were far from homogeneous in their ties to pro-Beijing or pro-British alliances. According to Lo, the first wave of Hong Kong's bourgeoisie cannot be regarded as synonymous with Hong Kong's new rich. Lo signposts the 1960s and 1970s as being the era in which a new capitalist class emerged, mainly from Shang-hai. The second generation of bourgeoisie in Hong Kong emerged in the

1970s and early 1980s, and it was this generation that gave rise to the four factions of Hong Kong rich, all supporting 'private property rights and capitalism' (ibid.: 165). These four factions were the local bourgeoisie, the local middle-class liberals, the local middle-class nationalists, and the mainland Chinese bourgeoisie. As Lo concludes:

> The rise of Hong Kong's new rich has had significant impact in the political, economic and social spheres. Politically, the emergence of the new rich has contributed to some degree of pluralism. Whether such limited pluralism will change remains to be observed. Economically, Hong Kong is becoming more dependent on China than ever before, and thus Hong Kong will definitely be affected by any change in China's political and economic circumstances. Socially, liberalization has been implemented to its permissible limits. But in the event that China's political system remains authoritarian, Hong Kong's liberalization may be curbed to some extent after 1997. Perhaps these uncertainties in Hong Kong's policy, economy and society may only be reduced or offset by an increase in the international concern over Hong Kong's future.
>
> (ibid.: 179)

Twelve years later, less change has been experienced than was predicted. The return to China was not as dramatic as inhabitants imagined. However, the economic crisis of 1997 made life precarious for the region's new and old rich, lending weight to a type of nostalgia that harked back to times of buoyancy and economic/social prosperity. Here we can see the mobile phone functioning firstly as a signifier of the new rich and secondly as an indicator for the user's alliances and rose-coloured nostalgia. In the first instance, we see the introduction of the first mobile phones in the early 1980s as a metaphor for the new businessperson and the rise of the yuppie. Unlike Britain, where the phone was seen as a pretentious prop in the vulgar public performativity of conspicuous new money (Agar 2003), Hong Kong seemed divided between these two different worlds of Britain and mainland China and attendant modes of public discourse. The previously mentioned *bus uncle* video demonstrated the way in which mobile phone etiquette in public spaces is deeply imbued by cultural and generational factors.

In Peter Waters's paper on 'Mobile competition: How many is too many?' presented at the *ITU Telecom Asia conference* in September 2004, he lamented that the competition was so strong in Hong Kong, it made it hard for companies to invest resources in innovation such as 3G content. In sharp contrast to the relatively self-contained and strongly regulated markets of Korea and Japan, which seemed to nurture monopolies that controlled service providing, hardware and software (i.e. DoCoMo in Japan and SK Telecommunications in Korea), Hong Kong's hyper-competitive market has hindered investment in technological innovation and the uptake of advancements such as 3G has been relatively slow to take off.

As Plant (2002) noted in her case study of nine different locations across the world, the local inflects what types of behaviour are accepted and what are viewed as rude. For the Chinese, voice calling in public spaces is a norm, while for British users there are types of voice calling that are acceptable in public and others modes that are frowned upon. For Hong Kong, the territory's hyphenated identity is demonstrated by the diversity of mobile phone etiquette in which differences occur across generational and cultural demographics.

In Hong Kong's current highly competitive market for handsets and service providers, we can see two dominant trends that highlight divergent modes of nostalgia through the politics of consuming and producing mobile phones. On the one hand, the new bourgeoisie are identifying with a Sino-British history by consuming brands that are popular in the UK such as BlackBerry and Nokia. On the other hand, there is a new wave of brands that project types of 'Asian' soft cultural capital such as Korean handsets (Samsung, LG). Many of these brands reenact forms of *wenqing zhuyi* so desired by particular demographics. The former group tends to be highly educated, on-the-go, aged 30–45 years old and usually expatriates. As noted in the sample study, these professionals prefer the ease and simple design of BlackBerries to the other jam-packed mobile multimedia devices. Younger respondents (aged 19–30) were divided between the Eurocentric classic of Nokia (which, as in Melbourne, has maintained a stronghold on the market until recently, when its lack of customisation and innovation saw it being edged out by savvier brands such as Motorola and Samsung) and Korean brands.

As Goggin observes in his chapter 'Txt msg: the rise and rise of messaging cultures', Hong Kong, unlike the Philippines or Japan, has been slow to join the upcoming 'thumb generations' (Plant 2002). As noted previously, text messages between server providers were initially often incompatible and with the cheap prices of voice calls, many Hong Kongers felt little inclination to text. Moreover, the preference of many Hong Kongers to utilise a mixture of Cantonese and English has seen a rise in particular hybrid forms of SMS textuality (Lin 2005). In keeping with the lack of strong nationalistic ties to either Britain or China (Ma 2005), the 'Hong-Kong mixed code' (Lin 2005) can be seen as a way in which Hong Kongers can struggle against a 'weak' (Ma 2005) or 'hyphenated' (Abbas 1997) national identity. One of the distinguishing features of mobile phone habits amongst Hong Kong inhabitants is the role of gender (Lin 2004; Lin 2005). According to Leung and Wei (2000), the gender differences noted in landline usage translated into mobile phone usage; females using the mobile phone for socio-emotional reasons, while males tended to focus on its instrumental and technological applications. For Angel Lin in her groundbreaking study on SMS usage amongst Hong Kong college students, differences emerge firstly in terms of gender and secondly in terms of bilingual linguistic practices.

As Goggin (2006) astutely elaborates, for immigrant workers the role and use of the mobile phone signals further differences. In the case of Hong

Kong, where a large proportion of the service workers – an underclass within the upward mobility of aspiring Hong Kong people – are Filipinos (the 1.6 million make up approximately 10 per cent of the population), the newcomers have imported the texting practices of their homeland. Text messaging allows immigrants an easily accessible and relatively cheap portal to their homeland – a fact that has not been lost on telecommunication companies, which have been quick to offer this diasporic market special deals for long-distance calls to the Philippines. Drawing from the work of Parreñas (2001) Goggin notes, 'in this way cell phones are deeply implicated in both the resolution yet also the intensification of social and gender contradictions that stem from the violence of the conditions of contemporary capitalism, its modes of forced as well as chosen mobility, that displace people . . . in their relationships as they try to survive' (2006: 140). He adds, 'the Hong Kong experience is a very interesting one because it questions the assumption that text messaging can be universally regarded as a "success", or that it has a particular trajectory' (ibid.: 141).

As aforementioned in Part I, McKay's (2007) observations of Filipino houseworkers in Hong Kong notes that the mobile phone helps to provide particular forms of intimacy that were previously not available. She implores us to reconsider the emotional and material dimensions of transnationalism (2007: 175) and to stop viewing migrants as passive victims of globalisation. However, the hyphenated modernity of Hong Kong *vis-à-vis* mobile media is undoubtedly present in the inequalities between migrants doing physical labour – such as Filipino houseworkers – and IP (intellectual property) capital workers. Here class, cultural capital and ethnicity are clearly demarcated, with the dividing lines clearly reflected in mobile media usage. For example, in a study of university students one is unlikely to encounter a Filipino – in sharp contrast to locations such as Korea and Australia.

For Lin, in her study on gender and bilingual text messaging practices by Hong Kong college students (expanding upon her 2004 study), the practices of texting can be read in terms of broader 'mixed, ambivalent feelings in national and sociocultural identification' (2005). In her 2004 quantitative study of 455 students at the City University of Hong Kong, Lin noted striking gender and linguistic differences in terms of application usage. As in the other sample locations, female respondents were more proficient with SMS. However, as Lin observes, the issue of gender and language is very much tied to notions of what its means to be 'Hong Kong-ese'. Reiterating Ma's discussion of Hong Kong's 'weak' national affinities with its British history and its Chinese present and future, Lin asserts:

> Given the special sociopolitical, historical context of Hong Kong, it seems that many Hong Kong people have not entirely accepted British colonial rule in the pre-1997 era and yet are equally ambivalent about Socialist Chinese domination in the post-1997 era. Such mixed, ambivalent feelings in national and sociocultural identification seem to correlate

with the freely intertwining of Cantonese and English words in the everyday public life of Hong Kong people, and these 'non-pure' bilingual linguistic practices seem to be playing an important role in marking out the Hong Kong identity – they seem to serve as distinctive linguistic and cultural markers of 'Hong-Kong-ness' and seem to constitute some defiant acts of identity. It is almost like saying: *We're Hong Kong-ese and I don't care whether I'm speaking 'pure Chinese/English' or not!*.

(ibid.: n.p.)

Lin draws comparisons between the Hong Kong linguistic mixing and the politics of 'Singlish' (a localised hybrid vernacular of English) in Singapore. The 'cultural defiance' phenomenon can be seen in the context of remediated mobile phone customisation practices in the region such as the 'kitten writing' (Kinsella 1995) in the 1970s and the growing use of ASCII and *emoji* by young Japanese, as acts of defiance to traditional overarching dogma. As Lin notes, there have been relatively few studies conducted in SMS practices in Asia apart from Ito and Okabe in Japan (2003, 2005) and Yeh in Taiwan (2004) – in contrast to the multitude of studies in Europe.

Hong Kong has been one of the places with highest penetration of mobile phone service in the world for many years. From 1998 to 2003 the number of mobile service subscribers had increased 1.5 times. The number reached 7.19 million by the end of 2003, representing a penetration rate of 106%. Despite this high mobile phone penetration, SMS is not as widespread as in other economically developed Asian societies such as Singapore, the Philippines or South Korea. The TNS Asia Telecoms Index shows that only 43% of Hong Kong cellphone users use SMS and the average number of messages sent per user is only 23 per month.

(Lin 2005: n.p.)

Migrating from Lin's quantitative study that discusses the difference in SMS usage dependent upon gender and age, I turn to my qualitative case study. Unlike Lin's study, my qualitative study hopes to explore some of the specific voices in the various usages (both material and symbolic) of the mobile phone, and how it operates for Hong Kong users in terms of gender performativity, self-expression, gift-giving practices, data collection, and as a repository for cultural capital and ideologies of *wenqing zhuyi*. I found that the role of gender played out in the types of pictures taken and the modes of sharing, as I will discuss shortly. Whereas camera phone practices in Korea are influenced mainly by whether the user is single or in part of a couple (in which case the mobile phone becomes a metaphoric engagement ring), and camera phone practices in Japan tend to be playful, with both male and female participants taking many 'funny' photos, in Hong Kong the role of gendered practices is pronounced with male respondents taking more typical snapshots of people or landscapes, whereas female respondents tended towards more quasi-professional genres of photo journalism such as paparazzi.

The use of camera phone images as wallpaper (screen-saver) decorations sees the media transformed from personally consumed to publicly accessible. Wallpaper transgresses the internal/external division of mobile media as often it is viewed by intimate strangers in public places; phone owners often leave their mobile phone on the table at a café, or have it hanging around their neck whilst walking in public spaces. Thus the wallpaper operates as window between the user and the public. To return to Goffman's (1969) analogy of front- and back-stage again, the wallpaper is the middle-stage: in between public and private spaces and performativity. The wallpaper operates like a hyphen between the public and private work an example of 'the interior being the new exterior'. They also function as a constant presence for the user, reminding him/her of significant others in the ambient co-present manner particular to mobile media. Moreover, the symbolic significance of the wallpaper to traverse both inside and outside, public and private modes of performativity, echoes Chow's identification of the site of the domestic as a struggle for Hong Kong-ese *wenqing zhuyi* (2007: 19). As a key icon of the embodied 'domestic', the domestic technology of mobile media is undoubtedly the site in which twenty-first century reterritorialisations are charted, explored and challenged. I will now examine some of the other issues that became evident in the case study.

Figure 6.5 Mobile phone customisation as a battlefield for cross-cultural consumption (Peko screen saver plus Disney cute objects hanging from the Nokia phone). (Photo: 2005)

6.4 Yearning for *wenqing zhuyi*: a case study among Hong Kong users

> In the Asia-Pacific region, the Hong Kong regulator, OFTA, is the first regula-
> tor to confront this issue [the expiring of 2G licenses] as the Hong Kong 2G
> licenses have shorter terms than typical. Disappointed with the low level of
> mobile data penetration in Hong Kong OFTA has opted to shake up the
> market by introducing a focused 3G operator.
>
> (Waters 2004: n.p.)

In the eyes of the multiple global telecommunication companies that are
fighting it out in one of the most competitive markets in the world, Hong
Kong could be seen as a type of yet-to-be colonised blank on the map. As a
hip and cutting-edge city, one would imagine that the mobile phone – as an
extension of the user's self in terms of fashion, self-expression and identification – would be central in the make-up of everyday life. Yet despite the
ubiquity of the mobile phone in Hong Kong, the market there has been a
source of frustration for the telecommunications industry. As already noted,
the uptake of SMS was slow due to in part industry incompatibility and the
fact that voice calls were cheaper. A compounding problem was that many
Hong Kongers speak a hybrid of English and Cantonese, a practice that lends
itself more easily to oral – rather than textual – representation. In contrast to
Lin's study, which noted hybrid SMS textuality, most of my respondents
chose to write in either 'proper' English or Chinese (Cantonese).

At the time of my first series of interviews and surveys in June 2004, the
content of most respondents' phones was still mainly in the same league as
2G customisation rather than the 3G–multimedia experience. The currency
of the Hong Kong nostalgia industry was very evident in my case study of
20 respondents. The case study consisted of half female, half male respondents, aged between 18 and 50. In June 2004 I surveyed 20 respondents and
followed up with in-depth interviews with ten respondents. I re-interviewed
ten respondents in June 2005 and, then again in February 2006. I sought
initially to interview users solely from Hong Kong City University, but having
managed to recruit only ten I resorted to a snowballing method. All the
respondents had at least undergraduate degrees from a variety of disciplines:
marketing, economics, English, media and communication, and film. Fewer
than half the respondents had a camera phone.

Respondents expressed themselves through their mobile phone primarily by
adorning the outside with post-production paraphernalia (such as characters
and straps attached to the phone). There was little internal customisation apart
from using camera phone images as wallpapers. This contrasted sharply with
the high innovation that was evident in Seoul and Tokyo. Parallels can be
drawn between Melbourne and Hong Kong users, but there were two major
differences, possibly attributable to the fact that Melbourne is the more multicultural of the two cities: namely the cultural capital of *wenqing zhuyi* (particularly through cute customisation) and the importance of linguistic hybridity
(mixing Cantonese and English) that Lin identified in her respondents.

Although a textual reading of SMS would be a wonderful way to elaborate on user-generated (as opposed to industry-created pre-production) customised content, my lack of proficiency in Cantonese precluded this. As aforementioned by Qiu (2008), the realm of *duanxin xieshou* is undoubtedly a new area of not only advertising but also creativity and income. In the case of my respondents, they seem to diligently deploy 'proper' English or Cantonese, with only two of the younger respondents noting that they occasionally wrote Cantolish as a joke. In this respect my findings differed from Lin's, probably because my demography contained a broader age bracket. In this study I was keen to focus on the interplay between personalised (user-created) and customised (industry-generated) content, and how respondents discussed the customisation codes inside and outside their phone as part of its symbolic communication processes.

Some of the younger respondents used QQ (IM). The term QQ can be used to denote an emoticon for crying, but in this context it refers to Tencent QQ (initiated in February 1999). Originally called OICQ, it was a free IM program that created a new breed of texters in mainland China and, to a lesser extent, Hong Kong. In mainland China, QQ has over 150 million users, with an estimated 9 million people online at any one time. However, QQ does not nurture the hybridity Lin (2005) spoke of, and this may explain why my respondents were resistant to using it. However, there was a marked difference between 2004 and 2005 – by 2005, respondents seemed more engaged in IM. Many respondents objected to the small charges for sending and receiving SMS, and for services such as ring tone downloads, and respondents over 27 years noted that 'they didn't have time' for such services.

As in Melbourne, Hong Kong does not have a dominant SNS. But unlike in Melbourne where SNS have continued to grow, Hong Kong oscillates between SNS and the more direct and individuated IM. As high percentages of postgraduate, and, to a lesser extent, undergraduate university students study abroad in Western countries such as the UK, US, Australia and NZ, many deploy 'Western' modes of SNS such as MySpace, Facebook and MSN Messenger to maintain the friendships that they made aboard. One female respondent stated:

> When I studied in Australia, there weren't Internet communities like there are today. The Internet was used for email. I would often either call or email friends and family back home. When I came back to Hong Kong, a friend in Australia suggested I try MySpace as she had just started using it and said it was fun. I used it for a little while but in the end I stopped as it took a lot of time to maintain. I still visit friends' sites and say hello but it just wasn't for me. Although, since then, some friends here are interested in joining an Internet community like MySpace or Friendster so that we can chat online while at work.

As this respondent notes, her attempts at maintaining an online presence were time consuming and she did not feel the remuneration was worth it. Moreover, the seamlessness between online and offline was, like Melbourne, related to technological infrastructure where broadband speeds were only incrementally more than dial-up. As Pamela Koch et al. (2008) noted in their study on the rise of QQ, entitled 'Beauty in the eye of the QQ beholder', Tencent has almost 500 million registered accounts with some 200 million of these actively used, and has approximately 18.4 million people logged in at peak times. In their study, Koch et al. argue that whilst their 347 respondents expressed the positive influences of QQ, QQ had its detractors who claimed that it was not a realistic image of Internet cultures and Chinese localisation of virtuality. As Koch et al. observe, the growth of Internet in China has often been associated with the decline in traditional morals, as symbolised by the sexual symbol of 'internet girls' (*wang-ji*). Here again, as with the Japanese *kôgyaru*, we see the female user being made the scapegoat for cultural shifts caused by socio-technological change.

The phone usage practices of 'modern' women, as symbols of both domesticity and sexuality, have been central to many of the discourses around domestic technologies (Na 2001; Yoshimi 1999). The rise of UCC can be seen as a double-edged sword: on the one hand, it has allowed emerging forms of intimate creativity such as mobile novels (authored and read by women) and, on the other hand, it has facilitated the continued exploitation of social and creative (feminised) labour. Unlike Japan where *keitai shôsetsu* is dominated by women as both writers and readers, *duanxin xieshou* is predominantly marked by class. However, it is significant to note that both male and female users are partaking in the feminised practice of UCC. Feminised customisation practices in Hong Kong sit in between those in Seoul, where male partners surrender their phones to the female partners, and those in Melbourne, where the phone is very much an extension of one's personal identity and is less likely to be shared or become a symbol of interconnectivity. As one female respondent noted:

> Both my boyfriend's and my phone are the same brand, 3. We bought them on a shared plan. I am often the one who ends up taking the photos when we are together or buying matching character phone straps. He pretends to take no interest. But I know when he looks at his phone he thinks of me. And then about us. I like the idea that the phone takes a ride with us. It means it makes it hard to upgrade the phone because when we look at them, we look at us.

From the 20 interviews, it became clear that gender and age played a significant role in how respondents used and customised their mobile phones. Female respondents seemed much more interested in using their phones to take pictures, which they sometimes shared with intimate friends in a face-to-face situation, and they also enjoyed more playful modes of texting. However, unlike the Melbourne respondents (discussed in the next chapter), Hong

Kong texters seemed less likely to play with language, often only customising the text with occasional abbreviations of 'u' for you. During the two years between the first interviews in 2004 and the follow-up studies in 2005/2006, one of the dominant changes was that respondents in general were texting more and taking more camera phone pictures. While half the users in 2004 did not have camera phones, by 2005, it seemed impossible to get a phone *without* a camera.

Most respondents used their phones for contacting friends rather than family or work colleagues. Male respondents seemed more enamored with the functions, and more likely to use services such as the Internet. One male respondent, aged 40, used his phone 50 per cent for voice calls, 45 per cent for texting and 5 per cent for internet/emails. On average he made about five voice calls and sent ten text messages per day. Most of his phones (15 over the last 15 years) had been Nokia, the rationale being that Nokia gave him the best deals. The respondent's main camera phone usage was for holiday photos, and he described his phone as a 'trusted tool and friend'. I asked him if he were to describe his phone as something else, what would it be? He replied 'a hot car'. Here the respondent was parodying masculinity and the role of both the car and the phone as status symbols, while at the same time acknowledging the way in which the phone can bridge distance and offer personal mobility, much as the car did in the twentieth century.

A female respondent aged 27 had had four phones over four years; the first three were Nokia and her current phone was Sony Ericsson. She noted, 'Mobile phones are becoming more integral to everyone's lives, expectations are higher and mobile phones have to be more than just a phone. It has to serve multiple purposes and become more indispensable to the user'. I inquired as to how she had noticed her relationship and feelings towards her mobile phone changing over the last couple of years and she replied, 'It's becoming more indispensable, and efforts are made to personalise it more'. On average she made about 30 voice calls per day and 14 text messages. Like the male respondent above, she noted that she did not like to abbreviate her text messages although sometimes she too used 'u' instead of 'you'.

For this respondent, the camera phone held greater personal significance in everyday practices than for her male counterpart. She stored about 40 images on her phone, mainly for her own viewing/recalling pleasure: unquestionably her mobile phone functioned as a repository for the *wenqing zhuyi*. Her main usage of the camera phone was to take pictures of her friends, which she called 'mug shots'. When asked to give three adjectives to describe her phone she replied, 'Indispensable, crucial, attached'. For both this female and the aforementioned male respondent, the importance of customising ringtones was vital in personalising the device. This customisation of the ring tone was not just the ring one hears when the phone rings, but more importantly (similar to Korea in this way) was the customising of what Koreans call the colouring (the colouring being the music the caller hears when phoning someone). This colouring was an important internal customisation mode that contextualised and spoke (sung) on behalf of the user's mood, emotion and personality.

One female respondent aged 30 noted that she had recently upgraded to a phone with camera functions just because 'the deal was good'. She had never had any interest in owning a camera phone, nor had any interest in photography per se. However, she had noticed that over the last couple of months the camera function had 'come in handy'. Having never owned a digital, or even analogue, camera, she had found that it was always up to her family and friends to document their times together. Now, with the camera phone, she was enjoying the ability to be an active participant in the making and thus re-calling of experiences.

She was initially quite embarrassed by her photographic skill, not helped by the poor quality of the 1.3 mega pixel camera. But through persistence she observed improvement and began to show her camera phone images to friends. Her empowerment in being the controller – rather than the subject – of the lens echoed research conducted by D.H. Lee (2005) on female usage of camera phones in Seoul. She found herself becoming more aware of advertising images, and even bought the application Photoshop to touch up and edit her photos. She noted that photography had become her new hobby and that she was now going to buy a stand-alone camera so that she could take more professional photographs. She also stated 'I also like the idea of being behind the camera, it means I'm less likely to be subject to someone else's bad photo of me!'

For one of the respondents, the camera phone was firstly about accessibility; it was always on hand (both materially and symbolically) and thus ready to shoot at any moment. As the male respondent noted, the camera phone was about documenting pleasurable experiences, particularly urban landscapes and images of friends. In this way, the camera phone reminded the user of places and friendships in the midst of his normally frenetic urban life traveling around Asia as a journalist. For this user, the camera phone was very much a device for re-inscribing geography and locality in the co-presence that is mobility today. In other words, the camera phone – far from the 'mobility' of the phone creating a sense of the placeless – was an important device for linking memories with place; in short, *reterritorialisation*. Pictures of urban landscapes are carried around in the phone, very much like postcards bought and stored for viewing later. Thus through the act of the *wenqing zhuyi*, reterritorialisation occurs.

The first two camera images (Figure 6.6) depict a night shot and day shot of Hong Kong and its families of towers. One image is ambient in its blurring – creating a sense of movement and action in its capturing a night-time outdoor cinema cradled by towers. The blurring, due to the camera's poor capabilities to deal with low light sources, gives the picture a feeling of movement that evokes the frenetic nature of a big metropolis such as Hong Kong. The other image is of Hong Kong during the day – a cityscape in which roads bifurcate the congregations of buildings. The two images operate as counterparts; the

Figure 6.6 Camera phone images taken by male respondent. (Photos: 2005)

Figure 6.7 Camera phone images taken by male respondent. (Photos: 2005–2006)

daytime one showing the normal and banal image of the heavily built up Hong Kong; the second alluding to a sense of joy and energy within the city's lights. The respondents' next two images (Figures 6.7a, 6.7b) are of friends; two lovers smile into the camera, and a friend's child smiles cheekily. The respondent took these photos to mark significant moments when he was having an enjoyable and relaxing time with friends. Whilst the place is obscured, for the respondent the image triggered remembrances of both the place and the people as part of the whole enjoyable *wenqing zhuyi* experience.

Although male respondents in general seemed less likely to share their images, preferring to save them for personal consumption by storing them on the phone or using them as wallpapers, the female respondents were more likely to share. Moreover the younger the respondents, the more likely they were to share images via predominantly face-to-face modes. Many felt apprehensive about 'forcing' them onto friends unless the friends were participants in the image, reflecting what Koskinen (2007) sees as the need for users to make their images 'newsworthy' in order to have enough relevance to send. Many female respondents preferred sharing in face-to-face situations such as sitting down at a café or bar, rather than sending them electronically. When asked about this preference, respondents noted that it was easier to contextualise the images, and that they enjoyed the way images could be used in simulating dialogue about certain topics such as family, friends, places experienced, etc.

Comparing my male and female respondents there were few differences in terms of content or composition, with many images following the 'banal snapshot' aesthetic. However, female respondents leaned towards more photo-journalism genres, particularly the mass media 'paparazzi' shot. Differences were however noted in terms of the sharing and distribution of the images. Younger female respondents were more likely to share their images rather than just store them for personal consumption. By circulating the images they become part of a collective unofficial nostalgia rather than an individual memento. These younger female respondents preferred to share their images face-to-face, much like the way in which analogue photo albums were shared around (but this time on a portable electronic screen, which was more likely than not viewed in public spaces such as cafés).

The younger female respondents tended to take photos of what could be deemed as cute subjects – cute animals, cute babies, and cute cartoon characters. The deployment of 'cute' is an obviously engagement with the *wenqing zhuyi*. Often it was the cute nature of the image that rendered the image worthy to show friends. The cuteness of these shots often lay not just in the subject matter but also in the way the images were taken – up-close and personal, to further emphasise the cute features. Seeing something cute in their travels would provoke the younger respondents to think of friends and their sharing of the cute; thus cute functioned as a co-present gift that would make friends wish they were there to experience it with them. By storing a picture on their camera phone until they next met face-to-face with the particular friends, the respondents could *defer* the moment and later relive it *collectively*.

One female respondent in her late twenties envisaged many of her camera phone images as potential wallpapers. Her choice of wallpapers was dependent on her mood as well as on her up-coming schedule. She noted that whilst the wallpapers were chosen for her own personal consumption, others often inadvertently viewed them. She noted that if she had certain work commitments, her wallpaper image would tend to not feature a friend but rather an object or something less telling about her personal life. Creating wallpapers was one of her main modes for customising her phone, and when she was with friends, the wallpaper often helped to show whose phone was whose. Many friends, particularly among the younger respondents, tended to buy similar models either because they heard by word of mouth that it was a good brand/model, or to facilitate compatibility. Although the SMS incompatibility issues in Hong Kong have long been resolved, having similar phones meant that respondents would be aware how their texts would be formatted and read on their friends' phones. Standardised viewer frames also helped to create a feeling of continuity in face-to-face sharing.

I have chosen five images by one female respondent in her mid-twenties

Figure 6.8 Camera phone images taken by a female respondent. (Photos: 2006)

to show some of the standard formats/genres that were consistent across male and female respondent images. Three images were taken specifically for use as wallpapers, the other two because the respondent wanted to document something and only had the camera phone available. These first three images (see Figure 6.8) are very much of personal content – a friend's baby, a young relative and the respondent's cat. In the first image of the baby the camera phone has been placed very close to the baby, an example of the way in which 'cute' subject matter is rendered cuter by close-up shots. In the second image we see a relative of the respondent looking curiously into the lens whilst being held by its mother traveling in a taxi. Again the shot is close-up and personal, emphasising the cute expression of the child that takes up almost the whole screen. The image of the respondent's cat is once again taken close-up and from above, to emphasise its cute face, small paws and watchful expression. This use of cute images of animals to warm up the coldness of new technologies can seen as a version of 'techno-cute' (McVeigh 2000), as discussed in Chapter 4. However, in the case of Hong Kong mobile media, the 'techno-cute' is part of the *wenqing zhuyi* affective mode of reterritorialisation.

Among Hong Kong users, the pervasiveness of 'cute power' to warm and humanise the technology occurs across various layers and levels of the phone. Although the most conspicuous and enduring form of cute customisation has been the dressing up of mobile phones with cute characters hanging from the phone, as users shift from 2G to 2.5G and 3G technologies, customisation becomes more internalised – both in terms of being inside the phone and also becoming much more, in the case of Hong Kong, about personal consumption, reflection and identification. Cute can make users laugh, cry or reflect; it is about picturing, visualising and representing emotions. These forms of cute are part of the semiotics of personalisation that I call emotology. Emotology can take various forms – from emoticon SMSs to cute camera phone images – reflecting both individual and collective *wenqing zhuyi*. It refers to the way in which the user grapples with their co-present representation on both an individual and collective level. In the following photos by the same female respondent we can see two very distinctively feminised genres – the cute and the mockumentary. On the one hand, users can mimic and find a voice in the reality presented by media; on the other, they can tap into the sublime moments and emotions that are 'on call' everyday though the ubiquitous and yet personalised genre of cute.

These images (Figure 6.9) were taken not as wallpaper but because the respondent wanted to document the moment and only had her camera phone. In the first image we see a band playing on a stage, ambient blurs giving a feeling of action and *mise-en-scène*. The photo is taken from a distance, equal amounts given to the audience, band and building in the backdrop. Unlike the cute images, this image does little to hold one's attention and is an individual's attempt at getting a souvenir of the moment that is likely to be reviewed and will not feature as a wallpaper.

Figure 6.9 Camera phone images taken by female respondents. These images are a good example of user-as-journalist (or paparazzi). (Photos: 2006)

The second photo is a comical image of a friend of the respondent being attacked by a Christmas tree. Amidst the dark ambient lighting (sprinkled with fairy lights) we can make out (if obscurely) a female with clenched eyes and a pained expression as she fights off the attacking Christmas tree. This image is an example of what the respondent called 'mug shots', in which the camera phone becomes the vehicle to transform the everyday mobile user into photo-journalist in the form of paparazzi. Often the camera phone image becomes a weapon that the user can employ to playfully blackmail friends with the threat of publicly showing the image – or to save it as wallpaper on the offended party's phone. These images mimic and parody mass-media images of celebrities; at any moment the mobile phone user can transcend the ordinary and become the 'produser', or journalist.

Mobile media not only operates as a site for fusing the public with the private through quasi-photo journalism, it also provides a space for personal nostalgia. Unlike in locations such as Korea where images are uploaded and shared online collectively, in Hong Kong images were predominantly shared via the face-to-face holding of the phone or, in some cases, through one-on-one mobile e-mailing. As vessels for nostalgia, the mobile phone operates to allow users moments to reflect – either by themselves or in face-to-face situations with others – upon the everyday frenetic nature that is Hong Kong urbanity; in sort, mobile media practices operate to 'reterritorialise' Hong Kong by forging personal ties between historical and geographical deter-ritorialisation. In its transformation into mobile media, we see Hong Kong customisation shifting from icons of the individual's cultural capital (i.e. the choice of phone brand and exterior adornment) to the maintenance of personal histories. These are intimate, individuated imaging communities that are embedded in reterritorialisation of place and personal historicity with contemporary Hong Kong life.

Imaging @ communities: *wenqing zhuyi* and cross-cultural consumption

As a symbol of converging fashion and technology, the mobile phone carries with it the weight of people's longing for more status and for times of more prosperity. As a device that traverses the public sphere and reinscribes notions of what it means to be individual and to practise the social, the mobile phone is also the perfect symbol for comprehending some of the transcultural consumption and production flows within the Asia-Pacific. In the case of Hong Kong, could the preoccupation with wallpapering as the ultimate mode of customisation be seen as a comment on identity formations in Hong Kong? Does the lack of sharing and the gendered preoccupation with cute genres of camera phone images speak of Hong Kong's relationship to the general flows of cute customisation initially coming from Japan? How does the *wenqing zhuyi* provide various pathways for reterritorialisation? Arguably one of the key differences is not so much the *what* (i.e. the genres of camera phone images) but rather whether or not, and by what method, these images are being shared. The modes of sharing (or not sharing) reflect meanings on both an individual and social level. A culture is the sum of its unofficial and official discourses, and, in the case of Hong Kong, we can see two paralleling imagined communities on a micro and macro level.

In particular, the use of camera phone images not as a shared experience (and thus not linked to the gift exchange logic of the mobile phone) but for personal consumption can be seen as an example of Wellman's 'networked individualism' (2002) or Ma's 'atomized individualism' (2005). The role of camera phone imagery – by mimicking the ubiquitous cute aesthetic or mass-media style (i.e. paparazzi) – can be viewed as a way for individuals to reflect upon, and participate in, the *official* circulation of nostalgic media. In Hong Kong, techno-cute modes of mobile media customisation reflect *wenqing zhuyi* modes that provide respondents with spaces in which to contemplate and rethink a sense of place and community. When people do share the images, the fact that they do so predominantly face-to-face rather than sending them via MMS or uploading them to a website, suggests a preference for more discursive and synchronous communication rather than for indirect and asynchronous communication (such as MMS). This has an influence on how stories are being told and the ways in which images are contextualised.

Mobile media practices such as camera phone imagery have become a dominant form of UCC. From the *bus uncle* instance to the case studies I have discussed in this chapter, mobile media UCC allows multiple forms of 'produser' that, in turn, reflect processes of hyphenated identity and nostalgia. From the deployment of IM and SNS – reflecting Hong Kong's position between Chinese and Western markets – along with the global smorgasbord of brands and customisation, Hong Kong mobile media is a hotbed of modernities. Mobile media UCC is growing strongly in Hong Kong, not only as a vehicle for representing private creativity and identity but also as a space for new forms of unofficial public media. From the 'performing' of paparazzi-style photojournalism to the recording of iconic 'real' events such as the *bus uncle*, the actual and symbolic dimensions of mobile media present proliferating imaging communities that unofficially reflect Hong Kong's own experience of post-modernity.

The mobile phone is not only a poignant symbol for post-modernity, it is also both an icon and a cultural practice integral to new forms of constructing and performing a sense of public culture and private intimacies, which in turn reflect cultural intimacies and constructions of the imagined community. The mobile device is a site where consumption, production and the politics of the everyday converge, making it an ideal symbol for rethinking the numerous contesting modernities and the resurgence of nostalgic transnational communities. Like the domestic and private act of consuming TV dramas in the home and then discussing them in forums such as online fan forums, the consumer communities of mobile phones are about types of privatised consumption in the public sphere. Activities such as using a theme song as a ring tone highlight to intimate strangers the user's tastes and interests, dividing public crowds into those who recognise and identify with that cultural product and those who do not.

The role of the mobile phone in signaling the user's tastes to others continued to feature prominently in interviews. One female respondent, aged 28, noted that often she felt 'judged' in a negative way by strangers or work colleagues by her decoration of her phone with Japanese 'cute' characters. This resulted in her stripping her phone in public and only dressing it when at home or going out with friends. When I asked her whether anyone had made a disparaging comment she replied, 'no, but I can sense the ways it could be misread'. The *wenqing zhuyi* is both a collective activity as well as a deeply private affective mode that manifests in different ways according to the age, gender, ethnicity and class of the respondent. Ironically, for the afore-mentioned respondent, the cute character was bought by her boyfriend of seven years and functioned like an engagement ring; they bought the characters together and customised both their phones. However, it was predominantly the female respondents that informed their male partners what to buy and what to customise their phone with; clearly demonstrating the role of female respondents as providing cultural capital whilst the male partners supplied economic capital. Thus social capital reflects types of cultural capital through specific modes of *wenqing zhuyi*.

What is curious about the mobile phone is that on one level the customising of the phone with pop cultural references is a process whereby the user is continuously updating the phone with current interests, so that it becomes a barometer for the never-ending processes of individual identity formation as 'one of unending layering and interaction of cultural knowledge acquisition' (Chua 2006: 35). On another level, the customisation is also about grounding and centring the individual within a social and cultural context – i.e. a community – as can be witnessed in the sharing of and reflecting upon camera phone images. We are reminded again of the way in which mobile phone practices recite earlier gift-giving rituals (Taylor & Harper 2002, 2003), which in turn reflect social nuances and power relations. In particular, the *wenqing zhuyi* is a key emotology in the reterritorialisation of mobile media subjectivities in Hong Kong.

For Hong Kong respondents, the mobile phone is a symbol of the cross-cultural flows of *wenqing zhuyi* – from Samsung phones with Hello Kitty cute characters hanging outside to ring tones from a Jackie Chan film and wall-papers of Korean TV drama stars (Yung 2005). Practices such as camera phone images are allowing respondents a space for individual and interior-ised *wenqing zhuyi* in the face of the overarching themes of nostalgia being activated by media not only within Hong Kong (as identified by Ma) but also within the Asia-Pacific. While from a cursory glance, the lack of sharing could be read as indicative of what Ma identifies as a decontextualised and 'atomized individualism', it is important to note that when images are shared it is in face-to-face situations and thus are part of collective modes of *wenqing zhuyi* as resistance to mainland reterritorialisation.

The capacity of the camera phone to provide a visual diary of experiences and feelings, which individuals can view by themselves or when face-to-face with intimates, highlights its importance as the device for collecting moments of the everyday and rendering them into poetic memories to be savoured. If Ma is correct in his assertion regarding the big business of nostalgia in Hong Kong media, and thus in the national 'imagined community' (in an Anderson sense) then it is the camera phone that is affording respondents with the ability to take control of their memories and to revisit them with intimates.

Nostalgia is not just an ideological construct on behalf of the government or media agencies; indeed, through the personal deployment of *wenqing zhuyi*, Hong Kong respondents are able to actively construct a sense of place and experience that differentiates their post-modernity from the mainland. Undoubtedly mobile phone practices in Hong Kong evidence the nostalgia that Ma (2005) speaks of while also highlighting Chow's (2007) characteristic *wenqing zhuyi*; for individual users, the affective politics of nostalgia and *wenqing zhuyi* are significant not only in the process of individual contextualisation but also in re-inscribing a sense of place and memory with intimates in face-to-face situations.

In sum, the particular emotologies of *wenqing zhuyi* reterritorialise upon both individual and collective levels, from the imaging communities of the respondent to the imagined community of Hong Kong post-modernity. The 'community of consumers' Chua (2006) identifies is, in the case of Hong Kong, a complex process negotiating micro and macro constructions of nostalgia. If the interior is the new exterior, then surely it is mobile media practices that are ensuring a sense of personal geopolitical re-territorialisation between 'hyphenated' histories and geographies.

Figure 6.10 'Communication is borderless.' This retro advertising lightbox found in Causeway Bay in 2004 is a great example of big business around nostalgia – even when dealing with new technologies and future generations. Photo: Hjorth.

Camera phone images afford Hong Kong people the ability to return to the past with intimates through face-to-face sharing, which suggests that the co-presence of mobility (here/there, virtual/actual) is made meaningful through recollections when face-to-face. Thus meeting at a café with friends and sharing images of the past is about providing discourse, a microcosm of an 'imagined community', whereby the personal is divulged in socioculturally specific ways that might be best characterised – in keeping with Koskinen's (2007) argument that camera phone practices 're-territorialize' – as imaging communities.

These cartographies of personalisation are marked by a sense of a longing and togetherness, a perpetual state of *wenqing zhuyi*. Through the emotologies of *wenqing zhuyi* camera phone images or saved SMS, Hong Kong respondents can grapple with a sense of place and reterritorialisation; a process forever filtered by a yearning of better bygone times that only existed in the specters of ideals. Mobile media practices give individuals a site for constructing their own *wenqing zhuyi* in the face of state projections of dislocated memories and ideologies. The cartographies of personalisation show that while China might well indeed be 'one country, two systems', Hong Kong people are being proactive in challenging a homogenous vision of a 'Global South'.

Through the symbolic and material dimensions of mobile media practices, Hong Kong respondents are able to negotiate the role of geographies and histories past, present and future that haunt Hong Kong's current postmodernity. From the transformation of SMS into a literary industry of *duanxin xieshou*, to the various ways in which place is inscribed with intimacy through gendered camera phone practices and the significant role of wallpaper constant co-presence, Hong Kong demonstrates that a sense of place has always involved multiple imaging communities and haunting of parallel historicities that are 'mobile, immobile and mooring' (Hannam et al. 2006). Through the symbolic and functional role of the wallpaper we can see practices of personal reterritorialisation that helps to locate Hong Kong's hyphenated modernity. Far from just being wallpapers or screen savers, these practices of personal consumption are re-enacting a sense of intimacy, affect and memory of urban spaces, these images are not mere backdrops to the movements of mobility and modernity: they reterritorialise the mobile – memories, subjectivities, geo-spaces and geo-imaginaries.

7 Postal presence

Persistence of the postal metaphor in Melbourne

> Alongside and interwoven into this global political and cultural economy of what used to be called the mobile phone, and now might more accurately be called 'mobile media' is the growing role and recognition of the user, including the cultural politics of the user-as-producer . . . In all of this with its imminently 'glocal' emergence the mobile phone is connecting various groups, communities and networks below, above, across nations.
>
> (Goggin 2006a: 12)

In the last case study in Part II, we turn our attention from the epicentres of mobile innovation, as noted in the Japanese and Korean case studies, via Hong Kong to Melbourne. Although the Japanese case study demonstrated high mobile Internet usage, the Korean case study showed a co-ordination between mobile Internet and personal computers (most notably in the case of the social space of *PC bangs*). Both Japan and Korea are examples of localised production and consumption of predominantly locally cultivated brands and service providers. By contrast, Hong Kong and Melbourne exhibit a smorgasbord of global brand names and service providers.

In each of these locations we can find different forms of gendered mobile media practices. Some transgress national boundaries, as the rise of the female mobile media 'produser' cited in the opening quote suggests. Other mobile media practices remain immobile and contained within socio-cultural enclaves and nation-state boundaries. In these various forms of mobility and immobility, gender features prominently. In Melbourne, we find mobile media practices of co-presence – particularly through SMS and, more recently, the rise of SNS such as Facebook – demonstrating the persistence of the postal metaphor. By postal metaphor, I am referring to ways in which new media practices rehearse older, remediated techniques such as SMS re-enacting the epistolary traditions evoked by nineteenth century letter writing traditions as well as the postcard. I am also referring to the role of time and space involved in mobile media. Far from mobile media destroying place and transcending temporal boundaries in a time–space compression characteristic of global ICTs, we can see that despite mobile media being heralded as the 'power of now' (Wilhelm et al. 2004), individual users are inserting the poetics of delay in order to actively construct spaces for contemplation and reflection.

These modes of practice are what I call 'waiting for immediacy' and are a pivotal part of Melbourne's specific cartographies of personalisation. In particular, 'waiting for immediacy' can be viewed specifically within three techniques – *delay, deference* and *decentring*.[1] These three Ds can be found from the feigning of sending a text 'immediately' to the re-enactment of a 'spontaneous' camera phone image that only much later is shared. The three Ds embody the postal presence underlying Melbournian imaging communities; the three Ds help to reterritorialise a sense of place and immobility in an epoch dominated by the rhetoric of mobility and 'death of geography'. In the case of Australia, the significance of geography and the sense of geo-spatial distance have far from been eroded; rather, as Papastergiadis observes, Australia is configured by the notion of 'South-South-South' (2003: 2). It is this 'South-South-South' – as a geo-political imaginary – that significantly shapes, and is shaped by, mobile media imaging communities marked by the three Ds.

As I argued in the Chapter 6, parallels can be made between Hong Kong and Melbourne in terms of the global array of mobile media, brands and service providers. This is demonstrated by the fact that brands such as Motorola use the same advertising images in both Hong Kong and Melbourne, and that Hong Kong and Melbourne have both been resistant to the usage of mobile Internet. Despite the emergence of 3G technologies in 2005, most users in these locations shy away from mobile Internet convergence, preferring to access the Internet via personal computers. However, there are marked differences that are influenced by industry regulation, socio-cultural and linguistic nuances.

Most notably, in the case of Melbourne, we can see the density of uptake and persistence of SMS as a central mode for users to customise mobile media. In this chapter we will explore the role of SMS and MMS, and how this is inflected by gender. I will begin by locating Australia, and specifically Melbourne, in the context of the Asia-Pacific. In the disparate construction of the region, Australia further illustrates the diversity and heterogeneity enveloped under the rubric of regionality. I then turn to my case studies and explore gendered mobile media practices and how these reflect emerging forms of 'produsers', in which resistance, as much as uptake, are key modes of expressing localised forms of intimate co-presence. In this study we see how cartographies of personalisation and the three D's are used to reterritorialise a sense of place and community for both local and international students. By investigating gendered mobile media in Melbourne we can gain insight into how Australia locates herself in both the post-modernity symbolised by mobile media and where she is situated in the region's twenty-first-century post-modernity, the 'Global South' (Dirlik 2007).

Reconnections: locating the practice of South

> As both a referent and a reference, 'going south' functions as a geographical distinction (that differentiates the Asian diaspora in Australia from Asian diasporas elsewhere) as well as a theoretical paradigm. Inscribed in a migratory movement of literal displacement and reoriented in the racialised landscape of a postcolonial settler Australia, the trajectory of 'going south' aligns itself with (Australia as) south of the West, (Australia as) south of Asia, and (both Australia and Asia as) south of the East and West. Implicit in the trajectory of 'going south' is an interrogation of how Australia, as 'south of the west' has also come to construct itself as specifically south of Asia.
>
> (Yue 2000: 192)

Australia has always sat uncomfortably within the Asia-Pacific. And yet Australia has, as a British colony, also found ambivalence in identifying with its distant colonial mother. While Hong Kong's vacillation between its British colonial history and its contemporary China identification is marked by a feeling of hyphenation that is negotiated through nostalgia media, Australia oscillates between its socio-geographic ties with its British colonial history and its geographic proximity to the Asian region. As Rao observes, Australia can be viewed as 'the west in Asia' (2004); a location constituted by an array of trans- and cross-cultural production and consumption.

As Yue identities in the opening citation, the geo-political imaginary of Australia is one that is marked by a perpetual de-centring of the West. The realignment of Australia's imagined community away from a colonial Eurocentric gaze towards a geo-cultural and postcolonial repositioning in Asia has seen Australia, for Yue, become both 'south of the West' and 'south of Asia'. As Yue notes, the 'south' is not just geographic but integral in the constitution of Australia's ambivalent imagined community. While Yue discusses the 'going south' trope in the context of 'a transnationalism that characterises the modernity of Asian diasporas in Australia' (2000: 193), for Papastergiadis the notion of 'South-South-South' refers to 'oscillations between deference and defiance' (2003: 2).

This notion of 'south' (of West and Asia) and 'South-South-South' – in a period marked by global flows, diaspora and decentring – suggests that in the new geo-cultural formations of multiple modernities, Australia is grappling with attempts to locate itself. It is this positioning and repositioning – defined by a sense of incompleteness and belatedness – that is part of Australia's geo-cultural struggle and that can be, in part, witnessed in the country's mobile phone industry. Initially mimicking Europe by consuming brands such as Nokia, 3 and Vodafone, Australia's market quickly became dominated by Asian companies such as Samsung, LG, and DoCoMo's i-mode. The fact that global companies such as Motorola and Nokia used the same advertising in Hong Kong and Melbourne suggests parallels in the ways global telecommunication companies view those markets.

Unlike Hong Kong's highly aural relationship to the mobile phone, the

Melbournian respondents' relationship is marked by a preference for textuality in the form of SMS. Yet there are marked similarities in the shift from externalised to internalised modes of customisation can be found. Interestingly, as Hong Kong respondents begin to give less time to customising the outside of their phones, Melbourne respondents are overcoming their Anglo-Saxon aversion to cute customisation and beginning to deploy overt forms of techno-cute customisation. Is this indicative of Australia's peripheral 'incompleteness' trying to relocate itself in the region? In one way, the co-presence of mobile communication and, in particular, the ways in which both industry and users try to locate the mobile can be seen as a metaphor for Australia's geo-cultural imaginary.

Like all forms of localisation and customisation in the face of global ICTs, Australia's own localisation techniques – in the form of industry (handsets, services and advertisements) and UCC customisation – speaks of Australia's particular relationship to modes of co-presence. As I have argued earlier in *Mobile Media in the Asia-Pacific*, the role of mobile media only further highlights the prevalence of place. For Michael Arnold (2003) mobile technologies are exemplary of Martin Heidegger's claim that the role of technology to overcome distance is paradoxically tied to the destruction of closeness. Mobile technologies complicate Australia's already existing sense of de-centredness. This is perhaps why applications such as SMS took off so successfully: the practice of SMS is one that overtly charts the genealogy of earlier co-present practices such as postcards. Moreover, the asynchronous mode of SMS means that gestures of intimacy are softened and deferred. SMS can allow for more consideration and delay, which in an age of immediate technologies affords an outlet for contemplation. I will further discuss the possibilities of SMS partaking in the poetics of delay in the case studies.

Australia's telecommunication service providers and handset producers present a market brimming with global products and a diverse cultural cross-pollination of customisation techniques (both industry created content and UCC). As in Hong Kong, it is not unusual to see a Nokia phone adorned with a cute hanging ornament from Asia and a personalised wallpaper of the user's pet or partner. These bricoleur techniques are part of broader UCC that seeks to personalise the communication device, operating to locate the user's identity at both an individual and social level.

As I have argued elsewhere in *Mobile Media in the Asia-Pacific*, today's imagined communities are increasingly being negotiated by localised customisation practices in the form of imaging communities. As Ito (2002) has argued, the practice of mobile technologies needs to be understood in terms of 'communities of co-presence'. In the case of Australia's 'South-South-South' imagined community, it is through examining 'communities of co-presence' that we can gain insight into the unofficial, micro and individual experiences and identifications. I argue that in the case of Melbourne, a sense of place and co-presence is determined by practices of postal presence that can take the form of SMS, MMS and other UCC customisation. It is this gendered mode of postal presence, informed by Australia's dislocation and incompleteness that makes delayed and deferred co-presence part of everyday life, which differentiates Melbournian practices from the other case studies. In the face of the 'power of now' (Wilhelm et al. 2004) rhetoric of mobile media practices, UCC cartographies of personalisation are increasingly becoming implicated in the poetics of delay that not only highlight the spectres from older, remediated forms of epistolary but also the everyday tactics at work to resist the compulsion of globalisation's so-called time–space compression.

The title of this chapter, Postal presence, operates not just on a micro, individual level but also on a macro collective level. By postal presence I mean the attempts to negotiate, most particularly, through the use of the mobile phone, the distance and difference Australia feels – as indicated by the multiple variations on 'south' – which manifest at the level of the everyday. Australia's involuntary experience of co-presence and attendant distance and difference upon numerous realms (geographic, spatial and temporal) inevitably informs the 'incompleteness' Papastergiadis refers to, and are now manifest in the micro-politics of mobile media co-presence. Unlike the 'personal, portable and pedestrian' qualities discussed by Ito et al. (2005), the role of co-presence in Australia is marked by a postal presence relationship to the everyday. The mobile phone only further instils the existing postal presence, and the related co-presence, in the form of three Ds: *delay, deference* and *decentring.*

These three Ds can be witnessed in behaviour around the dominant mobile media – SMS. In an age of immediacy, one of the interesting paradoxes in my case study was that respondents often delayed answering text messages, and often gave considerable thought to their seemingly off-the-cuff responses. This postponing, or deferral, allows respondents to gain control in the frequency of the messages, thus decentring the leash/freedom paradox of mobile media. This practice in texting can be seen as reflective of broader mobile media practices, in which respondents attempt to gain context and control in navigating co-presence and intimacy.

As I highlighted in Part I, mobile media is affording emerging modes of customisation and imaging communities, and these modes are remediated. This remediation, subject to the nuances of a cultural context, sees distinctive and divergent adaptations of mobile media. In the case of Melbourne, the remediated notion of postal presence is prevalent both in SMS and in camera phone practices. Practices such as SMS, whilst offering an affordable mode to keep in contact with intimates, have given rise to a new class of users, who see texting as an art form drawing on other, remediated forms of textual representation such as letter writing. SMS practices have also given rise to distinctive subcultural activities that deploy linguistic and emoticon variations. One example is the SMS practice of writing phonetically, which eschews conventions between textual and aural gestures. Through phonetic texting, respondents can toy with different accents that represent the user's class, age and ethnicity.

As Taylor and Harper have noted, mobile phone practices replay older rituals of gift-giving, whereby reciprocity and co-presence re-enact the custom of present giving (2002, 2003). Here the *present* transforms into a sense of *presence*. The aforementioned three Ds – *delay, deference* and *decentring* – are central to the temporal and spatial transformations of media such SMS. Rather than the three Ds being a prison sentence for all Australians, individuals can deploy them to regulate or, as Ling defines it, 'micro-coordinate' (2004) everyday temporality. The delay once intrinsic to all co-present correspondence now can be deployed by respondents in the form of SMS to delay and defer conversations between individuals.

In a land as large as Australia where most inhabit urban coastal areas, the idea of immediacy and intimacy is firmly located in the deferral notion of 'post'. Telecommunication has always been about trying to overcome geographic distance while navigating the desire to construct a sense of presence through co-presence. This idea of being able to circumnavigate a sense of distance and temporal dislocation through co-presence was most explicit in SMS; many respondents claimed to read SMSs immediately and yet reply or send after much deferral and reflection. Thus the experience of postal presence so integral to what it means to be mobile (in all senses of the term) is symptomatic of the micro-politics of Australia's perpetual location, dislocation and relocation. It is perhaps the idea of posting – which suggests an oscillation and movement between different places and time – that is wound up in the notion of belonging and home. Unlike Japan's 'communities of co-presence' (Ito 2002), Australia's co-presence is in the post.

It is the idea of the imagined community of the 'West' being located not geographically in the west but in the south that informs Australia's geo-political imaginary. Moreover, the fact that Australia is both a colonial settlement and also a multicultural country constituted by generations of immigrants from different cultural contexts, also informs its socio-cultural 'incompleteness'. Amidst the pressures of globalisation and the time–space reconfigurations epitomised by mobile phone consumption, Melbourne presents a curious constitution of media, capital and people. Unlike locations such as Korea where broadband is at its highest levels in the world, Australia's vast land mass has resulted in a new digital and mobile divide – urban versus rural dwellers. In contrast to the gated telecommunication industries of Japan and Korea, where local business was nurtured and thrived without outside threat (which then ensured their ability to launch globally), Melbourne's model of mobile telecommunications industry parallels Hong Kong's.

Both Hong Kong and Melbourne's telecommunication markets are saturated by companies from around the globe, all vying for supremacy. Far from being ruled by European companies such as Nokia, Asian brands are proliferating – epitomised by the adoption by Australia's dominant service provider, Telstra, of Japan's i-mode, in 2005. Korean brands such as Samsung and LG are growing in popularity. Even non-Asian brands such as Motorola re-imagine themselves as part of the Asia-Pacific. Motorola's RAZR V3i asserts its campaign on the cultural capital of Japan by using Japanese models in their advertising campaigns and slogans that exhort consumers to consume the cultural capital of Japanese techno-savvy. The Japanese/Swedish conglomerate Sony Ericsson has revisited the Sony Walkman with its Walkman mobile phone. Viewing the bombardment of mobile phone advertisements that grace trams, buildings and printed matter, one cannot but conclude that the Melbourne market – as symbolic of the Australian urban market – is being re-centred within the mobile cultures of the Asia-Pacific.

In an era when ICTs claim to reach users everywhere, being far from anywhere has meant that Australia has struggled with how it contextualises itself in the global market. The claims of the mobile industry only heighten the user's sense of profound isolation and a constant need to be connected. Despite having been a British colony, Australia's geographic distance has led to economic and cultural differences that isolated the colony from England. During the 1990s, before the economic crash of 1997, Prime Minister Paul Keating worked actively to re-imagine Australia in terms of its geographic proximity to Asia. As the country's former treasurer, Keating was mindful of the growing power of Asia in the global markets, and he established strategic geo-political and geo-economic imperatives that sought to link Australia to the region.

After the reign of Keating, Australia was governed for over a decade by an ultra conservative Liberal government that sought to reinstate the ideologies of the British monarchy under Prime Minister John Howard. During Howard's tenure, Australia followed the US and, to a lesser degree the UK, in its focus on terrorism, which saw various boundary making policies – such as the detention camps for refugees – erode Australia's standing in the region. This dark period in Australia's human rights record ended with the November 2007 election of the Labor government under Kevin Rudd. A fluent Mandarin speaker, Rudd seems likely to revisit some of the regional policies that were promoted under the Keating era.

During the Howard era, when much of this research described in this book was conducted, it is interesting to note that in contrast to the prevailing British monarchy ideologies, mobile phone practices – especially customisation – were more attuned with Asian trends. This was partly due to the advertising campaigns of global telecommunication companies, in which the relatively small market of Australia sat within the 'region', but it also had to do with users' own relationships to mobile phones as part of everyday practices. And while Australia has been slow, perhaps Anglo-Saxon, in its resistance to adorning and customising the outside of the mobile phone, the resistance appears to be starting to crack.

Postal cartographies: mobile media in Melbourne

In Melbourne, the second largest city in Australia, the mobile phone is a dominant form of everyday practice. The practices and experiences of mobile telephony are divergent and ubiquitous, marked by factors such as ethnicity, age, gender, class and sexuality. Reflecting its location in the Asia-Pacific, one can find many appropriations of customisation practices (such as cute characters and other fashion accessories). The streets are filled with a cacophony of polyphonic ring tones as users traverse the city with phone in hand. Due to economic restraints, the cheaper options of SMS-ing and instant messaging are the main forms of usage, marked notably by gendered inflections. Gender affects the motivations as well as the types of messages sent. Although cost may have been the initial main motivation for opting for SMS over voice calling, it is now a form of expression that is preferred by many Melbournians as *the* chosen medium for intimate co-presence.

With four main service providers – Telstra (the largest), Hutchison (Orange, 3), Optus and Vodafone – all vying for the Australian market, users have considerable choice. Telstra (formerly the government monopoly tele-communication service, Telecom) was set to adopt the BlackBerry phone – successful in UK and US markets – but the deal subsequently fell through. Instead Telstra has signed with NTT DoCoMo to take up i-mode, eight years after it was implemented in Japan in 1999. According to Telstra press releases, it was believed that one in 20 Australians would have i-mode in the next three years. The service failed spectacularly.

Although not the capital of Australia, Melbourne is arguably the capital for fashion and thus a good barometer for the uptake of mobile phone trends in urban Australia. With a population of three and a half million, Melbourne has a multicultural demographic comprising large numbers of Greeks, Italians, Chinese and Jews, among others, with residents coming from an estimated 140 different ethnic backgrounds. Settled in 1835, Melbourne has been subject to many influxes of immigration, most notably the 1850s' gold rush that saw the arrival of tens of thousands of immigrants. This colonisation was founded through horrendous crimes of humanity against the indigenous population. In contrast with the long and well-documented mobile phone histories of Japan, and the innovative mobile convergences of South Korea, Australia's mobile phone history has been relatively overlooked by researchers (Goggin 2006a). In the years since the launch of the first mobile phone service in Australia in 1981 and the first cellular mobile service in 1987, by 2006 Australia had acquired over 15 million mobile phone subscribers. As Goggin observes:

> . . . In 1987 Telecom established the first cellular mobile phone network, based on the USA Advanced Mobile Phone Systems (AMPS) . . . By June 1989, there were 100,000 AMPS services, rising to 290,000 users in

June 1991, and 500,000 in October 1992, 1 million in March 1994, and 2.6 million in 1996 (AMTA 2003).

(Goggin 2006a: 3)

For Goggin in 'Notes on the history of the mobile phone in Australia', while histories of media have their difficulties in terms of documenting the changes that are not just technological, this is even more the case for nascent media such as mobile phones. By outlining the history of the mobile phone in terms of the 'industrial, political economics and policy development', as well as 'representations of the mobile phone's adoption and use' (2006b: 7), Goggin explores some of the ways in which the mobile phone was received. Most notably, he explores the shifts in network choices in Australia that highlighted Australia's precarious position in aligning with the various global markets and networks in Europe and the US. The Australian Telecommunications Authority initially (in 1991) advised the government to adopt the European Global System for Mobiles (GSM), but GSM was best suited for dense urban populations and this proposal failed to consider Australia's rural population. Luckily for the government (and the rural community), Telstra, the country's major telecommunications provider, created a second network based on the North American CDMA (Code Division Multiple Access) system, and this was launched in September 1999. The decision was timely, because by the beginning of 2001 the number of mobile phones outweighed landlines.

In this rise of mobile telephony in Australia – parallelling other locations in the region such as Tokyo – the early adopters of mobile phones as social and creative media in the urban cities in Australia were predominantly young females. Researchers often seemed to focus on the 'youth' aspect, neglecting the key role gender played in this phenomenon, yet the industry advertising was clearly targeted at its main consumer, the young to middle-aged female. While scholars such as Wajcman and Gillard have been quick to characterise the role of mobile technology in Australia in terms of divides – principally gender and the urban/rural divide – it is surprising how little attention gendered mobile media in Australia has attracted until recently (Wajcman et al. 2008; Hjorth 2005b).

The role of gendered mobile phone production and consumption in Melbourne is unmistakable. Moreover, the growing importance and pervasiveness of Asian brands such as Korea's Samsung and LG, along with the Japanese/Swedish Sony Ericsson, in the configuring of types of femininity – and feminisation of technology – are also clearly evident. Here we become aware of what Ellen van Oost called 'gendered scripting' – that is, the way in which gender is inflected in the designing and marketing processes, imbuing technologies with gendered modes of practice. The role of gendered technologies has been particularly prevalent in the rise of domestic technologies such as fridges, televisions and now, through arguably the most gendered domestic technology in contemporary culture, the mobile phone. In the context of brands, the region is illustrative of the specific gendered production and consumption that is not reiterated in Europe or US. Interestingly, Australia, as a market, receives about half the images made for the 'Asia-Pacific' region and half aimed at the European market. However, much of the gendered scripting (designing with gendered usage in mind) is aimed at what Leslie Regan Shade argues is a 'feminizing of the mobile' (2007). For Shade, in her expansion of van Oost's work on gender scripting, North American mobile phone design's 'feminization' is all-pervasive.

Surveying advertisements in Melbourne from fashion magazines (*Vogue, Harper's Bazaar, Marie Clare*), tabloids (*Who Weekly, NW*) and newspapers (*The Age, The Australian*) three key factors can be identified: the increasing importance of gendered design and user interpellation (Samsung, LG, Nokia), the usage of cross-cultural Asia-Pacific capital (i.e. Motorola), and the selling of convergent mobile media as remediated (Sony Ericsson, Nokia). These factors form two tangents – female respondents are being interpellated through the image of the phone as fashion accessory, and male respondents are being targeted through the image of the mobile phone as convergent media (i.e. encompassing music and emails).

For example, Nokia has been careful to customise its ads separately for women and men by promoting phone models with gender-specific functions. After enjoying a strong reputation in the Australian market as a user-friendly device, deficiencies in its niche branding saw it losing its market dominance. Nokia's convergent Nseries have been marketed under the slogan 'Play, record, edit or listen while you send, share, print and download'. Expanding on Nokia's 'Connect people' slogan to encompass *connecting content*, Nseries proclaims 'See new. Hear new. Feel new.' Utilising the three 'Cs' – create, consume, connect – the ad features the four phone models in the Nseries with all but one graced by a male wallpaper. In this advertisement the obvious ideal user is a male who wants his mobile phone to be a multimedia device.

> Let your imagination take flight. The L'Amour Collection defines contemporary vintage – timeless elegance finished with modern touches. Exotic materials add glamour to innovation in a collection unlike any other. And with a range of exquisite accessories, the L'Amour Collection inspires the essential fashion indulgence.
>
> (Nokia Australia 2006)

Launched at the same time, Nokia 'L'Amour Collection' consists of three highly stylised phones with ornate detailing under the banner of 'Inspiring Imagination'. Unlike the Nseries, the L'Amour promotion neglected to entertain the technological features, as if the phone were a cultural artefact without purpose except to perform status and style. This is a clear example of how industry conceptualises the divide between male and female users: male users are interested in content and functional technology, while female users – represented in the advertisements by a female model dressed in retro clothing and looking off dreamily into the distance – are only concerned with the device as a fashion symbol.

The use of the mobile phone as remediated music player, competing with the popularity of MP3 players (most notably iPod), is explicitly addressed in the Sony Ericsson advertisements, in which advertising campaigns such as Sony Ericsson's W810i Walkman phone is clearly about remediating the success story of the Sony Walkman cassette player. Proclaiming, 'Your music. Your style', the ad features a hip urban young male enjoying his music on his phone whilst walking the streets. Here Sony Ericsson is not only selling the idea of Robert Luke's (2006) contemporary *flâneur* in the form of the *phoneur*, but also the idea that the phone is the embodiment of the user's lifestyle. The phone isn't just about personal consumption and enjoyment but about making use of it at the expense of sociality.

> Crank up your cool factor in stain black. Pump up the volume with crisp Mega Bass sound. Spin hit after hit using one-touch music control keys. Wherever you go, great sound and envious looks will follow.
>
> (Sony Ericsson W810i Walkman phone 2006)

Korean companies such as Samsung have clearly focussed on the female demographic through gendered handset design to functions aimed at the 'typical' female lifestyle. Samsung has been strategic in focusing on the phone not only as fashion accessory but also as state-of-the-art device for technological convergence. This has resulted in Samsung being deemed by *The Wall Street Journal* as 'the world's No. 1 purveyor of high-end cellphones and the third-largest seller overall' (Choi 2004). Samsung Z510, under the banner of 'imagine being 3G but ultra slim', is aimed at female users who not only want a phone to look good (with the phone being a metaphor for the 'slim' female user) but also to have multiple functions. In the ad we are met by the opened slim clam phone in the foreground, with the phone's shadow merging with that of a young, slim female in the background.

Samsung's E530 clam phone ad, titled 'Women's Life Phone', features functions to make shopping lists and count kilojoules. Here we are reminded of the way in which gendered scripting in design – and specifically mobile phone design – is about reinforcing traditional gendered modes of labour. The E530 advertising campaign reiterates the stereotype, not only that it is the role of females to do 'domestic' duties such as shopping, but also that women are obsessed by body image, reflecting the media's obsession with females' body image and weight issues.

The role of gender scripting by Korean company LG is more innocuous, the only gendered aspect of the model U880 clam phone being its pink colour. Motorola parodies stereotypical gender scripting with its clam phone V3X available in 'hot pink' and 'blokey blue'. Over the last three years campaigns have played on Japanese names combined with 'Moto' to create the impression that Motorola was a Japanese company. Utilising predominantly Japanese models, the US company has been keen to sell itself under an inverted glocal-ism in the guise of Japan's cultural capital. In Motorola's MotoRAZR V3i ads, featuring a Japanese female with short, hip hair and shirt and tie, the pre-dominantly black and white colouring toys with gendered stereotypes. The model assertively looks the viewer in the eye; she is hip, androgynous, and in control. So what is the reality for users beyond the industry imagery?

Quantitative data provides little insight into the various meanings and practices of the contemporary 'produser'. According to Nielsen Net ratings, Australia's mobile usage grew dramatically in 2005. MMS increased by 21 per cent, with 33 per cent of mobile owners over 16 years old having MMS capabilities, compared to 12 per cent in 2004. Of the 33 per cent, 25 per cent were able to send and receive photos (compared to 8 per cent in 2004), and 8 per cent had video capability (compared to 1 per cent in 2004). A survey conducted by LogicaCMG in September 2005 suggested that the potential worldwide market for mobile content will triple within 12 months. One-fifth of mobile phone owners worldwide have already downloaded content to their handsets, expected to rise to 60 per cent in the next 12 months.

The survey – covering Europe, Asia Pacific, North and South America – also revealed that the cost of the average monthly download per subscriber was $6.32, with more than 40 per cent of respondents expressing an expec-tation for the costs to rise next year. According to the US research group Gartner, globally users are buying handsets more frequently. In 2004, global handset sales reached 650 million (up from 520 million in 2003). Most not-able was the threat to Nokia's position as best seller, with Nokia taking 29.7 per cent (from 35.6 per cent in 2003) compared to Motorola's 15.8 per cent, Samsung's 12.1 per cent, Sony Ericsson's 6.6 per cent and LG's 6 per cent (IPC 2006).

According to ATMA (Australian Mobile Telecommunications Associ-ation), Australia has witnessed a dramatic growth in penetration rates from 64 per cent (2001–2002) and 72 per cent (2002–2003) to 80 per cent in 2004. Whilst post-paid dominated the market from 1998, by 2003 pre-paid plans accounted for around 40 per cent. In 1998, SMS made up 1 per cent of revenue for the carriers; by 2002–2003 it was up to 9 per cent with an estimated 3.95 billion messages (an average of 294 messages per mobile phone sub-scriber) sent (Australian Communications Authority, 2003). The Australian Mobile Phone Lifestyle Index survey conducted by AIMIA (Australian Interactive Media Industry Association) in 2006, found that 66 per cent of respondents had purchased mobile content in the last 12 months (up 50 per cent from 2005). The survey found that 73 per cent of women were likely to download content, compared to only 27 per cent of men. Eleven per cent of the respondents said they owned a 3G mobile phone, while 26 per cent didn't know if their phone was 3G. Twelve per cent said they would be upgrading to a 3G in the next six months because they needed to get a new phone and might as well get 3G.

These quantitative industry studies focus on *what* users consume from industry-created content, rather than on the more interesting question of *why* they make certain choices. They consequently fail to address what makes UCC so compelling. Indeed, there has been much mass-media hype about industry creating content for users for bigger revenues, without acknowledging what the history of mobile telecommunication has taught us – namely, that users enjoy generating their own content and customising techniques. This is highlighted by the history of the industry's series of 'killer applications' – a genealogy marred by failure ('discontinuous innovation').

At the same time, it must be conceded that some companies such as Melbourne's Ish Media are creating content for users not only to interact with but also to inspire creativity. A company dedicated to cross-platform media and specifically content made for the mobile such as the sitcom *Girl Friday*,

Figure 7.1 The star of mobile cross-platform mini-series, *Girl Friday*. The script, story, visual and aural aesthetics have been worked around the mobile phone as a symbol of contemporary everyday life.

Source: Ish Media.

Ish Media has been quick to innovate and to create content specifically for the mobile by operating upon different levels of audience engagement. Beyond the global industry hype, what is the reality for individual users in Melbourne?

Pixoleur: case study of sample group

Much of the advertising for mobile phones in Australia – from service providers such as Optus, 3 (Hutchison), Vodafone and Telstra to device providers such as Sony Ericsson, Nokia, Siemens, Motorola, LG and Samsung – reiterate the importance of being connected both literally and metaphorically. In the printed and electronic advertising media the phone is touted as a status symbol and an expression of the user's own identity and subjectivity. Increasingly, device providers are selling types of identity and status – from Nokia limited-edition designer phones (such as models 7260 and 7280) to the recent release of iPhone that has become the fetish for creative industry workers. A market once dominated by Nokia, the Australian market is now awash with various brands and associated consumer stereotypes. This was evident in my sample survey groups where, of the 20 respondents surveyed, only three owned a Nokia; the remainder had brands such as Siemens, LG and Samsung.

I conducted two case studies in Melbourne, one in 2004 and the other at the end of 2005/early 2006. Although I had hoped to work with the same respondents for the follow-up case study, I was only able to re-interview half the previous respondents and had to recruit more to keep the study size consistent. Over this time between the first and second studies I noted marked differences in what types of phones and functions the respondents were using. In particular, in 2004, few of the respondents had camera phone functions, but by the end of 2005 almost all of them did. To parallel the survey of camera phone usage that I conducted in Korea, I did a separate workshop with RMIT University students to gain a sense of their relationship to the camera phone and how they used it in everyday life. I will discuss these findings after I discuss the 2004 and 2005/2006 sample case studies.

In the 2004 sample survey I interviewed 20 students, administrators and staff – both male and female ranging from 20 to 50 years old – from the University of Melbourne.[2] The group consisted of an equal number of male and female respondents. I found that they viewed the mobile phone as a predominately personal and intimate device that reflected much about their relationships and correspondence. I conducted follow-up in-depth interviews with six of the respondents to gain a sense of the symbolic role of the phone and the gendered function of customisation. I asked respondents about the role of the mobile phone in their everyday rituals and social relations and how they employed customisation to personalise the device.

The aim of this sample study was to gain insights into some of the different ways respondents customised their mobile phones and what this said about representation, identification and co-presence. I was particularly motivated by a curiosity about prevailing gendered codes of intimacy, and the way in which the mobile phone helped maintain or transgress these codes. As in the other case study locations, the Melbourne case study was just that – a sample study. It did not seek to address all practices and attitudes of the Australian population, but rather to take a snapshot of the relationships and ambivalences that Melbourne users have towards mobile phones, co-presence and intimacy.

In the sample study, gender featured both in the practices and in attitudes to the role of mobile telephony in the everyday. In particular, stereotypes about gender and technology were continuously contested within individual responses. The stereotype of females not being adept with, and fluent in, the use of various applications proved to be inaccurate. Often the female respondents seemed more curious and adventurous in exploring and mastering applications than their male counterparts. However, female respondents tended to spend most of their mobile phone time contemplating, re-reading and editing text messages, unlike the male respondents who seemed less engaged in text messaging in general (often preferring voice calling).

Female respondents tended towards regular usage of their mobile phones and spoke eloquently about the importance of customisation, tending to view the mobile as a predominantly communication-based device. By contrast, male respondents tended to view it as partly a communication device and partly a data/content transferring technology. The usage of games on mobiles was overtly gendered – none of the female respondents had *ever* played the games, in comparison to most of the male respondents who had not only played the games once, but did so *regularly*. In my follow-up survey in late 2005, female respondents were not only beginning to play mobile games but were also more active in other applications such as their camera phone practices. However, the dominant mode of female usage was still SMS.

The younger the respondent, the less gendered differences were evident in the customising of applications such as SMS. Having said that, most of the older male respondents denied using any customisation despite the fact that they were doing so! By contrast, many female respondents were able to articulate different forms of SMS customisation corresponding to degrees of intimacy. When asked about their screen savers, ring tones and abbreviations for SMS-ing, the male respondents seemed indifferent to their customisation, often stating that it 'just came with the phone'. However, despite this denial in customising, male respondents did choose ring tones and screen savers – but they didn't view this as active customisation. In contrast, female respondents seemed much more attached to the customisation modes, likening the phone to a diary and describing it a significant mode of everyday communication in maintaining intimacy. In addition, female respondents seemed much more aware of the 'janus-faced' role of customising mobile phones as a grappling between modes of individualism and self-identification, and of the ways in which they were judged by these choices in public by intimate strangers.

In the sample survey in 2004, very few respondents used the relationship services, arguing that mobiles were more important in reinforcing *existing* – rather than establishing *new* – relationships. In addition, only two of the 20 respondents had downloaded their customisations from post-production mobile companies. Many preferred to either use their own images (mostly taken by camera phones) or choose from the images provided with the phone. Images used included places visited, Asian animations of cute characters, Betty Boop, the respondent's name and a flower. Some had tried the downloading services but had found them unsatisfying – too costly and often frustrating to use. While most respondents (70 per cent) selected ring tones and screensavers supplied with the phone, many claimed that they would do their own customisation if the phone had the capabilities (eg camera phone or Bluetooth).

Gender featured predominantly in discussions about customisation, with female respondents tending to be more decisive and opinionated about their selections, often downloading different screensavers and ring tones rather than using the generic (and unsatisfying) ones supplied by the manufacturer. In addition, female respondents were mindful of the ways in which people judged others by the types of mobile phone used and such features as ring tones. Desirable features for ring tones included factors such as being 'distinctive but not annoying'. As one female respondent noted:

> I have chosen Betty Boop (screensaver, face plate and doll hanging from the phone) because she is a bit of a role model of mine – she operates like a type of avatar or alter ego. There are some physical similarities such as we both have black curly hair. My ring tone is one of the Nokia ring tones supplied with the phone. It was chosen because it suits another alter ego of mine – so I felt it corresponded with that identity; it's like playing dress-ups.

When asked whether she saw customisation as an extension of the user's personality/identity she replied:

> I think so because I think you get judged by your ring tone when you are in public. When you hear someone's ring tone that is the same as yours you expect to find your doppelganger . . . It (customising) does become a fashion thing that you do get judged on.

Here, the respondent's discussion about the role of mobile phone's function as a form of identity and cultural capital is significant. Her discussion of the ring tone as a signifier of her own identity – the role of mobile phone customisation in the construction and maintenance of individualism – was exemplified by the respondent expecting to meet her doppelganger *vis-à-vis* the same customisation. Although the respondent was being humorous, this point signposted the performative elements involved in customising the phone as both a playful and thoughtful exercise. It is interesting to note that in locations such as Seoul and Hong Kong where users do have ring tones on public transport (as opposed to places such as Tokyo that do not), for respondents in those locations it was the internal colouring ring tone that held more meaning. For Melbournians, where features such as colourings are still yet to become a form of customisation, the external ring tone was one of the strongest signifiers of the user's identity in public places.

For this respondent – the youngest of the sample study – the performativity associated with mobile phone customisation was very significant in her everyday practices. She not only gave great consideration to the way in which she customised inside and outside her phone but also, as a regular SMSer (five to ten times a day), she spent much time personalising her text messages with emoticons and phonetic vernacular. As an essential everyday item, the mobile phone for this respondent was embroiled in paradoxes associated with identification and identity. On the one hand, she was aware of the ways in which she was judged in public by her phone's attire – from the hanging Betty Boop wearing a flashing light to the distinctive polyphonic ring tone – and thought this was often misread (i.e. that people did not understand her sense of humour). On the other hand, customising the mobile phone – from attachments to personalising SMS – was a site for gestures of intimacy amongst established friends.

The ring tone was one of the pervasive and public forms of cartographies of personalisation – operating upon both private and public levels. Ring tones highlight the fact that mobile phones are seen as both a literal and symbolic extension of the owner's personality, subjectivity, tastes, values and social capital. In her 2002 study, Plant discussed in great detail the pivotal role of the aural in performance of mobile media; her choice of bird analogies for mobile types only further reflecting the instrumental role of the ring tone has as an inversion of the Walkman's interiorised 'mobile privatization'.

Unlike much of mobile media that had increasingly become internalised, ring tones in Melbourne highlighted the performance of 'publicness of intimacy' (Berlant 1998). This is perhaps epitomised by the popular ring tone trend to deploy a person's voice – the epiphany of personalisation. Some, particularly male respondents, chose the 'anti-logo logo' in the form of the default ring while others chose ring tones such as beeps or landline ring to identify the mobile phone as a 'functional technology' that is an extension of previous, older technologies.

Ring tones function like slogan t-shirts – however, unlike t-shirts where one can just divert one's eye, music operates in an all-pervasive manner. Both music and mobile phones reflect contemporary struggle between what is public and what is private, what is individual and what is collective. Just as music is an important conveyer of one's tastes and values, so too is the mobile phone. Even the 'non-choice' default ring tone is a choice in the performance, construction and maintenance of the illusionism of individualism so important to contemporary lifestyle cultures. The ring tone reflects a different form of the three Ds from internalised media such as SMS and

MMS. On the one hand, ring tones reflect the user's tastes, values and attitudes whilst, on the other hand, participating in the *delaying, deferring* and *decentring* through various tactics such as switching to the silent mode or 'non-choice choice'. Even though both male and female respondents tended not to use the silent mode in public, they would often lower the volume of the ring tone and would answer calls briefly and quietly. Of the male and female respondents, only a couple of female respondents put their mobile on silent mode in public spaces such as transport. Female respondents, however, tended not to initiate phone calls on public transport, unless imperative. One female respondent stated:

> I think the correct mobile etiquette in public is brief and discrete. I use silence mode when I am in private, more than public. I usually don't have my phone on a loud ring mainly out of respect for other people's personal spaces. I don't think it should be banned; you should just act as you would normally – not talking loudly and making it brief . . . I don't think it is frowned upon to use your mobile in public but people do seem wary and self-conscious to use mobiles in public because – unless you're an extrovert – it is quite a self-conscious process as everyone can hear what you are saying and find out quite a bit about you (i.e. where you are going, where you have been).

Both male and female respondents predominantly used the mobile phone to contact friends (rather than family or work colleagues). Many of the female respondents preferred SMS as a means of communication, with 70 per cent of the respondents SMS-ing more than 80 per cent of the time. Both male and female respondents claimed that at least 80 per cent of their friends had mobiles; the only respondent (female) who did not SMS used her mobile mainly to contact family, particularly her elderly grandparents. Only 10 per cent of her friends had mobile phones. Given the large disparity between the costs of voice messaging and SMS in Australia, it would be easy to surmise that this was the main rationale for using SMS-ing over voice calling and MMS-ing. But while this was acknowledged, it was not the only reason. One male respondent stated:

> Most of my communication is SMS because it is cheaper. But I don't like telephone conversations; I think they are often misleading – there is not enough eye contact or body language to determine what they are really saying. But with SMS I can give much thought to the message, considering the way it could be interpreted. As a form of mediation, it is the most direct for me. So hence I prefer SMS-ing.

The same respondent noted a difference in his frequency of contact with people since his acquisition of a mobile only one year earlier. He became aware of the additional conversations people had via the mobile phone; the mobile phone afforded him greater forms of constant contact and higher sociability. He noted:

> Probably in a space of a week I keep in contact with just over a dozen people. It's very important – particularly with people I am close to – that I can communicate with them immediately when necessary. The mobile does reinforce relationships. I would take calls/messages from people at 2am; it is very unlikely that I would with the landline.

For one female respondent, SMS-ing was a new form of expression that she saw as an 'art form'. In weaving the spoken into the written, she viewed SMS as a very particular mode of communication that was not confronting (as in face-to-face) and operated as a form of reassurance. She stated:

> I see texting as a new form of expression; it's not necessarily destroying but a borrowing and reappropriation. It has a lot to do with compression, speed, and efficiency. The main form of writing I do is texting; I do see it as an art form. I enjoy making a funny message; and I appreciate receiving ones where the sender has put in time and thought by personalising and individualising it . . . A text message is like a book, each sentence can be compressed to become a chapter. If you have four different thoughts you can have four different sentences. I spend time editing texts, I really consider 'what is important' . . . 'what has to stay'. Often the original message is quite different from the one I end up sending; for example, if I am sending a long text message that goes over into two messages I will edit into one message. This is not because of the cost but more about the flow of the message; often it gets sent as two separate messages that hinders the message and its intentions. Recently I got a message from someone who sent six messages in a row; they were obviously not used to writing texts! She wasn't concise, it was literally as if she were talking!

Here the respondent identifies the role of customisation in the act of writing SMSs itself, to signify a type of performativity and self-presentation. As the respondent described the process of creating an SMS, there was nothing immediate about it. The editing and regulatory process was, as she stated, not just a matter of cost. Rather, it was about a type of conversion within a different genre – one that moulded the language of the user (just as writing a postcard moulds the language of the user). It was about flow and personalisation, not just efficiency and speed. Like all media and genres, SMS comes with often unspoken conventions and etiquette. As the female respondent conveys in her story of her friend's long-winded messages the medium has particular etiquette and conventions such as word compression. To borrow from McLuhan, SMS 'massages' the medium of textuality (1964).

When the above female respondent was asked about SMS and the function of language, she answered, 'It is a compressed form of writing and it does make you revalue words. Although it can be instant, it can also be very deliberate and premeditated'. Here the respondent seems in direct conflict with the stereotype that sees SMS as a spontaneous genre that erodes proper language. On the contrary, the very mediatory and co-present nature of language is highlighted in the role of SMS. This attests to the fact that language is continuously negotiated and 'butchered' through the specific practices of culture and place as denoted in Massey's notion of locality (1993a).

Another female respondent spoke of the gender divide in terms of the previously noted male respondents opting for predictive text and more direct conveying of data. The female respondents would often use certain terms for specific people – a type of intimacy that would be lost on the outsider. Yet, at times, predictive text can also become part of the expressive form. One male respondent played with the predictive function that converted his name 'brian' as 'asian' – now he uses 'asian' as his sign name with specific friends. Another female respondent commented:

> I'm not big on smiling faces; it's too generic. You want people to read the text like you would hear it – incorporating both the written and the spoken. When I read a text I read it in *their* voice. I try to make it a bit more personalised. Sometimes I put the generic kiss thing; I like it when people make strange faces or symbols. I don't like when people use predictive text; I never use that. Predictive text tends to choose wrongly . . . For example, 'go' becomes 'in'. I notice with my male texters there tends to a usage of 'in' when I think they mean, 'go'. I don't like it because I like people's personalities to come across, to express their sense of humour.

The gendering of mobile behaviour was noted by most respondents. Most also noted the influence of age and class. One male respondent stated, 'I don't know the difference. It seems as if women take more phone calls and text messages than men. That's something I have just noticed but I don't know if it's true.' Another female respondent stated:

> I do think gender has a role. I could agree with the myth that males use more voice calls and tend to be more to the point in their text messaging. I suppose young females text a lot, males tend to be more familiar with the games on the phone, whilst females don't care about the games. If I were to generalise I would say that males use the calling phone function more often, females send and receive more SMS. However, I do think it is subjective – it depends on the person.

In the interviews one year later between December 2005 and January 2006, some major changes could be noted – most particularly the acquisition of camera phones. Whilst in the earlier study, few had a camera phone and many projected a future scenario in which they would have a camera phone, by the second round of interviews all but two had camera phones. Most of the respondents had already owned at least four phones. Many of these respondents began with a Nokia phone for reasons of either price or easy functionality, but had then moved onto Motorola, Sony Ericsson or Samsung.

Many respondents tended to upgrade on average every one and a half years. Of the 20 respondents all but two had regularly changed brands; the two exceptions claimed to have owned only Nokias. Female respondents tended to buy phones similar to those of their friends and intimates. Some claimed this created a greater level of intimacy and co-presence. One female respondent noted, 'I like when I'm sending a text to someone that has the same phone as mine because I know the text will appear exactly as I wrote it on mine. It means I can be thoughtful in terms of layout and how the text reads'. Another female respondent complained about her phone's incompatibility with her two other friends. She stated:

> We like to send emoticons and animations to each other. One day we were at a pub sending each other messages and I suddenly realised that their phones had much more sophisticated imagery; I would send a smiley face and they would get a dancing character. We laughed at the idea that the phones had a mind of their own. But I couldn't help but feel disappointed that my phone was crudely translating gestures and thoughts.

Another female respondent noted:

> I purchased a phone the same as one of my friends'. It wasn't an issue of copying; it was just that she had told me how much she enjoyed the design and functions of the phone. She was also much more technologically savvy than me. That was really useful because she showed me how to use functions that I wouldn't have used otherwise. I now think of this phone as much more of a tool for expression. I take pictures. I play games. It lets me be creative in day-to-day life.

Most of the male respondents tended to not be mindful of their friend's phones and often chose phones for either price or functionality. For many of the male respondents, functions such as the camera phone were not viewed as devices for creativity but were used occasionally in social settings. Male respondents tended to spend less time considering which phone to buy and why, and tended to upgrade their phones mainly when the previous phone had been mislaid or had ceased to work properly.

When the respondents were asked about what changes they had noted in their mobile phone practices over the last two years, all respondents noted that they texted more than they used to with increases of up to ten per day. Over 70 per cent stated that they personalised more both outside and inside the phones, whilst the other 30 per cent noted how they customised less. Female respondents tended towards more customisation, especially internal, whereas male respondents had noted an increase in their proclivity towards outside customisation. When asked about why they thought these changes had occurred, respondents were united in citing the growing significance of SMS in their everyday lives. One female respondent noted, 'I text more as more people text me, but maybe they text me because I text them?' One male respondent stated, 'I never know what the charges for mobile phones are, but I always know texting is cheaper. I also like being a little poetic, which is easier via texting.' Another male respondent noted:

> Reducing cost has increased the use of SMS. It's now assumed that everyone has a mobile and those who don't are often scorned like those that did when mobiles first came about. In the early '90s anyone with a mobile was considered a 'wanker' nowadays those without are often considered 'self important anti-establishment freaks'. Ok maybe that's a bit harsh but think about it, if you ask someone for their mobile number and they reply with 'sorry I don't have a mobile', they usually also go into a long diatribe of why they don't have a mobile which more often than not is based around 'I don't see why, just because the rest of society want to be sheep, I should as well . . .'

Most of the respondents had used the mobile phone for romantic gestures, and thought that SMS lent itself to quick, fleeting flirting without the confrontation of face-to-face. One respondent noted, 'It's easier to flirt through SMS', while another stated, 'when I first met my partner we texted a lot between times that we saw each other. And now we text through the day to work stuff out for home or whatever.' Apart from the discussion about the growing importance of texting in everyday life and it helping create a *postal co-presence* between intimates, many noted the significance of the camera phone in not only documenting everyday objects and moments but in turn making the mobile phone into a more personal device. The deployment of multimedia and customisation operated on numerous levels of the three Ds, constructing a space for contemplation and reflection. The three Ds were very much part of the process of imaging communities and the inscriptions of intimacy. When I asked one female respondent how often she used her camera phone and for what purposes, she noted:

> Probably on average only once a day. More often than not it's to record an incident at work, like an OH&S issue or an example of something that others need to see. However my old phone (oh my glorious old phone) had some great shots of my 2-year-old niece drinking a giant chocolate milkshake!

Over half the respondents often transferred and archived their images to the computer. Some preferred to store them on the mobile phone claiming that their 'embarrassing' pictures of friends could be used as collateral and could be whipped out on the mobile at any moment. One respondent noted that she stored about 15 images, mainly of her friends or cats, on the phone because she didn't know how to get them off the phone. Eighty per cent of the respondents sent MMS, usually accompanied by a text message to contextualise the image. Of their rationale for taking pictures on the camera phone, many noted a need or desire to remember something, or to send to a friend and share a moment.

Among the respondents, there was a gender divide between the motivations for taking an image as well as how they would be shared. Male respondents tended to send social 'blackmail' images, while female respondents tended to send images to connote a feeling of 'thinking of you', or a wishful need for co-presence ('wish you were here'). As one female respondent suggested, the importance of the mobile phone is very much part of its everydayness, something highlighted through SMS and MMS exchange. She said, 'It is the everydayness, it's the immediacy, and it's the attachment to friends and family. It's also about a personal story being carried around on these devices. A way of documenting and sharing life around us.'

Many respondents noted that camera phones made them feel more creative in their communication modes; especially as they could put image and text together to give the receiver more of a 'picture' – creating an ambient co-presence. However, many found the cost curbed their desires for camera phone images as a form of reterritorialisation. As one female respondent noted:

> When I first got my camera phone I had such joy being able to take a photo at any time. A lot of my photos consisted of me drunk with my friends; it was fun to send them the next morning as evidence. But then I got back my first months' phone bill and decided to curb my sending tendencies.

For another respondent, finding ways to share camera phone images without the expense was a creative and important exercise to her. She stated 'as both my friends and I don't have much dispensable income it is important to find ways to still share images without the horrible costs.' For this respondent, Bluetooth-ing the images to the computer and then either sending via email or uploading them onto blogs was a great way of sharing. She also saw that camera phones were a great way of creating digital stories and noted that she was starting to use the images in her professional work. When asked about the problem of low resolution she stated, 'It was a problem initially but now I have found that it has freed me up and now I tend to make more playful compositions, so I like what it produces!'

Although five of the respondents claimed to use their camera phone around three times per week, mainly to document themselves with friends, little difference could be noted in SMS practices over the year apart from a general increase in usage. For all the SMS users, the acquisition of a camera phone had momentarily (for a couple of months) changed their usage of SMS (often opting for MMS), but they found that they were returning to their earlier SMS patterns. This was partly because many of their friends' phones were not compatible. As one respondent noted, 'It can be very frustrating when a friend doesn't have the same phone compatibility as you. The way to solve such a problem is to keep it simple, with no images or animations. Sometimes too much content can get in the way of communication.'

As noted by many of the respondents, SMS and MMS are not about a simple form of information dissemination or organising. Most (predominantly the female) respondents commented on the importance of SMS/MMS as a reassuring type of co-presence not unlike the gesture of the postcard to signal *wish you were here* or *thinking of you*. These customising practices reflect the micro-politics of the experience of place as something that is always negotiating presence and co-presence. As demonstrated by the case studies, SMS textuality helps to locate individuals in a sense of place, one that – in the case of Melbourne – is informed by postal presence. However, in the context of SMS phonetic textuality, wish you were *hear* might be more apt!

One-hour photo: a case study of camera phone practices

As demonstrated in the last section focusing on a case study of SMS, many respondents negotiated place and co-presence through a mode I define as postal presence. But this postal presence phenomenon, so particular to Melbourne and Australian notions of place and co-presence, also sees itself manifest in other forms of mobile media customisation such as camera phone practices. I will now turn to my case study on camera phone practices in Melbourne, conducted with a group of university students (aged between 18 and 29) at RMIT University to gain a sense of how gender is performed and reflected through types of co-present presentation and sharing.

During May and June 2006 I conducted a mobile media workshop at RMIT University similar to the one I had conducted with the Korean students at Hallym University (see Chapter 5). One of the obvious demographic differences is ethnicity. In my Korean case study all but two of the students were Korean (the other two were Filipino exchange students). In the Melbourne case study the group consisted of 70 per cent local students from a variety of backgrounds (Greek, Italian, South African, Irish, English, Slavic, and Chinese) and 30 per cent international students (Korea, Japan, Hong Kong, Thailand, Indonesia). Arguably, the most notable difference was between the local and international students' modes of sharing, with international students predominantly utilising virtual communities or the Internet to distribute their images to loved ones overseas. Interestingly, the types of subject matter taken did not differ, with both local and international students predominantly playing the role of 'tourist' in the everyday with their camera phones.

However, the role of gender and gender performativity was pronounced, with female respondents taking pictures of social scenes and 'cute' things while male respondents tended towards subject matter such as landscapes and social scenes. Female respondents tended to share only social-related images such as MMS-ing images of friends to friends, whereas male respondents tend to MMS objects of mutual interest such as guitars or cars. Here we see that female respondents from a variety of cultural backgrounds viewed camera phone practices as part of postal presence, since most of their images were about conveying and sharing a sense of co-presence. Overall, female respondents tended to share their images more often either in face-to-face situations (passing the phone around) or through MMS. The proclivity of female respondents to share camera phone images highlighted that such practices were primarily social rather than for personal consumption (Scifo 2005).

Unlike other locations such as Korea and Japan, the lack of easily accessible convergent mobile media in Australia is most evident in the divide between mobile telephony and SNS. Although 3G technologies were released at the beginning of 2006 with Telstra's i-mode, the notion of 'mobile with Internet' did not appeal to many users, who were used to accessing the Internet via PCs. Thus, unlike Japan's and Korea's seamless transition between mobile media and the Internet whereby users can document, edit and upload images, videos and text onto their SNS whilst on the run, for Australian users the co-presence between mobile and face-to-face and Internet and face-to-face are parallels that do not interconnect easily. Thus users would often send images by Bluetooth camera phone to their PC and then upload to the ever-growing SNS, MySpace (which was beginning to be challenged by Facebook) and to a lesser degree, Flickr, MSN Messenger and Gaia Online.

In particular, gendered camera phone practices could be noted in the way that predominantly female respondents embraced a photographic practice once deemed as a male profession, that of paparazzi. The surveillant and often intrusive role of the paparazzi was an important mode of capturing a social moment amongst friends and strangers. Conventions about the male gaze of traditional subject/object photographic techniques were deployed by female users in often an ironic way; female users took paparazzi pictures of friends to share and disseminate as a type of 'blackmail' mimicking the global media's usage of celebrities as commodities. Users made themselves and friends subject to media types of photography such as paparazzi techniques that provided them not only with different photographic subject positions (in relation to the gaze), but also a space for them to reflexively and critically parody such industries.

For international students, this type of still photography offers them a way to create a sense of social space that they can then upload to their various websites to share with friends and family back home. Only one male respondent tended towards paparazzi at times. He was Korean and like most of the Koreans in the Seoul study, he uploaded the images onto his Cyworld mini-hompy site so that he could keep his friends and family updated.

Recall that in the Seoul study, the gender issue was fused with age differences. As most male students tended to do their conscripted two-year military training before commencing university, the male students tended to be at least two years older than the female students. Moreover, the male students seemed more preoccupied with finding a partner than their female counterparts. This imbalance was not an issue in the Melbourne study. In the Seoul study, the most telling distinguishing factor was whether students were single or part of a couple. If they were part of a couple, the phone became the property of the female partner and was colonised with signs of being 'engaged'. This was not the case for Melbourne respondents; the overt multiculturalism informed various kinds of representing intimacy.

Unlike the relatively homogeneous ethnic demographies in Korea and Japan, the Melbourne study was marked by both the differences between ethnicities and whether respondents were local or international students. One of the dominant tropes was the various ways in which ethnicities informed underlying role of the three Ds – *delay, deference* and *decentring* – in representing modes of postal presence. Whether through camera phone practices or SMS, respondents using forms of mobile media performed the three Ds in an assortment of degrees. In particular, for international students, the geo-political imaging of mobile media to both secure a sense of place – reterritorialise – whilst also provide a 'connected present' with places and people elsewhere sees a particular socio-cultural deployment of the three Ds that differs slightly from local students. The camera phone often becomes a repository for memories that will be shared later with family and friends back home.

Gender differences could be noted in both studies. In Seoul, the gendering of both genres and modes of performing was overt, most particularly with females using the camera phone to represent fragments of particular inanimate objects, whereas male respondents were more likely to 'document' themselves in social contexts that often implied some type of hierarchy. Male respondents were also more likely than their female counterparts to take paparazzi or self-paparazzi shots, whereas in the Melbourne study the opposite was true.

One of the dominant differences between the two studies – apart from the single/couple distinction in Seoul – was the ways in which respondents shared, or didn't share, their images. For the case study group in Seoul, all but two (the Filipino students) respondents predominantly uploaded their images to their Cyworld mini-hompy (virtual community), so that the images could be instantly shared with friends. Korean respondents also liked to share in face-to-face situations such as at cafés. Melbourne respondents shared images via range of media (flickr, MSM, Gaia, MySpace, and more frequently Facebook) and took almost 40 per cent of their images for predominantly personal consumption. However international students shared almost

95 per cent of their images with family and friends abroad. MMS sending was relatively rare, with many respondents regarding the cost as the major determent. I will now detail some of the Melbournian examples to further demonstrate the role of postal presence through the maintenance of the three Ds – *delay, deference* and *decentring*.

For two international female students (one from Brunei, the other from Indonesia) working in a pair, camera phone practices could be divided into what they defined as primary essentials and personal essentials categories. For one of the students (see Figure 7.2), primary essentials were genres such as food, fashions and facades, while the personal essentials constituted friends. Under food, the student took pictures for various reasons such as sending images back to her family at home as well as for her own personal visual archiving in which she collected such images as if in an electronic cookbook journal.

Under the banner of 'facades', the student often used the camera phone to record the sights of Melbourne that were in sharp contrast to her hometown. For example, her pictures of buildings were taken to 'Create a memory in contrast with Brunei', where their tallest building is seven levels. (No buildings are allowed to be taller than the mosque.) She shared these images with siblings who hadn't been to Melbourne. The student often took pictures of fashion and shop displays as an example of the difference between what she dubbed 'global culture' (settings in real life where boundaries between cultures have broken) and 'virtual reality' (settings in a fantasised world constructed through popular culture references). The cross-cultural representations in Melbourne were, for the student, a source of interest and fascination.

When it came to her personal essential category, the student was a great example of a paparazzi photographer whereby, as Koskinen noted, 'senders have to make drama of the banalities of everyday life' (2007: 251). In her images she acknowledges that she makes her friends into celebrities for their group consumption, sometimes uploading her images to Friendster or MSN for public viewing after seeking approval from her friends. In speaking about what was going on in the images, the student noted that often it was 'just lame stuff', but that she felt it was important to store images of friends on her phone for personal consumption.

For this female respondent, the phone was not just a means of communication but an important tool in documenting the everyday and a source for archiving memories and moments with close friends. In this way, she is an example of Plant's identification of gender differences, whereby female respondents tended to value their mobile 'as a means of expression and social communication' (Plant 2002). Often the respondent alternated between the role of paparazzi photographer (of both friends and random strangers) and tourist. The alternation between a surveillant, all-controlling, predatory eye for media and being an unknowing 'tourist' echoed her feelings studying overseas, where sometimes she would feel in control, and at other times she would feel that she was overwhelmed by the unfamiliar.

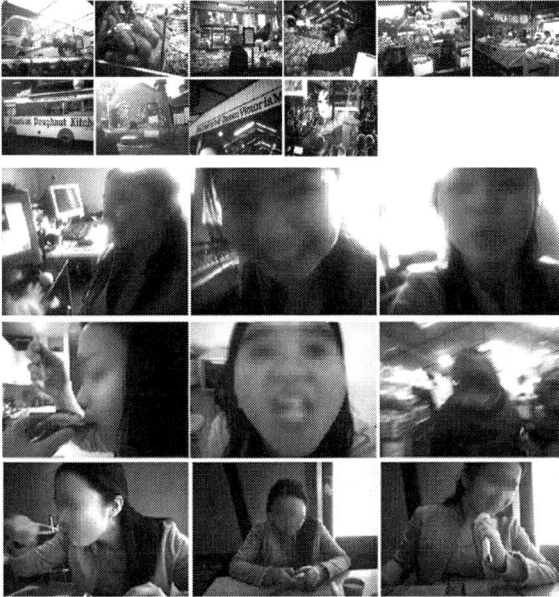

Figure 7.2 One female user's images illustrating genres ranging from paparazzi to tourist postcard. Here the international student utilises the camera phone to take tourist pictures for herself and family/friends at home whilst also providing her a space to perform her identity within an Melbournian context.

The student saw her camera phone practices as reinforcing her self-defined quasi-tourist position. She noted that she liked to play into media stereotypes about the 'tourist', continuously photographing while on holiday. In this way, she uses the camera phone to parody stereotypes about the culture of tourism and the tourism of culture. Here the camera phone is seen as an extension of earlier practices such as analogue photography in documenting experiences to represent to friends and family back home. In this way, camera phone images allowed her to remediate the tradition of the postcard, sending home images and texts of her experiences in a foreign land.

As an overseas student living in Australia for her studies, the camera phone allowed her to briefly slip into the role of tourist and then back again to a student. She noted that her circle of subjects and objects was limited, and saw the phone as a means for furthering sociality. Sharing the images amongst her close circle of intimates was very important to her because it functioned as part of a group memory exercise that reinforced existing social relationships. She noted that the function of self-presentation was minimal in comparison to the social benefits.

Another female international student (see Figure 7.3) divided her camera phone practices into five self-defined genres: personal space, social (expected), social (paparazzi), everyday poetics and mnemonics (camera phone as memory collector). For the personal space genre she took images of her dog M (Figure 7.3a). For her social images, she noted that not only did the pictures capture a fun time but also, due to the familiarity of mobile phones in day-to-day situations, reflected people's relationships and social order more appropriately than a staged conventional photograph. The casual and spontaneous nature of the camera phone meant that people often felt less on their guard (although still aware that they were being photographed). She argued that her paparazzi style photographs often represented a less biased visual memento, as people were not performing to the camera. The difference between the social expected and social paparazzi images can be seen in her pictures at a bowling alley (Figures 7.3b, 7.3c). Under the genre of everyday poetics she noted that thanks to the camera phone she was more likely to take a picture of something that caught her eye (and which she would not have been photographed with a normal camera).

She observed that the camera phone had made her a more active photographer and had given her a different sense of agency in everyday life. As with the car interior photo (Figure 7.3d), she said 'I randomly took this useless

Figure 7.3 Clockwise from top left: images classified by the respondent as personal space, social expected, social paparazzi and everyday poetics.

Figure 7.4 Images clockwise from top left: two shots of watches, a picture of respondent's toy for a wallpaper, a picture of the ANZ bank in Chinatown taken to win an argument with a friend, a shot of a scooter that the respondent had considered buying (female respondent).

photo for no reason', but she found herself using it as wallpaper (screen saver) later. The camera phone was also a repository for nostalgia whereby she often took photos that recalled something from the past. She shared many of her images with friends through MSN messenger or her website, as she 'liked to remind people of times they had shared'.

For a female local student whose parents had migrated from Hong Kong (see Figure 7.4), the camera phone was a tool for documenting objects of desire from scooters to watches. An anomaly in comparison to the other female students in her group who laid great importance on the camera phone as a tool for socialising, this student's motivations were wound up in pragmatic goals such as comparing watch styles when buying a watch, or proving herself correct when she was arguing with a friend as to whether an ANZ bank was in Chinatown. A self-confessed tomboy, she arguably used the camera phone in a very masculine manner as a tool for recording pragmatic details rather than, as noted with most of the female respondents, as a tool for expression and sociality. In this way, this female student demonstrates that masculine and feminine usage of the mobile phone is not necessarily attached to specific genders.

For male students, the main differences were between international and local students with international students tending to share more often, especially with friends and family at home. One local male student (Figure 7.5) used his camera phone mainly to reflect his family and social life, and especially for sharing images with his girlfriend. He did not think of the camera phone as inferior to the stand-alone camera but rather as a useful way to capture pleasurable and thought-provoking moments. He would share his images of friends with friends, but kept thought-provoking images for his (and often his girlfriend's) personal consumption. For example, his picture of

Figure 7.5 Images clockwise from top left: image of girlfriend, with mates, with a friend, vegemite-brother metaphor, visage in the coffee, cigarette burning.

the two pieces of toast, one bathing in vegemite, the other lightly dusted with vegemite could be seen as a metaphor for himself and his brother. He said, 'When I realised that my brother has hardly any vegemite and he is skinny and I have loads of vegemite and I'm tubby, I thought it was a nice symbol to keep and make reference to.'

Another local male student (Figure 7.6) used his camera phone for collecting images of his family and objects of great significance. Whilst he used the images he took of his family as ID caller images, the other images were strictly divided into those shared and those for personal consumption. The image of his dog and the sepia toned image of his house from the gate were

Figure 7.6 Images clockwise from top left: gate to house, pet dog, mother, father, sister, hockey game, favourite object (guitar).

Figure 7.7 Images clockwise from top left: graffiti, cat, Korean food, empty tram, friends playing soccer video game, mini-hompy page.

stored for personal consumption both on the phone and on his computer. Images of a hockey game and his guitar were freely shared amongst friends via MSN messenger; these images were 'relevant' to his friend's interests and thus didn't need to be justified when sent.

One Korean international male student (see Figure 7.7) used his camera phone similarly to international female respondents except that he tended not to take paparazzi shots of friends, preferring staged ones. He also took pictures that operated as a point of contrast from his homeland, such as the shot from his balcony overlooking a pool, an empty tram, and Korean food in Australia. This student saw the mobile phone as a way of collecting data for his Cyworld mini-hompy page, which kept him in constant contact with friends and family at home as well as operating as an archive for his experiences. Images such as his aunt with his cousins and his Korean friends playing a soccer video game in the lead-up to the World Cup were important images that he constantly referred back to, giving him a sense of being both home and away – one of the integral components to mobile media co-presence.

For another international male student from Hong Kong (Figure 7.8), the camera phone operated much like a stand-alone camera in its role of documenting social and personal moments. Whilst the aforementioned Korean student saved images to desktop and uploaded to his mini-hompy page, the Hong Kong student tended to share images via email. Both students had an

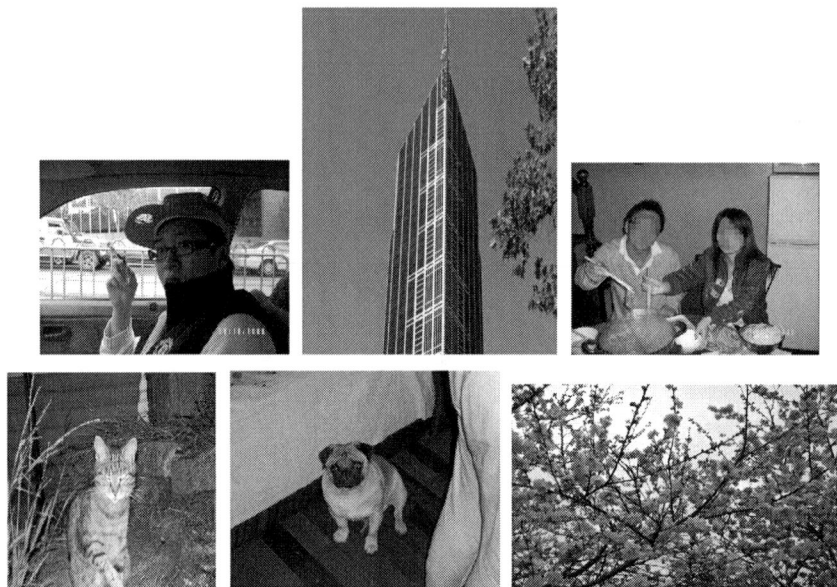

Figure 7.8 Images clockwise from top left: mug shot of friend, Melbourne building, eating out with friends, blossoms, 'funny' dog and cat.

interest in taking pictures of animals and nature that became mementoes for their experiences in Australia. These students rarely took paparazzi shots, instead favouring what they called 'artistic' photos of standard photographic genres such as portraits and landscapes. Like the standard stand-alone camera, they used the camera phone occasionally and did not deviate from the types of pictures they would normally take with a stand-alone camera.

This difference could also be witnessed in the work of another male international student from Hong Kong (Figure 7.9). His genres consisted of landscape (for personal consumption) and people/occasions (operating on a social level where he would share the images). His landscape images tended to reflect an artistic approach that was about self-discovery and self-expression. For his pictures of people and food, he adopted a photographer persona whereby he took it as his own responsibility to be the memory collector such as at his parents' anniversary. Used in this way, the camera phone becomes an extension of the Kodak camera in discourses of constructing family and private memorialisation. As Bourdieu notes, 'As a private technique, photography manufactures private images of private life . . . photographers see the recording of family life as the primary function of photography' (Bourdieu 1990: 30). This student was more interested in surroundings rather than constructions of self-presentation, demonstrating a traditional method of camera use with no silly or cute shots taken.

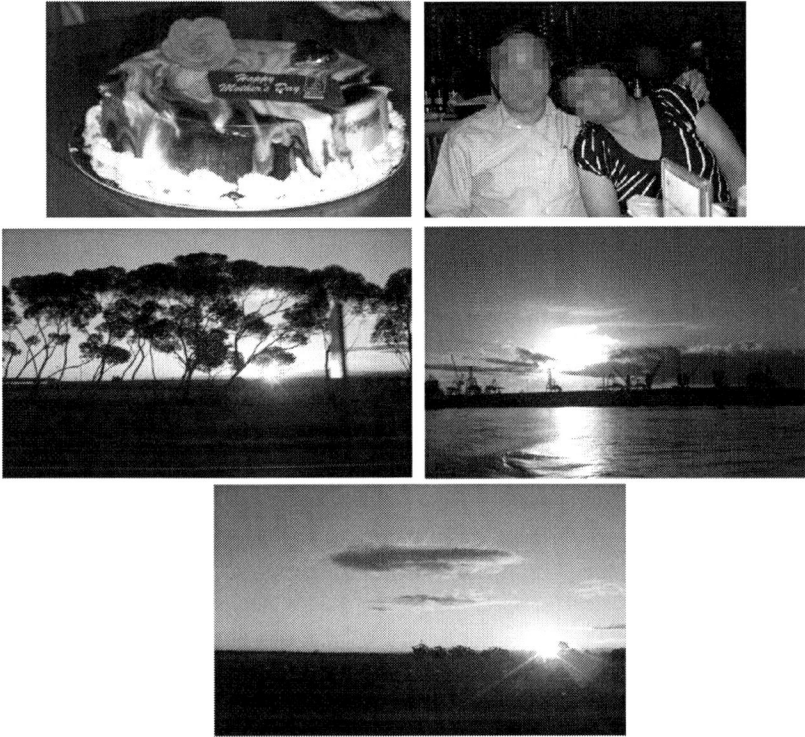

Figure 7.9 These images, which exemplify the classic photographer style, from top left (going clockwise): mother's day cake, mum and dad's anniversary, and three shots of sunsets in Australia.

In the workshop, female respondents tended to be versatile in their approaches to the subject matter and in their intimacy towards the photographic subject. There were also gender differences in the ways in which images were shared, with female respondents tending to take more photographs for both personal consumption and sharing. Moreover, conventions about masculine and feminine styles of photography as denoted by the amateur versus professional divide were blurring, with female respondents sometimes adopting masculine (i.e. pragmatic) usages and male respondents adopting feminine approaches (eg taking pictures of cute subjects such as animals or toys). Ethnic differences were most apparent in the way camera phone images operated as a form of tourist voyeurism, and also in the way in which images were shared, especially with friends and family overseas.

For permanent residents, Australia's 'South-South-South' imagined community is richly metaphoric as individuals and communities deploy the three Ds to gain a sense of reterritorialisation via the camera phones (Koskinen 2007). Unlike Hong Kong, where camera phones are deployed as resistance to the reterritorialisation of mainland China through the rehearsing of the trope of nostalgia or, Korea, where camera phones are a site for challenging patriarchal constructions of the *kukmin kajok*, Melbourne's treatment of camera phone reterritorialisation techniques are marked by the three Ds central in locating a sense of 'south' in a century of the 'Global South'. Given Australia's multicultural demographic, many citizens grapple with some notion of diaspora in their distant or recent history: a type of yearning and displacement underscoring a sense of belonging. These experiences are rehearsed through the three Ds – from contemplating text messages that are seemingly 'immediate' to the use of camera phone images to re-enact a place. It is a reterritorialisation that is governed by postal presence.

So how do these types of mobile phone practices reflect Australia's own grappling with national identity and ideas of self and sociality? Unlike locations such as Korea and Japan where almost all respondents had a shared nationality and ethnicity, Australia's multicultural demographic, along with its proximity to Asia and the Asian mobile market, demonstrates an assortment of modes of self-presentation, sociality and exchange. As noted by one female respondent, the use of the camera phone was a way for her to parody stereotypes about tourist culture and cultures of tourism. Rather than the mobile phone inferring a sense of deterritorialisation or 'incompleteness' through its everywhere-ness, mobile phone customisation – from inside (camera phone, SMS) to outside – reflects the ways in which individuals interconnect with groups and locate – reterritorialise – themselves within a sense of 'south'.

Practices such as camera phone imagery or SMS provide people with ways in which to comprehend the role of distance and difference in Australia. As can be witnessed by the international students' usage, many deployed the camera phone as a way of practicing place for both themselves and family and friends back home. This was particularly the case with the female student from Brunei, who used the camera phone to slip in and out of being a tourist and being an Australian. She used it to 'perform' being a tourist for her co-present family and friends. However, it is not just the international students that are 'performing' place through camera phone practices, since examples could be found among both international and local students. Through these imaging community processes, respondents can centre, decentre and re-centre themselves in the 'south'.

The 'imaging communities' techniques of camera phone practices are inflected by gender, class, age and ethnicity and, in the case of Melbourne respondents, they could be viewed as a snapshot of the micro-politics in motion within the macro workings of Australia's emerging imagined (and imaging) community. These imaging communities are particular to the Australia's notion of place and the practice and maintenance of co-presence. In an age of the tyranny of immediacy, Melbournian gendered mobile media are demonstrating forms of remediated and emerging *postal presence*.

Wish you were hear: conclusion

As we can see in these sample studies, Melbournian respondents are customising as much as their techno-savvy counterparts, but in different ways. In particular, the persistence of SMS highlights some characteristics of mobile media in Australia as underscored by the three Ds – *delay, deference* and *decentring*. This can be witnessed in SMS practices as well as emerging other forms of mobile media representation such as camera phone imagery. Moreover, these three Ds can be viewed as part of an ongoing maintenance of intimate co-presence through the postal presence metaphor. As we can see in both the SMS and now camera phone practices, the metaphors of earlier forms of propinquity such as letter writing and postcard writing are evoked and re-enacted.

There are marked differences in camera phone storing; most of the local students and sample survey respondents tended to use images for personal consumption rather than networked sharing. In this way, Melbourne's modes of camera phone practices are similar to that found in the Hong Kong case study. However, as I have argued throughout this chapter, one of the distinguishing features of Melbournian mobile media is the role of postal presence. Such practices as camera phone image-taking and sharing not only re-enact earlier, remediated co-present technologies such as postcards and analogue photography, but also highlight emerging modes of negotiating co-presence, mobility and place in the incompleteness that is 'South-South-South'. The significance of phonetic SMS so prevalent in the case study not only demonstrated gendered differences but also class distinctions. Particular accents signified specific forms of cultural, social and economic capital. This is reminiscent of Pertierra's observations in the Philippines, whereby mobile media practices such as SMS can reflect social inequalities at the same time as expressing inner modes of subjectivity (2006: 100–101).

Camera phone practices, as a particular mode of cartographies of personalisation, are clearly about localised tactics of reterritorialisation. For international students, camera phones are used to both defer and delay their feelings of displacement ('decentring') and modes of resettlement. Sometimes they take the role of 'tourist', other times the ethnographer, charting and documenting a sense of place for both themselves and friends and family at home. One can find multiple camera phone practices – from self-paparazzi genres to the storage of the images (of family back home) on the phone. These are the emerging forms of imaging communities that are reflecting the everyday lives of international students in Australia. But it is not only international students that deploy the camera phone to reterritorialise: often local students perform similar emotological techniques being simultaneously the tourist and the guide. However, the relationship towards the three Ds is obviously more metaphoric for local students and more explicit and direct for international students.

Customising is also clearly occurring differently among groups. In this regard, issues such as age, class and ethnicity underscored the role of gender in defining modes of mobile telephony in Melbourne. Applications such as SMS are continuing to grow, especially among female respondents, and the burgeoning textual possibilities are being further explored and exploited. The emerging SMS and camera phone practices demonstrate differing cultural, social and economic capital even within this small sample study. While from the outset Melbournians seem less embroiled in the cute character customisation frenzy seen in such places as Tokyo and Hong Kong, this is not to say that less customisation is occurring. Rather, much of the personalising of mobile media is through the (internalised) genre and conventions of SMS-ing. The play with the vernacular, colloquial and dialect through phonetic textuality is as vast as the city is multicultural. Here, through implicit modes of SMS customising, individuals can denote types of similarities and differences that extend beyond gender into class, age and ethnic distinctions. Having said that, the fact that female respondents not only SMS more but also see it more as a form of creativity and self-expression than the male respondents highlights the significance of gender in shaping the cartographies of personalisation.

Moreover, unlike the other locations, the multiculturalism of the demographic, as well as the fact that some respondents were international and others local, undoubtedly affected the trends of mobile phone usage. Hong Kong students studying in Melbourne noted that they were SMS-ing more. The international students in general tended to use their camera phone in place of the stand-alone camera, and often sent their images back to friends and family at home via SNS such as Flickr, MySpace, MSNmessenger. Overall, female respondents were much more aware and articulate about the textual and visual possibilities of mobile media and seemed, in general, to be embracing of the multimedia elements of the mobile phone. Male respondents, and particularly local students, tended not to view the mobile phone as a creative tool and were less playful with SMS textuality and MMS visuality. One of the strikingly features was that many chose asynchronous forms of mobile media over than voice calling. Only a few male respondents viewed the mobile phone solely as a voice-calling medium.

Australia's co-presence is always in negotiation with the 'incompleteness' of place that is marked by a sense of distance. By engaging with the persistence of the postal metaphor that haunts mobile media practices, localised cartographies of personalisation – in the form of three Ds – are providing methods for both visiting and permanently based residents of Melbourne to negotiate and reterritorialise the 'South-South-South' that is Australia. Rather than new technologies overcoming this distance, mobile media respondents are deploying predominantly co-present and asynchronous forms of visuality and textuality to highlight the paradoxes of closeness and 'un-distance'. These asynchronous forms remediate earlier practices of co-presence range from the postcard (in the form of camera phone images) to letter writing (in the form of SMS). In resistance to the tyranny of immediate technologies promising the impossible wish of overcoming all forms of distance, Melbourne gendered mobile media deploys postal presence. Melbournian SMS and camera phone practices can be defined as the art of being *hear* ... where going south is always in the *delay, deferral and decentring* that is postal presence.

8 Domesticating cartographies
Gendered mobile media in the region

The *keitai* culture occurring in Japan has spread to females in other Asian countries, particularly those of a younger age. In South Korea, mobile service providers have started to concentrate on the female market by introducing distinct handsets, rate plans, and special service packages appealing to women . . . Meanwhile, Chinese female users, especially those in the white-collar class, are known for their preference of red clamshell designs with ornaments made of synthetic or real diamonds. A number of handset manufacturers consequently began to produce such cell phones to meet the needs of this market segment.

(Castells et al. 2007: 52)

One of the most enduring images that haunt contemporary global media is the image of the young female mobile phone user. In the case of the Asia-Pacific, it is the dominant and enduring image. Thus, to say that mobile media customisation is a highly gendered activity seems to state the obvious. Yet despite the pivotal role that gender has played in the rise of mobile media and the produser, the significance of the female user has often been overshadowed by the conflation of new technologies with youth cultures. And although the mobile phone has in some ways liberated the female user it has also become her controller, relegating her to the status of consumer and reinforcing the traditional female role of reproduction labourer.

The fact that the global rise of mobile media as a social tool in everyday practice owes much to its adaptation by female users has reinforced the derogative view of women's cultures as the lesser, 'hyperfeminine' form of unofficial 'gossip' (Martin 1991a, 1991b; Moyal 1992), in contrast to 'hyper-masculine' official culture. Here the rise of 'hyperfeminine' languages in new media spaces – which I have called emotology – reflects the broader global emergence of a 'hyperfemininity' that is affecting both males and females in different ways. Mobile phones, with their capacity to facilitate socio-emotive practices, are no longer just 'women's business', but are part of a global trend of the 'feminising' of technology (Shade 2007; Fortunati 2005a) and increasing full-time intimacy. The politics of the personal and the 'publicness of the intimate' (Berlant 1998) have returned as the contemporary *modus operandi*. This is epitomised in the body politic of the mobile phone as an icon of intimacy and hyperfemininity, as encapsulated by Fortunati who posits that 'of all the mobile technologies, the mobile phone is the one most intimately close to the body, extending, enhancing and transforming many of our sensory experiences as well as influencing the 'traditional management of intimacy and distance' (2002b: 48). This is most apparent in the rise of the female mobile media user.

The pivotal role of the female user in the rise of mobile technologies seems to purport that one of the primary functions of mobile technology is to elicit increasing feminisation and that this is why women have been so quick to adopt mobile technology. It can also be argued that the 'feminisation' of the phone from a business tool into a social device only further entrenched women into various forms of 'care cultures', which are part of an increased 'full-time intimacy' phenomenon. Technologies such as mobile media only further enslave users into the exploitative politics of hyperfeminisation, validating and substantiating the increasing global rise of precarious labour. In short, the mobile phone is a vehicle of hyperfeminisation that exploits women as both consumers and producers in the hypermasculine logic of post-industrial capitalism. This argument is persuasive, and we could take it as indicative of why society and culture today are eroding. But it doesn't explain the whole picture, because the twin hypergenderisation has not evenly affected all women in the Asia-Pacific, nor within any one location.

Here we are reminded of McKay's (2007) discussion about Filipino global care workers and the importance of not consigning them to the role of victim. If hyperfeminisation is one of the key traits of this phenomenon then in order to understand it we must call upon post-structuralist models of subjectivity to see the various ways in which gender is localised. Moreover, if this hyperfeminisation is global and affecting all areas of labour, why aren't men also practising modes of hyperfeminisation? How do the dual influences of hyperfemininity and hypermasculinity differ between the genders, and how are they informed by various forms of locality and subjectivity? And what can the conspicuous forms of mobile media teach us about the relationship between the gender practices of consumption and production in the region?

It is this last question that *Mobile Media in the Asia-Pacific* has endeavoured to explore in all its complexity through detailed case studies in four locations. This section of *Mobile Media in the Asia-Pacific* has argued that we can gain greater acuity on the current forms of modernity and mobility in the region by looking at the various forms of gendered consumption and production around mobile technologies. As I have demonstrated, one of the reasons it has been difficult to isolate emerging forms of hyperfeminisation is the problematic way in which gender has been sidelined in mobile communication literature. This oversight is partly a result of the way gender is conceptualised in the crossroads of lifestyle, consumption and everyday life. Since mobile technologies are increasingly viewed as symbols of lifestyle, they are decreasingly critiqued as technologies. This is despite the numerous and compelling feminist debates that have surrounded socio-technologies – particularly in the context of cyberspace – that could provide great insight in discussions around propinquity and co-presence (virtual and actual, here and there, work and leisure).

One of the main problems is that much of the research to date has looked at women and mobile phones through the lenses of youth and fashion. Although there is a vast body of literature on the politics of fashion as a barometer for culture and power relations (McRobbie 1999, 2000, 1996), these areas of fashion, women and technology rarely cross-pollinate. A key point that has been overlooked in mobile/gender research up till now is that the gendered practices of mobile media play a deeply political role that links into notions of the local, national and transnational. That is, the mobile phone is part of the construction of everyday subjectivities and emotional grammars, simultaneously reflecting gendered forms of paid and unpaid labour. The perpetual emphasis of industry and advertising on young females as the prime consumer and the conflation of the mobile phone with fashion obscure the very real and complex formations of labour, gender and mobility.

This bias has limited the possible lines of inquiry, and it has also tended to perpetuate and naturalise gender inequalities around technology and labour. To counteract this bias, researchers such as Matsuda (2007) have shifted their research frame away from the conspicuous and thus much cited phenomenon of youth and new technology, towards the function of the *keitai* in the relationship between mothers and children. Matsuda's research has revealed how the *keitai* serves to create a full-time intimacy in which the mother's material role, and labour, is extended. The added degrees of micro-coordination – Matsuda has dubbed the phenomenon 'mom in the pocket' (ibid.) – are arguably unnecessary, since everyday life for Japanese children has changed little, with after-school studies timetables and train schedules remaining rigid and punctual. Similarly, the demand for the *keitai* as a 'security' device reflects national stereotypes about increasing crime, although there is no evidence to support such fears (ibid.). Moreover the rise of mobile media as a 'wireless leash' (Qiu 2007), facilitating and entrenching hyperfeminine modes of labour, is not isolated to Japan.

By observing the gendered customisation of mobile media, we can gain insight into emerging patterns of work and leisure in the region. These patterns of post-modernity reflect both literal and metaphoric forms of gendered mobility and labour practices. Through gendered mobile media we can begin to see new forms of visuality, textuality and virtuality that demonstrate the shifting relationship between post-industrialism and hypermasculinity, and the emergence of increasing social capital in the form of hyperfemininity. Mobile media – as a nexus between communication technologies and new media – becomes a site in which we can reflect upon changing modes of labour, media literacy and expression; a space in which gendered differences continue, but in which the power relations between male/hypermasculine and female/ hyperfeminine labour are undergoing change. This is what makes mobile media such a fertile ground in which to rethink modernity in the region.

Hypermobility @ gender: the politics of gendered labour and media

> What exactly are the new technics of mobility? Can we even pin them down? Are mobile media media technics in the old sense, like film or TV (with their own disciplines, their own established forms of mediation)? Or do we need to rethink the whole question of mediation? How do we map the new technologies and techniques of mobility, along with the social processes, individuation of collectivities or subjectivities they make possible? What modes of living – of work and loving – come into being in a mobile world?
>
> (Murphie 2007: n.p.)

As Murphie (2007) notes in his discussion on the impact of mobility on practices of work and life, the role of media usage can provide a lens for understanding some of the enduring and emerging forms of these practices – in short, a lens onto what it means to live today. These 'new technics of mobility', which I call cartographies of personalisation, are imbued by the remediated spectres of older media as much as they are exemplary of new media. As Morley observes, the mobile phone 'currently carries the heaviest symbolic burden' (2007: 301) of our century as one of the icons of the new. Rather than dismissing mobile media as synonymous with new media, we need to recast the rubric for what new media means within twenty-first century post-modernity. This, in turn, necessitates that we reflect upon emerging patterns of life, work and love.

The multivalent dimensions of mobility – labour, technology, capital and people – means that the mobile phone bears the brunt of debates around globalisation, post-modernity and post-industrialism. Indeed, the mobile phone is such a poignant symbol that the cultures and practices of mobile media can often be neglected under the weight of such urgent and critical issues. In particular, mobile media provides a vivid lens onto one of the prevailing issues for twenty-first century post-modernity namely the increasing feminisation of labour (Fortunati 2008). As the leader in the production and consumption of global ICTs, the Asia-Pacific is a key example of this phenomenon, characterised by Ling (1999) as the hypermasculinity of post-industrialism relying on the exploitation of hyperfeminity. Indeed, the structures of hypermasculinity rely for their existence on the exploitation of social, reproductive and creative labour – that is, the hyperfeminine.

The region's cartographies of personalisation challenge conventional boundaries between 'territorial' and 'sedentary' tropes underlining twentieth century social sciences (Hannam et al. 2006). Through the multiple imaging

230 Mobile media cultures

communities we can see the various localised immobilities and mobilities across symbolic and material histories and geo-imaginaries that are forming the region in the twenty-first century. Rather than mobile media signalling the much-heralded 'death of geography', place and re-territorisation becomes increasingly prevalent. In the face of immediacy rhetoric, we see that cartographies of personalisation are being deployed through various localised UCC tactics to provide contemplation, reflection and re-inscribe a sense of place. That is, waiting for immediacy. Central to these tactics and re-inscriptions is the way in which female users have increasingly become 'produsers'; and, in the case of Tokyo, female 'produsers' are not only constructing new forms of media genres but also new areas of creative industry.

However, with the expansion and convergence of mobile media into a multimedia device integral in everyday life, a paradox has emerged. As work and leisure patterns increasingly blur, so too the gender-scripted labour blueprints overlap. In developed countries, both male and female users deploy techniques of labour feminisation, as social labour is no longer just the prerogative of the female. Rather, both male and female users actively become part of the labour feminisation by which social capital – exemplified by the rise of Web 2.0 and SNS – becomes a newly valuable commodity. But in view of Rupert Murdoch's acquisition of MySpace in 2006, and the importance of emerging cartographies of personalisation such as *emoji* in Japan, the question arises: how does one put a price on these new forms of mobile labour in a system dominated by hypermasculine modes of assessment? This question will continue to haunt the region as social labour is increasingly deployed amid multiple forms of mobility.

The co-dependent relationship between hypermasculinity and hyperfemininity in developed countries such as Japan and Korea is in sharp contrast to the overriding exploitation of the hyperfeminine labour of developing countries – specifically the Philippines – in developed countries such as Hong Kong. Through the rise of the global care cultures, mobility and immobility take on a new agenda. Although mindful of these 'slaves of globalisation', *Mobile Media in the Asia-Pacific* has focused upon developed countries in the region in order to expose some of the politics of gendered production and consumption behind the image of global technological innovation. I have endeavoured to contextualise the technological rise in the context of the region's sociocultural and historical issues in which gender changes function as an indicator of other, broader issues.

As I have signposted in the four very different case studies, we can see various ways in which gender operates within the every-expanding discourse of mobile media in twenty-first-century everyday life. In Japan, we can see that creative labour around mobile media, drawing from the history of gendered genres of expression such as *gyaru-emoji* and kitten writing, has transformed into a multifaceted industry of popular *keitai shôsetsu* (portable novels) and is now being adapted into other media such as film. Here we see that mobile media as new media are undoubtedly remediated with *keitai shôsetsu* feeding back into older media such as printed novels, *manga* and film. Through Web 2.0 media such as 2ch and mixi, community storytelling is taking on new value again as exemplified by the *Densha Otoko* story, which features female directors, creators and producers.

One of the interesting aspects of the rise in social labour and cartographies of personalisation within *keitai* new media discourses is that they have attracted as many fans as dismissive so-called arbiters of taste. These forms of creative mobile media invite us to question and reconsider notions of cultural, economic and social capital; indeed I would describe this particular milieu as mobile capital. While over one million *keitai shôsetsu* are currently in circulation, and five of the top ten printed novels last year in Japan were adapted *keitai shôsetsu,* the new media debates still rears its head. For example, Jane Sullivan discussed the *keitai shôsetsu* phenomenon as 'pulp fiction', positioning the authors and audience as 'teenage girls and women in their 20s' (2008: 28), and cited the phenomenon as an example of how 'democratic' media – while 'spontaneous, fertile, exciting' – suffers from 'quality control' that fails to lead readers 'into new territory' (ibid.: 28). This type of argument is far from new, and one could easily substitute *keitai shôsetsu* for new media in another epoch. This polemic and conservative argument neglects to engage with the new forms of media literacy and expression being fostered by *keitai shôsetsu,* and overlooks the ways in which these hyperfeminine forms of new media are providing new paradigms for author and reader relationships that are predominantly orchestrated by women for women. One can't help but wonder: If Jane Austen were alive today, would she be writing *keitai shôsetsu?*

In Seoul we can see that the conflation of family with technology in the national culture – *kukmin kajok* – has clearly been scripted through the symbolic function of mobile media. As Hong (2007) noted, through the deployment of discursive strategies on behalf of industry and government, the hypermasculine *Hyoja* became symbolic of the national mobile phone industry. Meanwhile, the mobile phone, and specifically Samsung's Anycall, became symbolic of Korean twenty-first century post-modernity. However, despite the attempt by the industry to ensure that the technology conformed to techno-nationalist agendas – i.e. to the hypermasculine – the ways in which couples use mobile technologies tells a completely different story. Here female users are not only active creators of mobile media, but also act as gatekeepers for their male partner's device. As a vessel for the storage of future family memories the device becomes a symbolic engagement ring, and it is the female who chooses and often dictates the customisation – internal and external – of her male partner's phone.

Moreover, the use of mobile media such as camera phone imagery is affording Korean women with new forms of self-expression, representation and identity (D.H. Lee 2005; Hjorth 2007a, 2007b). Beyond the image of *sel-ca* with its associated narcissism, female users are creating new forms of storytelling and sharing their stories through SNS such as Cyworld mini-hompy. These visual storytelling modes are providing new forms of documenting, articulating and managing the fast acceleration of Korea into twenty-first-century post-modernity. As D.H. Lee remarks in her analysis of the role of camera phones and Cyworld mini-hompy, these practices are informing 'emerging geospatial imagery' (2008: forthcoming). Thus, these new forms of mobile media help users in negotiating new urban spaces that are no longer the site for the *flâneur* but rather for the co-present 'phoneur'.

In contrast to the two mobile 'centres' for technological innovation, Tokyo and Seoul, Melbourne and Hong Kong, although less technologically advanced, demonstrate other paradoxes of hypermasculinity and hyperfemininity that are being performed through twenty-first-century mobile media. In Melbourne, we see media such as SMS functioning as 'artforms' not dissimilar to the emergence of *emoji*, kitten writing and *keitai shôsetsu* in Tokyo. These art forms are deployed in creative ways predominantly by female users and the industry has responded by producing both female and male forms of innovative mobile media. Ish media's *Girl Friday* is a key example: a *Sex and the City*-type sitcom revolving around the story of a girl who finds a mobile on a tram. Ish media creative and technical team is almost all female.

Girl Friday's cross-platform program of 2-minute episodes has been carefully contextualised for the mobile frame. Registered readers receive SMS and emails from the heroine, 'Girl Friday' – a great example of the deployment of mobile media's specific cartographies of personalisation, hyperfeminine labour – to create intimacy and new forms of reader–author–producer relationships. *Girl Friday* has developed various interactive tangents that enable readers to become enveloped in the *Girl Friday* world, such as live gigs (localised to where readers are) and parallel stories about the other actors. Readers become 'produsers', able to control how much of the worlds of *Girl Friday* they enter. Examples of mobile media aimed at male users, such as *Three Day Growth*, explore the on and off air dynamics of two radio presenters. However, in general, female 'produsers' seem more active in mobile media content – displaying multiple cartographies of personalisation. While male 'produsers' were more implicit in personalisation modes, often opting for anti-logo logo such as the default ring tone that emphasised mobile media's functionality rather than creativity.

In Hong Kong, mobile media functions as a vehicle for negotiating the personal within the context of Hong Kong's hyphenated identity. Here the role of mobile technologies as a lens for conceptualising contemporary mobility is played out through mobile media as a site for personal consumption (as opposed to heavily shared aspects found in the Seoul and Tokyo case study), as well as a symbol for contemporary postmodernity (with its associated dilemmas as the *bus uncle* mobile movie demonstrated). In Hong Kong, mobile media are used to stitch together a sense of hyphenated mobility within personal reterritorialisations of place. Extending upon the filmic images that Chow (2007, 2000) identifies as displacement techniques against the mainland, Hong Kong camera phone practices have become a site for a

Figure 8.1 Girl Friday, mobile media movies that feature stories revolving around a mobile phone. The innovative cross-platform content includes regular SMS from *Girl Friday* to the registered user as well as parallel interactive content via Web 2.0 formats.

further re-enactment of nostalgia as resistance towards mainland reterritorialisation.

Images of people and places are stored in the mobile, creating a portable pocket of memories to draw upon at any time. Female users are most active in their deployment of mobile media, particularly by reappropriating mass media genres such as paparazzi as a form of expression and creativity. The increased deployment of media such as SNS, along with IM and camera phone practices, allows users a space in which to monumentalise their relationships with memories of place. By customising both the inside and the outside of the phone, users grapple with their hyphenated identity both on an individual level (using mobile media as predominantly as a site for personal consumption) and on a collective level (the 'atomized' national culture Ma spoke of as being performed through individuated 'satellite' mobile media).

In each of the locations, different forms of imaging communities can be found, mostly amongst the various forms of media visualities, textualities and virtualities that I call emotologies. Emotology, like hyperfemininity, is no longer the prerogative of female users, just as social capital increasingly becomes a form of marketable capital. Undoubtedly, in these early years of mobile media we can see that the shape of the industry is changing and evolving, with women graduating from consumers, to 'produsers', to producers. As producers and creators of content – exemplified by the *keitai shôsetsu* in Tokyo and *Girl Friday* in Melbourne – women seem to be part of a new generation of media practitioners that takes a playful and innovative approach to author–audience paradigms.

In each of the case studies I have highlighted some of the emerging imaging communities' practices that are reflecting, on the one hand, forms of intimacy, labour and community, and, on the other, materialising forms of national and transnational modes of post-modernity and post-industrialism. One of the ways for conceptualising these emerging practices is through the rubric of cartographies of personalisation – a process that is marked by localised gendered formations. In each location, different forms of imaging communities operate, reflecting the particular role the female mobile 'produser' occupies. In Japan, the user has become a professional creative and generated a new, but remediated genre of mobile media. In Korea, the user has re-defined the role of the family and technology in the nation-state. In Hong Kong, the user has produced new modes of writing personal history that operates to reterritorialise the hyphenated identity. In Melbourne, the user draws on the three Ds – *delay, deference* and *decentring* – to re-enact the post presence metaphor apparent in both new and old media.

Although it is still early days in the epoch of mobile media, its rise has been marked not only by female consumption and production but also by a dialectic between post-industrialism hypermasculinity and hyperfemininity. Examples such as *keitai shôsetsu* and *Girl Friday* give hope that the hyperfemininity will gain value in the market place in terms of financial remuneration, instead of remaining unpaid 'women's business' that multinational media corporations continue to exploit and profit from. As my case studies have demonstrated, the region's various forms of post-modernity and mobility – as seen through the lens of mobile media – are characterised by a shift in gendered power relations. Whether these new forms of media literacy, expression and creativity become a vehicle for further female empowerment and agency, is a subject for future research.

Mobile phones can extend current modes of intimacy and facilitate social 'hyperfeminine' labour through various forms of localised customisation and gender performativity that I refer to as cartographies of personalisation. From emotologies such as *kaomoji* and other UCC such as 'imaging community' camera phone practices, to the fashioning and adorning of the device with various forms of presentations of self, the mobile phone is both a playground and battlefield for emerging politics of hyperfeminine capital in the twenty-first century. It is its symbolic dimensions representing both postmodernity and intimacy that make mobile media such as fascinating site for reflecting upon the region's shifting gendered landscape. By investigating burgeoning localised modes of gendered mobile media customisation from a user level, we can gain acuity into some of the shifts in gendered deployment and consumption of technologies in the region.

Re-imaging communities: rethinking the region

The role of globalisation in the region, as reflected in gendered patterns of mobile media use, has been one of the central tenors of *Mobile Media in the Asia-Pacific*. From camera phone images to the coordination and customisation of SNS, mobile media plays an increasingly significant role in articulating both the unofficial and official images of a nation's, and of the region's, imagined community. These forms of mobile media emotologies create imaging communities, in which users become active producers of images and texts of cultural vernaculars. Imaging communities are demonstrating new and remediated forms of media literacy, expression and creativity that alleviate yet, paradoxically, at the same time amplify the disparities of twenty-first-century consumption and production. Thus mobile media encapsulates the *zeitgeist* of contemporary labour practices and the concomitant role of gender in geo-spatial mappings of belonging and yearning. The symbolic and literal dimensions of gendered lifestyle cultures are, I suspect, a phenomenon that will increasingly come to dominate in a global, mobile 'networked' society.

Just as history can be viewed as the sum of the images that are left when personal memories fade, how and what we document of an experience can have great importance to how not only participants but also non-participants reflect on it afterwards. Arguably, parallels can be made between the introduction of new media such as film and radio at the beginning of the twentieth century and the current state of mobile media. Media such as film and television became the embodiment of twentieth-century modernity and post-industrialism, reflecting shifting modes of temporal and geo-spatial imaginaries. These cartographies were charted by an ocular-centric focus, in which visuality was the portal between mobility and immobility, here and there, global and local. In the twenty-first century, the visual is re-ordered as mobile media reprioritises other sensory experiences such as textualities, virtualities and haptics in the practice of place and co-presence. It is the individual and collective micro-narratives of arising mobile media visualities, textualities and virtualities that both reproduce, and are the product of, the region's post-modernity.

These micro-narratives can be read as indicative of some of the ways in which consumption, production and representations of individualism and community are changing in the region. While the region's accelerated rise to twenty-first-century post-modernity has been inextricably linked with its central role in producing and manufacturing global ICTs, it is the role of ICT consumption that offers more insight through its rapid convergence with new modes of user–producer creativity and labour such as UCC. During this time, the region's economic and technological prowess globally has soon transformed into a significant form of twenty-first-century cultural capital. As an icon of mobile media in the region the young Asian female consumer is also changing, reflecting the emerging and residual forms of labour, capital and agency.

When I began my research in 2000, the role of the female consumer was only beginning to translate into new forms of creativity and expression. It was still unclear how this would translate into actual viability in the market place. However, as increasingly as the region's hypermasculinity accelerated through the production of ICTs and the corresponding force of hyperfemininity in the form of mobile media consumption blossomed, mobile media came less and less to exemplify hypermasculinity and increasingly to typify hyperfemininity as it became a site for 'produser' practices. This is not to say that exploitation disappeared through lauding the sovereign consumer, rather, mobile media has become a space in which new forms of labour and creativity can be performed. Some of these forms exploit social or reproductive labour, but many provide some form of remuneration to the user, from various metaphoric gift-giving practices to actual financial gain.

This transformation at the level of gendered consumption and production also reflects metaphoric gender shifts in the region. Over the last twenty years, and most notably in the last ten years in the case of locations such as South Korea, the rise of twenty-first-century post-modernity has been concurrent with the increasing production of innovative global ICTs, and most particularly, mobile media. This phenomenon has seen the region shift from techno-logical and economic power to 'soft' capital prowess: a transformation in which the region has re-orientated its particular forms of capitalism and modernity into localised forms, as exemplified by the 'Confucius capitalism' period at the turn of the century. The region demonstrated its hypermascu-linity on a global scale during this period. However, paradoxically, the tech-nology that most clearly symbolised the region's hypermasculinity, mobile media, has now become the vehicle for new and empowered forms of hyper-femininity, as emerging emotologies and female-orientated and -run indus-tries such as *keitai shôsetsu* begin to take hold.

In *Mobile Media in the Asia-Pacific* I have argued that the semiotics of UCC (rather than industry-created) customisation, and the dominance of feminised modes, follow cartographies of personalisation. These are the ways in which new technologies come to be imbued with emotion and identity as an extension of their user. They are also imbued by a sense of community, especially insofar as they are deployed as a form of imaging community. As the currency of social labour becomes increasingly significant in an age dom-inated by hypermasculinity global ICTs, we can gain insight into contempor-ary modes of labour, intimacy, community, locality and transnationalism by studying the evolving role of mobile media. In particular, mobile customisa-tion in the Asia-Pacific is part of an emotional and mimetic geography that sees the mobile phone taking on new cultural meanings.

Mobile media, as indicative of new media, is remediated and yet emer-ging. If we listen to the micro-narratives of imaging communities of users we can gain insight into how the region is negotiating globalisation, con-sumption and individualism, while maintaining strong localised notions of place and community. Through the lens of gendered mobile media we can begin to re-conceptualise the role of post-industrial technologies in the region and how they reflect emerging forms of lifestyle, consumption and identity politics. As an integral part of the practices, cultures and societies of twenty-first-century post-modernity, it is undoubtedly mobile media that will facilitate, foster and inform the personal and geo-spatial narratives of everyday life. This is but the beginning of an epoch marked by *Mobile Media in the Asia-Pacific*.

Having surveyed mobile media as *societies* and *cultures* in Part I and II, the book now migrates to Part III, where we contemplate the paradigm of emerging *Mobile media practices* as part of twenty-first-century new media.

Figure 8.2 A *keitai shôsetsu* author at home.

Source: CNet (Japan).

These case studies are not meant to be indicative of all mobile media as new media, but, rather, some sample studies to contemplate a couple of the many directions mobile media as new media are moving. Let us concentrate upon the 'art' within the art of being mobile.

Part III
Mobile media practices

9 Domesticating new media

A discussion on locating mobile media

As convergence leaves its mark on this century, the ultimate alibi in the convergence rhetoric seems to be the mobile device. Convergence can occur across various levels such as technological, economic, industrial and cultural. As Jenkins (2005) observed, in the growth of the mobile phone into converging various forms of multimedia – into the ambiguous and yet ubiquitous mobile media – one could almost forget that mobile media arose from an extension of the landline telephony.

Now, the twenty-first century's equivalent to the Swiss Army Knife (Boyd 2005), mobile media encompasses multiple forms of media including camera, gaming platform, MP3 player and Internet portal. As we begin to chart the burgeoning phenomenon of mobile media, we must re-assess the methodologies and frameworks being used. How do we grapple with mobile media's interdisciplinary background? Should mobile media be framed in terms of the mobile communication and material cultures traditions, fathered by Silverstone, that have contextualised the socio-cultural processes of media technologies in terms of the domestic technologies approach? Or should mobile media be framed by the creative theories and practices of new media?

The rise of the mobile phone into mobile media has attracted scholars from various disciplines such as media studies, gender studies, cultural studies, media sociology, virtual ethnography and new media, all bringing with them a wealth of traditions, methodologies and approaches. One of the dominant and highly successful approaches in the field of studying mobile phone cultures is, undoubtedly, the domestic technologies approach.

As discussed in the introduction of *Mobile Media in the Asia-Pacific*, the interdisciplinary framework of the domestic technologies approach draws from anthropology, cultural studies and consumption studies. A significant part of its lineage lies in anthropology and its commitment to analysing the processes of material cultures in everyday life. Undoubtedly, the seduction of the domestic technologies approach is that it focuses on the symbolic dimensions of technologies in everyday life. In particular, the domestic technologies approach focuses on meanings that individuals and cultural contexts give to their technologies, extrapolating on the ways in which users perceive them.

However, as the mobile phone expands into a multimedia device, how can the dimensions of social and reproductive labour – addressed by domestic technologies approaches – be incorporated into the growing realm of mobile media as new media? Domestic technologies approaches seem to fail in grasping the role of creative labour associated with mobile media beyond social and reproductive labour paradigms. In turn, new media approaches to mobile media seem unequipped to address the political dimensions of social and reproductive labour. Since both approaches have been useful in addressing the dynamic, social, creative and procedural nature of mobile media, it seems fitting to discuss these two enveloping traditions in the context of locating mobile media within a 'domesticating new media' approach.

In this chapter I will explore the marriage between the two traditions – on the one hand, the domestic technologies approach, on the other hand, the new media remediation approach – in order to conceptualise some of the paradoxes found in mobile media in terms of earlier, ongoing processes. I will outline some of the key attributes and paradoxes that have plagued both traditions' examination of mobile media. Through the example of mobile location-aware gaming, I will draw upon current discourses around mobile media and its co-habitation in both domestic technologies and new media discourses. As this chapter will argue, through mobile media we can gain insight into some of the recurring paradoxes that run across disciplines and boundaries, continuing to haunt and limit interdisciplinary approaches to twenty-first century new media practices. In particular, I argue that the emphasis upon visuality and screen-centric views have neglected to address one of the most important aspects to mobile media, the haptic.

Mobile @ convergence: divergence and convergence

Mobile media is a strange animal to tame. Part domestic technology, part new media, the phenomenon has attached much stargazing and posturing about the future. Through the portal of mobile media, we have witnessed mobility becoming conflated with futurism. The rise of the mobile phone has been marked by its shifting symbolism, usages and adaptations (Agar 2003). When mobile phones first graced mainstream media in the 1980s, they were associated with yuppies and conspicuous displays of wealth as demonstrated in the iconic 1980s film *Wall Street*. Then, as mobile phones were adopted and adapted by youth cultures, the phone shrunk into a complex creature adorned by user-generated customisation from phone straps to sticker faceplates and screen savers. Then, as the phone became more than *just* a phone and started to emulate this century's Swiss army knife, it expanded in size both physically and psychologically to become an integral component in visual, textual and aural practices in contemporary everyday life.

It is with this size change that we moved into an epoch of mobile multimodality that became synonymous with contemporary mobility. The rise of mobile media as multimedia par excellence has also been accompanied by corporate smoke-and-mirrors around the so-called 'empowered' user by way of UCC and 'prosumer' agency. In this climate of optimistic futurism, mobile media promised a further democratisation of media. But as Koskinen (2007) notes, this accessibility of multimedia often resulted in the aesthetics of banality; images and media rehearse well-known genres and themes. Within the so-called banality are normalised power relations inscribed at the level of everyday practice; thus mobile media serves to remind us of the growing significance of place.

Much work has been conducted around the 'banality' of mobile media practices in terms of camera phone visual and distribution characteristics, with many theorists pointing to how the content of mobile media rehearses the banality of earlier media (i.e. camera phone images re-enacting analogue genres), but the context in which they are shared (or not) provide much signification (Scifo 2005). However, it seems that the haptic economies, so particular to mobile media, are in need of re-evaluation. While Ito and Okabe's (2005a, 2005b, 2005c) three 'Ss' – sharing, storing and saving – noted some of the particulars, we need to examine the politics of 'waiting for immediacy' just outside the frame/screen. In other words, what are some of the haptic workouts occurring just outside the frame that undoubtedly affect what is inside the frame?

So what do I mean by haptic? Just one glance at the current models of mobile media such as iPhone and LG Prada and we can see that the screen is no longer about visuality; it is about haptics – haptic screens, to be precise. The engagement of mobile media is not ocular in the case of the gaze or the glance, but rather akin to what Chesher (2004) characterises as the 'glaze', as discussed in Chapter 1. To reiterate his argument, Chesher draws on console games cultures to identify three types of glaze spaces – the glazed-over, sticky and identity-reflective. For Chesher, these three 'dimensions' of the glaze move beyond a visual economy, deploying the filters of the other senses such as aural and haptic.

The haptic has often been under-theorised in mobile communication discourses, often left up to new media practitioners to grapple with in such projects as location-aware gaming. In the growth of mobile media discourses, much has been discussed in terms of media such as camera phone practices and the associated sharing and distribution methods. However, much of the rhetoric around mobile media and convergence has been focused upon the frame and visuality – as such concepts as 'cross-platforming' entail. These models have discussed media in terms of twentieth-century preoccupations with the visual and the screen, neglecting to re-orientate frameworks around what makes mobile media so particular; whether being mobile or immobile, the logic is of the haptic. It is the touch of the device, the intimacy of the object, that makes it so meaningful.

For new media artists such as Rafael Lazano Hemmer and his relational architecture projects, it is this very oscillation of the haptic and the cerebral that partakes in mobile media co-presence, that makes it such a particular vehicle for twenty-first century new media practice. In urban spaces, it is not so much the camera phone images that are transforming the spaces, but, rather, the haptic workouts of the everyday. Much of the discussion of mobile media has encircled the important role of mobile media co-presence, and yet the integral notion of the haptic, apart from the hype around SMS thumb cultures, has been largely ignored.

However, the critique of normalised everyday practices and the haptic workouts outside the frame can be found in the various upsurges of experimental new media projects such as location-aware gaming, mobile gaming or 'big games'. Location-aware or pervasive games often involve the use of GPS (geographic positioning systems) that allow games to be played simultaneously online and offline. As Finnish theorist (and director of DiGRA) Frans Mäyrä (2003) notes, gaming has always involved place and mobility and yet this is precisely what is missing in current games, especially single-player genres. Mäyrä points to the possibilities of pervasive (location-aware) gaming as not only testing our imagination and creativity but also questioning our ideas of what constitutes reality and what it means to be co-present and virtual.

The notion of 'big games' does not so much relate to the gadget's gluttonous size but rather it has more to do with the role of people and the gravity of place in the navigation of co-presence. These projects served to remind us of the importance of locality and its relationship to practices of co-presence. The potentiality of 'big games' to expose and comment on the politics of co-presence – traversing virtual and actual, here and there – in contemporary media cultures has gained much attention. They highlight some of the key paradoxes of everyday life that have been exemplified in mobile media projects such as location-aware gaming. The paradoxes include: virtual and actual, online and offline, cerebral and haptic, delay and immediacy.

As Frank Lantz (2006), a New York-based game designer who has been involved in such pivotal projects as PacManhattan notes, the importance of location-aware mobile gaming – or 'big games' – definitely plays an important role in the future of gaming. Citing examples such as PacManhattan, UK's blast theory, Geocaching and Mogi, Lantz emphasises the importance of these projects in testing the notion of reality as mediation. As Lantz observes, the precursors to Big Games and the 1970s New Games Movement were undoubtedly the art movements of the 1960s, such as happenings (impromptu art events) and the Situationist International (SI) tactics of Guy Debord such as *détournement* which operated to interrupt/disrupt everyday practices and the increasingly role of media and commodification. In this way, this can be paralleled with the trend in contemporary art from 1990s that French curator and critic Nicolas Bourriaud (2002) called 'relational aesthetics'. As Bourriaud observed, 'relational aesthetics' dominated the international art scene from the 1990s onwards, building from an emphasis upon locality and the de-institutionalisation of installation, and the 'international' in favour of the vernacular and local.

Locative mobile gaming illustrates the paradoxes of mobile media as part of the cyclic and dynamic processes of technology. For example, in an age of so-called immediate technologies, such projects enlighten us to the conundrum of instantaneity, that is, the inevitable poetics of delay. They highlight the price of mobility and its oscillation between freedom and leash (Arnold 2003; Qiu 2007), in which work and leisure boundaries are increasingly blurred (Wajcman et al. 2008; Gregg 2007). Locative-mobile gaming also emphasises other paradoxes apart from the aforementioned immediacy/delay temporal conundrum. These projects highlight the way in which mobile media can often interfere with, rather than help, face-to-face connections. For example, the tyranny of mobile media's creative labour and democratising of media dimensions, as epitomised by UCC, sees users becoming more enslaved to the technology rather than it freeing up time to spend with intimates. Locative-mobile gaming projects afford us one way in which to reflect and meditate on the paradoxes of contemporary mobile media.

Moreover, locative mobile gaming illustrates that in the face of democratising of media, new media is still far from the insights and interests of the everyday person. It also reflects new media artists' fears, and yet curiosity, about mobile media's ultimate creative conundrum: is it the rise of democratised media and the mainstreaming of new media, or does the 'banality' represent the dominant pedestrianisation of new media? Can mobile media reconfigure new media discourses to connect the 'shock of the new' of art history? Moreover, can mobile media's aesthetics of both 'delay' and the 'banal' revise new media?

As a new conflation of many techniques, traditions and media histories, it is no easy task to outline the nebulous terrain of mobile media. In this chapter I argue that one way in which we can understand mobile media is *vis-à-vis* its borrowing from, and adaptation of, various sociological and new media traditions. In the following section I will address two traditions – domestic technologies and remediational, new media approaches – in the cartography of mobile media. I argue that many parallels can be found in the two traditions and that by incorporating the two genealogies we could gain much insight into mobile media.

Just as mobile media needs the rigour of domestic technologies approaches to comprehend the social dimensions of new media, it also needs the innovative approaches of new media theory in order to reconceptualise the conflations between creative and social labour in mobile media's fusion between media communication and new media practices. By combining the two separate approaches – domestic technologies and new media – we can begin to conceptualise mobile media no longer as a third 'screen' but as a third 'space'.

Domesticating new media: two examples of the multi-traditions of mobile media

The rise of mobile media could be read as nascent. However such a belief, propagated in global media's lauding of the new mobile revolution in consumer agency (in the form of the prosumer and Web 2.0), neglects to address the dynamic dimensions of technology as a socio-technological process. In the case of domestic technologies approaches, in which domestication is always an ongoing and never-completed process, the dynamics of mobile media extends already existing cyclical models. So too, in the tradition of new media, in which old and new have had a dialectical and dynamic relationship that disrupts any linear or casual notion of time.

As mobile communication and media industries converge, the all-pervasive futurist rhetoric becomes stifling. And yet, if the twin histories of new media and mobile communication have taught us anything, the 'new' is always remediated and mediated. Each 'new' technology deploys techniques of the older technology, which in turn revises the earlier media. This cuts to the core of all communication and cultural practices implicated in intimacy. As discussed in Chapter 1, for Bolter and Grusin (1999) new media is remediated with older media into a dynamic ongoing process that disrupts any causal or linear notion of old and new technologies. As Morse (1998) concisely notes in the case of the Internet, all forms of intimacy are mediated – by language, gestures, and memories. Emerging forms of visual, textual and haptic mobile genres such as SMS and camera phone practices – re-enacting earlier rituals such as nineteenth-century letter writing, postcards and gift-giving customs – have only served to highlight the remediated nature of the rise of mobile media.

There is much to be learnt from understanding the parallels between new media theory on remediation and mobile communication's usage of the domestic technologies approach. Like the domestic technologies approach, the study of new media through the lens of remediation echoes a similar philosophical stance. As mentioned in Chapter 1, influential theorist in the field of media-archaeology, Huhtamo (1997) has argued that the cyclical phenomena of media tends to transcend historical contexts, often implicating a process of paradoxical re-enactment and re-enchantment with what is deemed as 'new'. As Parikka and Suominen note, the procedural nature of media archaeology approaches means 'new media is always situated within continuous histories of media production, distribution and usage – as part of a longer duration of experience' (2006: n.p).

Citing an example of the launch of Nintendo DS that heralded a new and 'unique' experience for twenty-first century entertainment, Parikka and Suominen note that much contemporary post-industrial digital media culture is inundated by futurism that seeks to break with the past. Parikka and Suominen note that this 'creates the impression that, in the new media discourse, the past functions solely as something worse or less sophisticated, something that has to be left behind and practically forgotten' (2006; n.p).

When McLuhan (1964) identified the content of new media as that of the previous technology, he highlighted the non-linear and dynamic role of new technology imbued by the specters of old technology. In short, that the 'new' is far from superseding or breaking with the old as modernist mythologies would have it. The fact that the notion of 'new' in new media has been continuously challenged and demonstrated as a fallacy echoes the way in which technology has been approached by many mobile communication scholars (from predominantly sociological and urban anthropological traditions) through the domestic technologies approach. Moreover, as Agar (2003) notes, the fact that the mobile phone is the 'icon of the new' further galvanises its position within the always-contentious terrain of new media.

In the picture painted by the domestic technologies approach, domestic technologies such as the radio, TV and mobile phone are seen as part of the cyclical and ongoing process of consumption in everyday contemporary life. Much of the literature analysing mobile communication has utilised the domestic technologies approach to identify adoption and adaptation of technologies as always ongoing and never completed. Like the cultures that they inhabit, domestic technologies are always in flux. Domestication is ongoing and dynamic, and through customisation practices we can domesticate domestic technologies as much as they domesticate us, in a productive tension. In the case of the mobile phone, while the domestic technology device may have *physically* left the home, it *psychologically* resonates what it means to be at home and local, no matter where it is located.

As Morley has noted, the mobile phone has often been cited as a key example of domestic technologies par excellence (2003; 2007). As a key scholar in the field, Haddon (2003) provided robust methodologies to comprehend the dynamic and enduring processes of the domestic technologies. Often users' relationships to their domestic technologies can wax and wane, drawing feelings of ambivalence through an inability to escape mobile media's increasingly present part of everyday urbanity.

As an approach, the domestic technologies method sees the process of engagement with technologies undergoing various stages or nodes of a cycle that include: 'imagination, appropriation, objectification, incorporation, and conversion' (Ling 2004: 28). Consumption is seen as an ongoing process that is perpetually negotiated, way after the actual point of sales. As Ling notes, 'our consumption becomes a part of our own social identity. Further, others' consumption is a type of lens through which we see them and through which we interpret their social position' (ibid.: 27).

Mobile media represents a meeting of the crossroads between the genealogy of domestic technologies and media archaeologies of new media. In both these traditions, we see mobile media remediating and re-enacting previous media cultures and modes of domestic regimes. Reminding us of our forgetting whilst harnessing the inevitable amnesia that accompanies any notion of 'new', mobile media represents the conundrum of new technologies. In new media discourses we can find many examples of the content or spectres of the older media. Like the domestic technologies approach, the study of new media through the lens of remediation echoes a similar philosophical stance.

According to Kopomaa (2000), the mobile phone is an extension of nineteenth century media. For Kopomaa, mobile media creates a new 'third' space in between public and private space. On the one hand, the project of examining mobile media entails observing the remediated nature of new technologies and thus conceptualising them in terms of media archaeologies. On the other hand, mobile media's re-enactment of earlier technologies is indicative of its domestic technologies tradition that extends and rehearses the processes of precursors such as radio and TV. It is the fact that Kopomaa draws our attention to mobile media as a third space, rather than third screen, which is significant.

Both traditions – the domestic technologies and new media remediation approaches – emphasise the cyclic and dynamic process of media technologies that cannot be simplistically divided between old and new, or inside and outside the screen. Rather, the cartography of mobile media is one imbued by paradoxes. In the case of camera phone practices – whether still or moving – mobile media demonstrates two distinctive paradoxes, that of the *reel* in the real, and the inherent poetics of *delay* in the practice of immediacy in navigating of offline and online co-presence. As noted in Chapter 1, Manovich (2003) identified contemporary new media and digital practice as all consumed by fetishing the real through the lens of the reel – that is, the texture and skin of the analogue.

For Manovich, the way to understand the remediated emerging digital cultures and the haunting by the ghost of the analogue is through a series of paradoxes. These sets of paradoxes are located around the relationship between the real and the reel. As Manovich identifies, while the analogue may disappear, it will continue to haunt the digital in the form of the analogue's particular realism, the 'reel'. This is evident in the way in which camera phone practices echo previous analogue norms and that, in turn, make mobile media, according to Koskinen, characterised by 'banality'.

However, one of the most compelling examples of the real/reel phenomenon, where the tactile process of the analogue is fully felt both metaphorically and actually, is the rise of screen cultures in mobile media. In particular, the rise of such mobile media devices as iPhone, LG Prada, and Samsung Armani phone – to name a few – all incorporate one key feature, the haptic screen. Here the reel/real paradox is played out in the haptic versus visual, in which the haptic is undoubtedly the more meaningful factor that 'domesticates' the device into the user's everyday life. Much of the specters of the analogue reel are more about the tactical experience of image processing; and while these processes have been deleted in the rise of the digital, it is the legacy of the haptic – which has moved from the filmic developing process to the actual politics of the touch screen – that continues unabated. However, in the language set of twentieth-century media cultures, much discussion was given to visuality rather than the increasing role of the haptic.

While location aware projects are invaluable in geo-caching (such as GPS) and demonstrating the importance of place and specificity in a period of global technologies, they also served to highlight one of the greatest residual paradoxes of mobile media as a metaphor for socio-technologies: that is, the paradoxical politics of co-presence. One example can be found in the aims of twentieth-century technology to overcome difference and distance from geographical and physical to cultural and psychological. This attempt to overcome distance and difference sees the opposite result, the overcoming of closeness. Practices of co-present intimacies become fetishised, through a proclivity of contemporary lifestyle technologies to further fuel full-time intimacy. This recites what Arnold (2003) identified as the janus-faced nature of mobile media that operates to push and pull us, setting us free to roam and yet attaching us to a perpetual leash.

As Arnold notes, the janus-faced phenomenon is symptomatic of what Heidegger characterised as 'un-distance'. The role of technology in the twentieth century has always been to overcome some form of distance – whether geographic, physical, social, cultural, temporal, or spatial. But herein lies the paradox. The more we try to overcome distance, the more we overcome closeness. This is the kernel of un-distance and its temporal and spatial tenor. Un-distance can be seen today in the practice of mobile media, particularly pervasive location-aware projects that rely on the so-called immediacy or instantaneity of the networked.

However, one could argue that un-distance has been perpetuated by the ocular-centricism of twentieth-century 'tele'-media, a phenomenon that has been disrupted by mobile media's emphasis on the haptic. For Richardson, mobile media needs to harness the importance of the haptic. Conducting a small ethno-phenomenological study on the use of phone-game hybrids, Richardson disavows the ocular-centricism prevalent in 'new media screen technologies' to focus on 'the spatial, perceptual and ontic effects of mobile devices as nascent new media forms' (2007: 205). As she persuasively observes:

> In order to grasp the epistemic, ontic and phenomenological status of screen media it is important to trace their ocularcentric legacy; by understanding this history we can then interpret how mobile screens in particular work to bewilder classical notions of visual perception, agency and knowing.
>
> (ibid: 208)

Indeed, one of the compelling factors to arise from mobile media, and this links back to its fusion of remediation and domestic genealogies, is the persistence of the ontology of the reel. However, unlike the twentieth-century 'reel'– in the form of the aural modes of address embroiled in 'screen-ness' – the mobile reel, and thus possible creative worlds and realities, is undoubtedly governed by the haptic. The game of mobile media – whether it be partaking in camera phone imagery and the haptic exercises outside the screen, or mobile gaming, in which interactivity and engagement are navigated by haptic mobility and immobility rather than visualities of screen cultures – is undoubtedly changing how we are thinking about domestic technologies and new media. Through the lens of paradoxes that encompass virtual and actual, online and offline, haptic and visual and delay and immediacy, some lessons about twentieth-century media practice can be learnt.

For anyone that has participated in a mobile pervasive game, they will quickly identify the lack of coherence between online and offline co-presence. The more we try to partake in the *politics of immediacy*, the more we succumb to the *poetics of delay*. This paradox extends beyond just mobile gaming and can be found in many of the multimedia possibilities of mobile media – from camera phone imagery and MMS to moblogging and SMS.

In the case of the growing interest in urban screen cultures as an analogy for the twenty-first century, one could argue that it is indeed the very eruption of the twentieth century's obsession with visualities into the twenty-first century's politics of the haptic that dominates the canvas of mobile media. From the haptic screen interfaces to the various multimedia tactics such as mobile gaming, which disavow the screen for the haptic and audio, mobile media revises the perceived hierarchies of the senses which, in turn, could breath new life into new media and domestic technologies approaches.

Time after time: the never-ending concluding beginning

In modernism, the role of originality was celebrated. For the modernist avant-gardists, such vehicles as technology served as a decisive break from the past in what art critic Hughes characterised as 'the shock of the new' (1981). In contemporary post-industrial digital cultures, the 'new' promised by mobile media is in fact 'banal' and located in nostalgic politics such as the real/reel paradigm. In this chapter, I have assembled two traditions – domestic technologies and remediation – in order to show the similar cyclic debates operating across disciplines. I argue that perhaps mobile media needs to be conceptualised as the 'shock of the banal', that is, its paradoxes – online and offline, virtual and actual, delayed and immediate, haptic and visual – are far from new and can be traced through various disciplinary traditions.

In my ethnographic studies into camera phone practices in Seoul, Tokyo, Hong Kong and Melbourne detailed in Part II, one of the increasingly clear features of the tyranny of the full-time intimacy of mobile media customisation is the use of immediacy to camouflage delay. Many respondents spoke of their pretending not to see the SMS or MMS so that they could savour and think – the poetics of delay – which immediate technologies seem to demand. Moreover, the persistence of the reel in much of the camera phone images, genres and mobile movies was significant. In particular, for many respondents, mobile media making was less about visual economies and more about aural and haptic modes of address, akin to earlier 'reel' domestic technologies such as the TV and radio. Thus, the conflation between domestic technologies and new media approaches could further address one of the greatest paradoxes of shifts from twentieth- to twenty-first-century media; that is, rather than it being a history predicated on such visualities as the 'screen-ness' would entail, it is a history of the rise of the audio and the haptic that are becoming the key indicators and characteristics of mobile media.

I have chosen to focus on two traditions – one draws from media and communication and material cultures (anthropology) in the form of the domestic technologies approach; the other calls upon new media approaches of remediation and media archaeologies approaches. In these two traditions we can see various similarities – the focus on the dynamic, socio-cultural processes of mobile media. While the former allows for more insight into social and reproductive labour debates, the latter affords us acuity into shifting modes of accessing creative labour and everyday life. In the case of mobile media projects such as location-aware gaming, I argue we need to draw upon the two models, incorporating them into a new framework for evaluating dimensions of mobile media, and twenty-first-century screen cultures, in terms of key attributes such as the haptic. The important factor here is against the seductive and simplistic futurism prevailing in much discussion around mobile media: we need to recognise that mobile media, like new media, is inevitably involved in the politics of the banal and nostalgia.

Much of the futurist posturing accompanying mobile media discussions in global media have celebrated the potential democratisation of new media. With the rise of the produser (Bruns & Jacobs 2006), much of the media of late has celebrated UCC and the prosumer as part of the Web 2.0 enterprises. But behind this rhetoric is the pivotal role mobile media has played in creating and re-enacting debates about technology, labour and creativity that have long accompanied new media and domestic technology discourses. Rather than just domestic and artistic labour having little or no remuneration in the general community, now the UCC associated with mobile media could see the everyday person subject to the injustices of industry convergence, whereby corporations buy and exploit social and creative labour in the form of Web 2.0 media such as SNS.

The conundrum of new mobile technologies is that they are *supposed* to free us up and yet, as a good existential crisis would have it, the freedom is a leash. Work becomes mobile; labour is on a perpetual drip. We are supposed to be available at all times, perpetually connected. Rather than free us, the 'immediacy' logic of mobile technologies makes us feel like we must be quicker and must achieve more. Rather than saving time, applications such as camera phone image making – and the attendant customising and modes of sharing/distribution – mean users spend a lot of time sharing and editing the so-called immediate. The present gets put on hold. However, one of the features that becomes apparent in mobile media is the need to move beyond the screen-centric and ocular-centricism of twentieth-century media and re-connect with the very reason the mobile phone has grown into mobile media, its importance at the level of everyday haptics.

By engaging in the significance of the haptic in mobile media we can grasp some of the paradoxes at play. It is important to recognise that this conundrum of delay and immediacy is not new with the rise of mobile media. Rather, these paradoxes have been central in the emphasis upon screen cultures in the face of the importance of the haptic in making sense of mobile media. As I have attempted to discuss, mobile media represents some interesting paradoxes about contemporary media and consumer cultures. In this chapter I have tried to show the ambivalences surrounding mobile media from both new media and domestic technologies approaches in order to re-conceptualise the philosophical and phenomenological dimensions of mobile media, to socialise the creative media dimensions and to innovate the social, domestic dimensions.

In order to grapple with the burgeoning field of mobile media we need to comprehend the twin histories – that is, the domestic technologies and remediation approach – to fully grasp the histories, contemporary and future paradoxical permutations of mobile media and not just fetishise the 'new' by futurist posturings. Mobile media is undoubtedly a project involving the domesticating of new media in which old boundaries between art and life, production and consumption perpetually change and shift, repeat and pause. But it is time that we moved away from twentieth-century preoccupations with visual cultures and screen-ness that deems to view mobile media as a (advertising) *third screen* and instead acknowledge its genealogy as a *third space* that is governed by the politics and aesthetics of haptics.

10 The big bang

An example of mobile media as new media

The notion of mobility, like play, is inflected by the local. As a region, the Asia-Pacific is marked by diverse penetration rates of gaming, mobile and broadband technologies, subject to local cultural and socio-economic nuances. This makes the region a compelling case study for both gaming and mobile technologies. Both media are entrenched in the social and local. In this convergence, discourses around gaming and mobile communication could learn a lot from each other.

In the world of mobile gaming, contesting definitions and beliefs prevail. This has to do with the fact that in an age of globalisation, localisation is a tenuous force. As social, cultural, economic and technological convergence conflates with globalisation, processes of localised disjuncture are inevitable (Jenkins 2005). The diversity of the region is clearly demonstrated by the bipartisan definition of mobile gaming in Seoul and Tokyo. As two defining locations they are seen as both 'mobile centres' and 'gaming centres' to which the world looks towards as examples of the future-in-the-present. Unlike Japan, which pioneered the *keitai* IT revolution and mobile consoles such as PlayStation2, Korea – the most broadbanded country in the world – has become a centre for MMOs (online massively multiplayer) games played predominantly in the social space of *PC bangs*.

Adorned with over 20,000 *PC bangs* in Seoul alone and with professional players (Pro-leagues) making over a million US dollars per year, locations such as Korea have been lauded as an example of gaming as a mainstream social activity. In a period marked by convergent technologies, Korea and Japan represent two opposing directions for gaming – Korea emphasises online MMO games played on stationary PCs in social spaces (*PC bangs*) whilst Japan pioneers the mobile (privatised) convergent devices. These two distinct examples, with histories embroiled in conflict and imperialism, clearly demonstrate the importance of locality in the uptake of specific games and game play. As Brian Sutton-Smith (1997) identified, game spaces are social spaces. These social spaces have histories that are imbued by the local which in turn inform how we conceptualise and practice various forms of mobile and immobility.

One of the compelling features of mobile gaming is that it represents the bipartisan of views on mobility (especially electronic) and its impact on place. Casual mobile gaming suggests an extension of what Williams characterised as 'mobile privatization' (1974); while the potential corporeal and electronic synergies of pervasive mobile gaming suggest the indefatigable significance of place. Much of current location aware gaming highlights for users that co-presence is undoubtedly disjunctive with delay often winning out against the desired immediacy. As I have argued elsewhere (2007b), pervasive gaming is the practice of one of the enduring paradoxes of mobility – the acrimonious marriage between immediacy and delay. That is, the wish for immediacy to have seamless co-presence is disrupted by the actual practice of delay through the haptic participation of place. In order to understand this phenomenon we need to reexamine what mobility and place means today.

In one way, the rise of mobile media parallels the rise of the webcam (Koskela 2004) by affording everyday users with the ability to document and edit their stories, however, mobile media promises more – the portal to new arising forms of distribution such as MySpace, Facebook, Cyworld mini-hompy, mixi, 2ch (*ni-channeru*), YouTube, etc. Much of the innovative mobile media art work has been conducted around hybrid reality and location-based mobile gaming (De Souza e Silva 2006a, 2006b, 2004; Davis 2005) that aim to challenge the role of co-presence and everyday life – forging questions around boundaries between the virtual and actual, online and offline, haptic and cerebral, delay and immediacy. Examples include the PacManhattan (US), Proboscis' *Urban Tapestries* (UK), Blast theory (UK), aware (FIN), mogi game (JP), and INP *Urban vibe* (SK). In these projects we can see the forging of various traditions and disciplines, often unable to be contained by games studies and new media discourses, apart from exceptions like the 'relational architecture' of new media artist Hemmer.

As discussed in Chapter 2, in debates about global media and the role of the local, two dominant arguments have persisted. One argument sees the global and global media such as the Internet and ICTs part of a broader 'mobility turn' (Urry 2000a, 2000b) in which cosmopolitanism, transnationalism and 'flows' of networks have prevailed. Mobility, *vis-à-vis* globalisation, can take the form of the movement of objects, people and capital (Parreñas 2001). Yet for others, this mobility is undermined by residual immobilities that are part of what it means to practice a sense of the local. As Turner observes, the contemporary milieu characterised by mobility is creating more borders and 'us vs. them' feelings in what he calls 'enclave societies' (2007). Through the lens of mobility and immobility we see localised forms of social capital, community and a sense of cultural context. In short, various types of mobility inform contemporary forms of what it means to be intimate and what it means to be public.

There are various lens for examining the multiple dimensions and implications of what its means to be mobile today. Through the role of location-based gaming in different contexts, we can gain a sense of contemporary localities. In each location, mobile gaming connotes different associations – from hybrid reality and location-based games to casual games played on mobile phones. In the rise of the ubiquitous rubric of mobile media, many forms of practices have emerged: micro movies (that is, movies made for the mobile devices), pocket films (movies made by the mobile device to be screened either on the mobile device or other screens including the cinema), casual games, location-based games, *keitai shôsetsu* and camera phone practices. The socio-cultural dimensions of mobile media were central to the earlier projects of mobile media as new media (de Souza e Silva 2006a; Davis 2005).

In the interdisciplinary rise of mobile media we must ask what role it can play in bringing new challenges to new media practice and theory. How can mobile media proffer insights into new media practice? How can mobile media's history as both a socio-cultural artefact and form of communication provide a lens onto emerging forms of new media politics in an age of global media? In order to address the potentiality of mobile media as new media, I will begin by discussing a case study of one example of mobile media as new media in Korea. I will briefly introduce the main new media organisation, Nabi, and then discuss the interdisciplinary mobile media groups, INP (Interactive and Practice). Art Center Nabi, Seoul's new media centre (funded by South Korean giant, SK Telecommunication), has been instrumental in establishing mobile media projects that attempt to question the possibilities and potentiality of mobile media (Chung 2003).

In particular, INP has been prescient in forging new ground for mobile media as new media in Korea, so much so that they have moved out from the neat new media context afforded by Nabi. In their recent 'mobile hacker' (2007) project called *Dotplay*, INP shifted the paradigm of mobile media as new media away from location aware games and towards a more risky project that not only sought to critique the hegemonic role of mobile technologies in South Korea, but also, its relationship to North Korea. Throughout the process of the project, one is continuously reminded that to contact a North Korean is illegal and yet such an event was increasingly possible as *Dotplay* hacked and hacked. I argue that through this type of hactivist project we can begin to reflect upon the art (and politics) to being immobile and mobile in an age of global UCC 'flows'.

The inertia of mobility

> If the wireless experience is basically a street culture thing, lived by youth expressing themselves and communicating by any means available, including changing language by merging visual and text messages, for example, should we – those who are in the art field – feel threatened or enlightened? Maybe what we are seeing is the beginning of a new epoch in which the conventional meanings of the terms 'artist' and 'audience' are losing significance, not in a theoretical sense, but based on real situations in an everyday

context. The potential for wireless creativity and 'art' being a critical and creative engagement with the intimate and the everyday context is here today.

(Chung 2003: n.p.)

As Eunhye Grace Chung (coordinator of the Nabi's Resfest's Wireless Art Competition) notes in her article on Korean wireless experience, the potentialities of mobile media to challenge conventional relationships between artist and audience, user and producer are endless. Over the last few years, with the shift from 2G to 3G, there has been much focus on the possibilities of mobile media. As a remediated (borrowing from other media traditions) and yet emerging medium that is now intertwined with the growth of Web 2.0 and social software, we are seeing UCC becoming increasingly determined by context.

Discourse on the possibilities for experimentation have seen many artists and theorists orientate themselves around the role of mobile media as more than a miniature and mobile version of the conventional gallery space. Thephone-book Limited have explored the emergent genres of SMS, MMS and ring tones to highlight the conventions and codes (compression, immediacy, intimacy) of these remediated and vernacular-driven discourses – for example, SMS poems being poems restricted to the formats of SMS compression (i.e. 160 characters). In Proboscis' *Urban Tapestries* project, a section of London is navigated and reorientated through mobile location devices, making one recognise that mobile media help reinforce place rather than destroy it.

Here, we are confronted with this century's *flâneur* in the form of what Luke (2006) calls the 'phoneur'; a postmodern *flâneur* who strolls with mobile phone in hand 'whilst stalked by corporate hunters'. As Luke notes, the lines between creativity and corporation have blurred in the global mixing pot of the phoneur phenomenon. And yet, the socio-cultural nuances of locality seek to undermine any homogenising of context. South Korea, as the most broadbanded country in the world, is a key demonstration of this by the fact that government and corporations are pouring money into the digital, yet users are still very mindful of the importance of contact over connection.

In the Resfest's Wireless Art Competition, Nabi sought to get various international new media artists to make work for mobiles which resulted in little more than screen savers due to the current generation of phones at that time (2G). In 2005, Nabi had a collaborative group INP – consisting of artists, engineers and media theorists – working to produce various mobile media projects such as *Urban vibe* in October 2005. In 2006, Nabi conducted a 'mobile Asia' competition to acquire mobile media (content made by or for the mobile) and pervasive projects. Whilst pervasive (location-aware) projects are invaluable in geo-caching (such as GPS – geographic positioning systems) and in demonstrating the importance of place and specificity in a period of global technologies, they also serve to highlight one of the greatest residual paradoxes of mobile media as a metaphor for socio-technologies.

What became significant about INP's various projects around mobile media – from location-based projects conducted by INP to mobile movie making – has been the re-occurrence of the haptic/cerebral, online/offline and delay/immediacy paradoxes. The experience of the virtual is always at play in the offline space of the haptic; far from them becoming seamless, the politics of immediacy is always the poetics of delay. It is INP's projects outside of the 'new media' confines of Nabi, as in *Dotplay*, that we begin to see a type of new media practice that incorporates the 'everyday media' of mobile media along with hactivists characteristics. It is in a location such as Seoul, a city synonymous with innovative mobile and broadband technologies, that the politics of mobile media as new media activism takes on new dimensions.

Through the role of mobile hactivism we may begin to reflect upon the politics of increasing full-time intimacy whereby the role of social and creative labour can come into question. In an age where work and leisure practices are being blurred by the 'wireless leash', how can we reflect upon the value and capital associated with endless expenditure of socio-creative labour on behalf of the user? Surrounding the neo-liberal rhetoric enveloping new mobile technologies and Web 2.0 UCC, another picture is painted – are users being renumerated (paid for their content)? Can these necessities of everyday life, become 'soft' capital for users to express their dissatisfaction and criticisms? In order to contemplate some of these notions I will turn to contextualising the socio-technological scape of Seoul and then discuss the mobile hactivist project, *Dotplay*.

The big bang: contextualising the socio-technological space of Seoul

As noted in Chapter 5, Korea's rise to the most broadbanded country in the world (OECD 2006) – and easily one of the dominant examples of twenty-first-century post-modernity – has been relatively nascent. I will briefly recap on some of the key features of Seoul's socio-technological scape. Having experienced a tumultuous twentieth century of Japanese rule and the UN imposed thirty-eighth parallel, to then being bailed out by the IMF after the Asian economic crash of 1997, South Korea has grown to be the ninth largest GDP in the world and the third strongest economy in the region. Part of its rapid rise has been thanks to the governmental and industry focus on technologies. The IT policies of Korea have been noted as the best in the world (West 2006) and the IT industry in Korea represents over 15 per cent of GDP (MIC 2006).

In Seoul one can find two types of youth sociality predicted around technologies – that of the mobile phone being used to contact friends and family (Yoon 2003) and the Internet through online communities such as mini-hompys (Hjorth & Kim 2005) and online multiplayer games (Chee 2005) mostly played in the very social space of *PC bangs* (PC rooms). The use of technologies in Korea is a key example of socio-technologies – that is, the contemporary urban culture is permutated by technologies that are social in nature. As Yoon's ethnographic study of young people's use of mobile phones (hand phones) noted, the mobile phone helps to reinforce physical contact and exchange (Yoon 2003).

In my aforementioned ethnographic study with Heewon Kim (2005) on youth using Cyworld's mini-hompy community, it was found that virtual connecting was always about the need and desires to be connected on various levels and never about substituting for the real thing (actual contact); thus the co-presence between virtual and actual was inevitably about corporeal relations and connections (Hjorth & Kim 2005). In Chee's ethnography on *PC bangs*, these spaces are social spaces that are viewed as 'third spaces' between home and work (2005). For youth of Korea – where most still live at home before getting married – these third spaces operate as spaces to connect with other like-minded people. In these examples, we see that the underlying logic of technological uptake is about the practice of socio-technologies.

Unlike the console-driven industry of Japan, Korea has been focused on online multiplayer games accessed from PCs mostly in PC *bangs*. These PC *bangs* are social spaces, participating in simultaneously online and offline 'communities of presence' (Ito, 2002) and co-presence. These PC *bangs* offer more than just a place for online gaming. For Chee (2005) Korean *PC bangs* operate as 'third places' – places in between work and home spaces that offer psychological comfort and support. Drawing on Sutton-Smith's (1997) *The Ambiguity of Play*, Chee argues that social play needs to be understood in terms of 'play culture'.

As Chee notes in her ethnographic study, the *PC bangs* not only offer a place to play games but also a space for comfort and sociality. Gaming culture has become a meaningful part of everyday Korean life. Pro league gamers (professional gamers) are treated like celebrities; while pro-gamer tournaments are social events akin to boy-band concerts full of screaming youthful energy. With many young Koreans still living at home (usually until one is married), *PC bangs* offer a social space away from the watchful eyes of parents. As Huhh (2008) observes, the *PC bang* ensures the success of online games in Korea by nurturing both the culture and the business side of the industry; thus the online game is seen as synonymous with the *PC bang*.

The symbolic dimensions of mobile technologies as integral to the projection of Korea as a twenty-first-century centre for post-modernity cannot be understated. As mentioned in Chapter 5 and Chapter 8, Hong (2007) notes that through the vessel of the mobile phone, the naturalisation of South Korea as a 'global leader' has been all pervasive. Since the 1997 IMF financial bail-out, South Korea has constructed a well-orchestrated production on various levels from the techno-national to the personal, to locate the mobile phone – and specifically Samsung's Anycall – as symbolic of South Korea's twenty-first-century post-modernity. As Hong identifies, the symbolic dimensions of mobile technology in Korea cannot be underestimated, a scenario

Figure 10.1 Dotplay workshop hactivists at work.

that is undoubtedly unravelled in the hactivist practices of South Korea's *Dotplay*.

Mobile play: *Dotplay* and the politics of being mobile

As a ubiquitous part of everyday life, mobile technologies can provide much insight into local customs and rituals. Projects such as location-based mobile gaming have the ability to challenge the normalisations around a sense of place and space (De Souza a Silva 2006a). However, the power of mobile media to provide a vehicle for critiquing conventions, particularly within new media, is still relatively under-explored. This is, in part, because mobile media is so integral to everyday life, it makes it a difficult media to separate and create a critical space; rather, much of the way in which mobile media as new media gets discussed is via the once all-pervasive art trend of Bourriaud's 'relational aesthetics' (2002).

However, the distinctive characteristics of mobile media – what Ito et al. call the 'personal, portable and pedestrian' (2005) – undoubtedly serve not only to comment on the increased exploitation of full-time intimacy and socio-creative labour practices, but also upon the arising techno-nationalist neo-liberal policies underscoring much of the UCC rhetoric. As Palmer persuasively argues in his paper, 'Mobile art', contemporary media culture can be broadly defined as 'participatory'; in particular through the dominant 'modes of address' that 'function to blur the line between the production and consumption of imagery' (2005: 4). He notes, 'that all forms of media participation need to be considered in relation to defining characteristics of contemporary capitalism – namely its user-focused, customized and individuated orientation' (ibid.: 4).

In INP's 2005 location-based project, *Urban Vibe*, the deployment of the project back into the new media space of Nabi did little to challenge norms about new media. One of the frustrating elements of mobile media, when contextualised as 'new media', is the way in which it is separate from the everyday socialising context that makes mobile media so exciting. This, to me, seems to defeat the point. Why aren't artists thinking of creative ways in which to disseminate and interact beyond the safe context of the arts institution? How can locative projects make users, actors and the general public reflect upon the localised and political nature of play? And when the art is taken outside the context, as was the case in the *Urban Vibe* project, why is it then documented and exhibited in the gallery space? Why does legitimation continue to plague and thus limit the possibilities for playful forms of contextuality?

This is where the significance of mobile 'hactivism' lies, as demonstrated by INP's recent confirmation as mobile 'hactivists', *Dotplay*. Conducted in December 2007 (www.dotplay.org), the series of workshops fused pedagogy with politics, re-energising the possibilities of play in challenging the role of technology. According to *Dotplay's* manifesto-sounding outline of the organisation:

- Dotplay is a network of local and international media artists, engineers, cultural researchers, and participants.
- Dotplay maintains critical perspectives on media technology, and delivers creative alternative to the mobile culture;
- Dotplay explores and challenges various forms of happenings, participatory workshops, open source manuals, exhibitions, performances, and online publications;
- Dotplay prefers process-oriented creation, rather than final outcome;
- Dotplay welcomes sponsorship and collaboration with art institutions and corporations.

This manifesto is followed by a list of what *Dotplay* isn't:

- Dotplay is not a research center on media technology, nor mobile contents design firm;
- Dotplay does not work for art institutions or corporations;
- Dotplay deliberately displaces utopian fantasies based on technology;
- Dotplay disagrees with media art that just uses technology or imitates other art.

From the outset, *Dotplay* is undoubtedly political. The idea of hacking mobile technologies, in a country in which mobile technologies have figured greatly in techno-nationalist agendas and to which the symbol of the mobile phone is a motif for national post-modernity, is deeply political. The rhetoric of *Dotplay*'s quasi-manifesto is reminiscent of avant-garde tactics – a hybrid between dadaism, surrealism and the SI. The appropriation of the notion of organisation not only parodies the heavily corporate world of mobile technologies but also is a nod to such pivotal new media artists as Young Hae Chang Heavy industries.

Dotplay as the birth child of INP emerged in December 2006 when INP became interested in developing their autonomy apart from Nabi; the receipt of a grant from Arts Council Korea allowed them to begin with a 'mobile hacking workshop'. By April 2007, a mobile technology workshop was held under the title Understanding Mobile Hardware (conducted by Sungmin Huh from Mobion Inc); and by May the group had renamed itself *Dotplay*. By July, *Dotplay* was participating in symposia such as *Dislocate 07* (an *International Symposium on Art, Technology, and Locality*) and *DAUM YouthVoice Media Conference*. At both the *DAUM* conference and later in August, *Dotplay* expanded its repertoire via various mobile hacking workshops. As both a symbol and a technology, the mobile phone comes under scrutiny in the ever-growing group of *Dotplay*.

In short, *Dotplay* is a network of media artists, technologists and cultural researchers. *Dotplay* is part of 'Dotplay Telecom' – a virtual (fictional) mobile telecommunication organisation consisting of Yangachi, Jaekyung Shim, Miyoun Kim, Soni Park, Saye Min, Hyeri Rhee, and Taeyoon Choi – all of whom provides real free services. According to the website, '*Dotplay* Workshop provides access to hack mobile phones – physically and conceptually – to intervene into mobile device and environment in a physical, social, cultural, political manner in order to creatively re-intervene the mobile.'

Dotplay is a great example of teasing out one of the central tenors of new media debates – what role technology should play in constructions of society, culture and art. As one of the most intimate and personal technologies, it is often hard to construct a critical space in which people can reflect upon the role of mobile technologies in everyday life. It is often easier to return to older, remediated technologies such as the landline in order to comment and transcend the hypnotising role new technologies can occupy. As a group dedicated to the analysis, critique and play with mobile media, INP – now *Dotplay* – has held a series of Mobile Hacking workshops that seek to free the 'end-user' from their bottom position in the circuit of global technology companies.

By December 2007, *Dotplay* had managed to have a list of workshops and growing participants in a new form of UCC not encouraged by the multinationals. Arguably, a new technology becomes fully democratised when it creates its first breed of 'hactivists' that seek to challenge corporatism – for example, new media group e-toy's clash with and destruction of mega company E-toy's website business. More recently, the hacking of celebrity 'mobile-in-every-photo' Paris Hilton's mobile phone – and the dissemination of her personal addresses and details – sent a message that the ephemeral and precarious nature of the intimate, convenient and ubiquitous mobile phone could just as easily be transformed into a weapon in the wrong hands.

While much of the rhetoric around UCC and mobile convergent technologies seems to suggest media for 'the people', the reality of this 'democratic' media – comparable to the webcam revolution – is notably different. Examples of 'people power' are repeated constantly – from the camera journalists of the London bombings to the texting political revolutions of Korea (Kim 2003) and the Philippines (Rafael 2003) – and yet the realities of everyday life and increasingly precarious labour, along with trends towards full-time intimacy, provide little space for mobile media literacy and political action. In the case of South Korea, the hacking of mobile phones – and particularly holding workshops to teach budding 'hactivists' – sends a message that is not just symbolic. The hacking of software could result in accidentally calling a North Korean number – inadvertently facilitating a practice that is not only illegal but could also result in bringing the caller before a court martial. Ironically the very technology that symbolises South Korea's large embrace into democracy and post-modernity, the mobile phone, can become the weapon that bridges the two worlds of North and South, rather than representing the economic and technological distance between them.

The 'hactivism' on the level of hardware provides some extraordinarily beautiful results, highlighting the interesting cultural and material dimensions of technologies. This reminds us that in the everyday – symbolised by the ubiquitous mobile phone – we can find the 'terrible sublime', the meeting of the trivial and the sublime, familiar and yet unfamiliar. From transformer like toys to sculptures reminiscent of Brancusci, the work of *Dotplay* demonstrates that there is much more to mobile media as new media than is currently being deployed. For *Dotplay*, the 'end user' is far from the user–agency fantasy of a 'prosumer' or 'produser' (Bruns & Jacobs 2006). Where telecommunication companies still exert a trickle-down model of industry, it is important for *Dotplay* to reinstate a bubble-up, grass-roots model for mobile media as new media.

Dotplay aims to provide an 'ideal and imaginative' telecom service in which priority is given to engineering a 'politically correct and culturally free service' rather than 'traditional emphasis on economic efficiency (investment wise)'. For *Dotplay*, it is important to recognise, deviate and revolt from the capitalist and 'technocentric fantasy' by defining new infrastructures, different kinds of services for different kinds of users. Working towards an official opening and beta service by 2009, *Dotplay's* group is growing as a creative commons for mobile media. The ideas generated in the workshop will later be the business strategy of *Dotplay* telecom.

Dotplay telecom grew out of *Dotplay's* struggle to use the mobile network to create mobile art (such as access to free network and black box devices), and to abuse the network. It is a virtual platform for democraticising the mobile network, so that individuals can decide their own method of

Figure 10.2 Dotplay's (hardware and software) art of mobile hacking.

communication and create art works. *Dotplay* re-connects new media to the socio-political dimensions that made such an important discourse in the first place; reminding us that technology is as much cultural, social and thus political as it is a functional tool. *Dotplay* demonstrates that far from the mobile setting us free, it has further entrapped us into various erosions between work and leisure. But, maybe, through setting the mobile technology free, we can, in turn, mobilise new media. That is, unless you are left hanging on the phone, waiting.

Mobile but not free: conclusion

Mobile media have, as the most ubiquitous and pervasive technologies in everyday life, a formidable capacity to comment upon both the social and creative dimensions of contemporary practice. Far from the mobile setting us free it highlights various limitations and invisible leashes (Arnold 2003). The mobile phone communicates on various levels – both literally in the form of visual, textual and aural mobile media as well as symbolically as a cultural artefact reflecting the user's identity and social and cultural capital. Mobile media can be read as symbolic of global technologies and the growing importance of the level of the local. This makes it a plentiful object for rethinking new media discourses among other things.

Through mobile media projects such as *Dotplay*, we can further explore the dimensions of new media as well as investigating the overlap between social and creative labour practices so prevalent in contemporary life. Mobile media as new media can help to provide a space in which to critique the paradoxical dimensions of socio-technologies of everyday life. For example, in an age of so-called immediate technologies, such projects enlighten us to the conundrum of instantaneity, that is, the inevitable poetics of delay. They highlight the price of mobility and its oscillation between freedom and leash (Arnold 2003) in which work and leisure boundaries are increasingly blurred (Wajcman et al. 2008; Gregg 2007). We are left to contemplate that perhaps the object closest to you can expose even more conundrums and paradoxes within everyday life that not only reflect upon redefining new media but, also, new forms of creativity, labour and soft wars mobility and immobility. These are but a few examples of mobile media as new media in an age marked by both mobility and 'the art of being mobile'.

11 On hold

Reflections on mobile media in the Asia-Pacific

There is a secret bond between slowness and memory, between speed and forgetting. Consider this utterly commonplace situation: a man is walking down the street. At a certain moment, he tries to recall something, but the recollection escapes him. Automatically, he slows down. Meanwhile, a person who wants to forget a disagreeable incident he has just lived through starts unconsciously to speed up his pace, as if he were trying to distance himself from a thing still too close to him in time.

(Kundera 1995: 34, cited in Cho 2000: 67)

In an age marked by 'Mobilities, immobilities and moorings' (Hannam et al. 2006), the motif of mobile media provides a poignant symbol for the various temporal and geo-spatial transformations occurring at the onset of twenty-first-century post-modernity. However, far from the promises of ICTs as vehicles for twenty-first-century globalisation – to overcome distance and difference through a time–space compression whereby immediacy becomes the constant – intricate forms of localised resistance and delay continue unabated.

Indeed, contemporary everyday life seems dominated by new forms of exercise and martial art gestures ... in the form of waiting ... *Waiting for immediacy*. These poses and gestures are part of the emerging practice of mobile media practice. In the flux of contemporary mobility and immobility, the space of contemplation has been replaced by the increasing need to document – to an audience in the theatre of life. However, the performance is not just in the images we take, share and send to each other, rather, it is the haptic space in which we perform co-presence. The moves we make, the decisions we rapidly deliver – this is the dance of *waiting for immediacy*. Clicking, eating the world as junk food, clicking, deleting, clicking, saving and sending, clicking, memory full. Download. Click. Waiting.

Brimming with immediate 'saved' and 'delete' moments, the everyday has arguably become a series of sequences rather than moments. Snap, pause, delete, snap, save and share have become the cultural repertoire for youth cultures. We quickly snap at moments, voraciously consuming the fragments of meaning in a rapidly accelerating life. But is it so pessimistic (or so simple)?

Is it all about disjunctures of co-presence whereby work and life struggle in an acrimonious embrace? In this rush, what are we saving and what are we deleting? Or is there something, slightly *outside the frame*, that allows us space to wait and reflect upon the tyranny of immediacy? This is the dance of *waiting . . . for immediacy*.

From the delayed and deferring practice of SMS in Melbourne that engages with a sense of postal presence, to the role of camera phone images to reterritorialise and rescript a sense of place and memory in Hong Kong, to the feminised customisation of male partners' phones as a revision of the *kukmin kajok* in Korea and the emergence of *keitai shôsetsu* in Japan (paralleling the nineteenth-century rise of female writers in the West) mobile media practices are presenting new avenues for expression, creativity and labour. As much as they partake in the new, mobile media practices are also remediating and rehearsing earlier modes of expression. In the two case studies of this section, I have considered mobile media within the context of new media practice. I have argued that mobile media can teach new media about the complex localised socio-cultural dimensions of technologies.

In Chapter 9, I discussed the parallels and intersections between domestic technologies approaches and new media models. Through the case study of context-aware mobile gaming that deploys both the virtual and actual spaces, we can see the limitations and need for revision of both approaches in order to chart the temporal and geo-spatial paradoxes of mobile media. These processes are ordered by two dominant tropes – 'waiting for immediacy' and the 'poetics of delay'. In Chapter 10, I explored the case study of Dotplay, which, as a mock Korean telecommunication corporation, integrates the symbolic and material dimensions of mobile media as new media. Drawing on the genealogy of Korea's pivotal new media artists such as Nam June Paik and Young Hae Chang Heavy Industries, Dotplay utilises mobile media in order to question not only contemporary Korean technocultures but also the ways in which mobile media as an art form has many possibilities.

I began this concluding chapter, which is more of a reflection on the journey of gendered mobile media in the region, with a quote from Milan Kundera. In *On Slowness*, Kundera proffers multiple counter narratives of slowness in the face of the need for speed presented by globalisation; as we have witnessed, those with enough economic and cultural capital are choosing to opt out of the fast-food life and instead become part of the slow food (and life) movement. However, for many, the luxury of slowness as a constant isn't a reality. Slowness is a new form of capital that only a few can afford. But slowness can be deployed in the everyday realm, in which people insert a moment of delay and slowness against the growing proclivity towards immediacy. Hence the need for the creation of minuscule everyday practices that envelope slowness, *the poetics of delay*.

Interestingly, this Kundera quote was cited by the aforementioned Korean scholar Cho (2000), in her discussion of Korean post-modernity and the role of male–family–nation conflation encapsulated by *kukmin kajok*.

Paradoxically, in an age that seems to embody the fast-walking man sketched by Kundera (although a contemporary cipher would be talking, texting or 'filling-in time' on the mobile), there are multiple forms of walking slowly. Through the rise of cartographies of personalisation, we can see two distinctive directions for users. One is the direction signposted by sovereign consumer rhetoric, in which the user-as-produser seems an empowered, and yet misunderstood, phenomenon. Here the rise of mobile media is viewed as part of the increasingly convergent media (and media conglomerates), a democratisation of media whereby those once unable to gain access to such resources and capital are granted entry and agency. Fears about the demise of the amateur artist challenging the position of the professional begin to surface, as do discussions about the on-going indivisibility between work and leisure within the increasingly all-pervasive role of social labour. In the other direction, we see cartographies of personalisation – evoked through the various imaging community mobile media practices – both liberate and exploit the language of social labour and grammars of intimacy. Mobile media becomes the 'wireless leash' (Qiu 2007) that further serves to exploit social labour, or what Ling (1999) defines as 'hyperfeminine' practices.

It is indeed difficult to have a 'last word' on the dynamic set of cultures, societies and practices that define 'mobile media'. In his conclusion to *Cell Phone Culture*, Goggin (2006a) mulls over the possibilities, realities and challenges surrounding the nascent mobile media and its relationship to user agency, media politics and convergent media. As he notes, it is important for mobile media to reconcile its relationship with both old and new media and its place within 'a broad knowledge of communications, cultural and media history and theory' (ibid.: 211). He argues that 'the cell phone's metamorphosis into media par excellence' involves 'not only to follow, however, but also to participate actively and knowingly in the opening up and shaping of such media and technology' (ibid.: 211).

Undeniably, in the face of much of the futurist hype about mobile media and, particularly the way in which geo-political imaginary of the region as both representing potential mobile futures as well as becoming this century's new 'Global South', the dynamic formation of mobile media is haunted as much by the past as it is by spectral projections about the future. The role of mobile media – at both material and symbolic levels – to reflect salient and emerging forms of subjectivity and intimacy as well as national and transnational communities makes it a compelling emblem for examining the region. Mobile media both liberates and reinforces traditional modes of localised gender performativity; it affords new forms of access to multimedia, distribution and sharing, as it does provide some spaces and modes for expressing different forms of subjectivity and storytelling.

We can learn a lot about twenty-first-century forms of cultural practice through the nascent but remediated mobile media. We can reflect upon new forms of creativity and labour, in turn, reconsidering new dynamic paradigms for author and audience relationships in which previously demarcated spaces become blurred. According to Goggin, some of the crucial issues surrounding mobile media and media politics include 'what cultural forms will survive and be supported . . . how audiences will be imagined and served . . . how the new publics will be understood . . . in whose interest will policies be arbitrated . . . and, crucially, who will be involved in these arenas and decisions that will shape mobile media' (2006a: 209). In conclusion, Goggin presents the rubric of mobiles as media (ibid.).

As an integral part of everyday life, mobile media has indeed become the emblem for post-modernity. It epitomises both the trivial (banal) and sublime – echoing the arguments (particularly media effects approaches) that previous domestic technologies have attracted. For example, consider the role that television has occupied throughout the second part of the twentieth century as a symbol of popular culture and post-modernism (Collins 1989). Consider how much the television – like mobile phones today – functioned both at a material and symbolic level in contemporary culture, and how much of the arguments around dependency and mediation issues are similar. But unlike television's symbolic mobility, mobile media partakes in actual material mobility that invariably shapes the place of mobile media not as a 'third' screen, but, rather, a haptic third space.

Slow spectres of immediacy

Last year a friend of mine went to Ars Electronic. He informed me that there were some mobile media works there – but none *seemed* to be working. This scenario of 'new media' being denoted by abject technologies that fail to behave in professional situations – is a familiar one. But, in particular, the specific visual economies of mobile media seem to work against any version of formal aesthetics – the images are low resolution and miniaturised, the modes of storing are precarious and ephemeral, and the content is personal and immediate.

With the emergence of mobile gaming, we are reminded that play – like mobility – is informed by the socio-cultural. In a period of global ICTs, the time–space compression has not occurred; rather, localities have utilised mobile media to reinforce everyday practices that are marked by specific temporalities and geo-spatial narratives. For anyone who has participated in a mobile context-aware game in which an 'immediate response' is savoured, delayed and even feigned, they would be mindful that the more we try to partake in the *politics of immediacy*, the more we succumb to the *poetics of delay*.

As new media practitioners try to grapple with the anomaly that is mobile media art, it seems that everyday users are left with another set of incongruities. As Goggin notes, within the 'assertive noisiness of individual consumers . . . it remains unclear how multitudes of users will come together to steer change in their own interests.' (2006a: 209). In other words, how will the emerging 'imaging communities' take on power and agency at national and transnational levels?

Let us return to the example of *bus uncle* video in Hong Kong, in which a mobile movie ('pocket film') records an everyday situation, such that the phone becomes part of a bigger disjuncture between different generations and understandings of modernity. In this video, we see the debate around mobile media actually being transformed into a debate about conflicting notions of modernity and attendant value systems. Earlier domestic technologies such as television have engendered much debate and heated discussion; Jim Collins argues that this is because television is both the material and symbolic icon for postmodernism and popular culture (1989).

Now, it seems that the mobile phone has taken a leading role as the harbinger for these debates about post-modernity, particularly, as I have shown, in the region. This makes mobile media, like its predecessor television, a symbol for some of the most familiar and yet unfamiliar, sublime and yet trivial moments. Indeed, as Koskinen noted (2007), one of the key aesthetics of mobile media that makes it so compelling – adding realism and authenticity – is its 'banality'. Koskinen's apt observations rehearse much of the earlier theorisation of the everyday as a source for paired dualities such as the banal and sublime. Mobile media also provides three tropes: shifts in networks and modes for identification, mobile-specific modes of digital storytelling and virtuality and the emergence of new political intensities. These tropes could be broadly defined under the rubric of cartographies of personalisation and the attendant imaging communities that, in turn, circulate into localised and transnational imagined communities.

As highlighted in my two case studies in Chapter 9 and 10, the immediacy of mobile media is engaged in the poetics of delay, which is also marked by the geo-spatial; thus far from 'being mobile' disintegrating a sense of place it operates to reterritorialise it (Koskinen 2007). This is the inertia of mobility that will undoubtedly inform the emerging forms of creativity, expression and labour – imaging communities – in the region. Behind much of the smoke and mirrors of UCC are stories of emerging subjectivities and communities that could reconnect us to questions about the various attendant paradoxes – the delayed and immediate, mobile and immobile, banal and sublime.

Mobile Media in the Asia-Pacific has attempted to engage in the ways gendered mobile media can provide us with cartographies to conceptualise and understand the various dynamic changes occurring in the region. Through the rubric of 'cultures', 'societies' and 'practices', I have endeavoured to intersect and highlight some of the ways we can rethink both micro and macro systems and rituals – from the localised and emerging forms of gender performativity, creativity, labour and lifestyle patterns to the ways in which they reflect transpiring localised, national and transnational processes of the region. Over the next decade, it will be fascinating to see how these gendered mobile media practices play out in the symbolic economy of femininity and feminisation – as well as within the material conditions of women

– in the region. Continuing longitudinal studies in the region is pivotal if we are to understand the dynamic symbolic and material dimensions of mobile media as a recomposition of old and new technologies and socio-cultural forms.

Rewind, fast-forward, replay

The role occupied by mobile media impacts upon how we conceptualise convergence, new media and multimedia in the twenty-first century. As a set of cultural practices, mobile media's various forms of visual, aural, textual and haptic modes – along with its recombinant rise with Web 2.0 and the attendant new types of distribution and context – are providing emerging forms of self-expression, representation and identity. In this way mobile media is not only a material and symbolic product of the Asia-Pacific's rising NICs but, also, actively part of documenting the new forms of storytelling, agencies and subjectivities as they emerge into twenty-first-century post-modernity.

Let us return once more to the opening Kundera quote. We could argue that the region's rise in the late twentieth century has been akin to the fast-walking man. Like the man, the region didn't want to continuously remember the tumultuous wars, violence and conflict; instead it focused upon transforming western forms of capitalism into one of the most exemplary models of post-industrialism globally. As the region unevenly gained technological and economic power, the multitudes of localities adapted this currency into cultural 'soft power' capital – all-the-while, like the fast-walking man, incrementally increasing the pace, and leaving the spectres of the twentieth century behind. He, representative of post-industrialism in the Asia-Pacific, ran into the twenty-first century while most of the world – strolling like a *flâneur* – remained in the previous century.

However, now, we can see that it is no longer a fast-paced man; instead the motif has become a female *phoneur*. She strolls between online and offline spaces, partaking in cartographies of personalisation that actively engage and reflect on the stories of the everyday and the intimate. She does not run, instead, she pauses and greets the immediacy of ICTs with her own poetics of delay. She savours the SMS before sending what appears to be an immediate response. She writes mobile novels in keeping with nineteenth-century epistolary traditions. She shares camera phone images on her mini-hompy, images that rehearse earlier analogue traditions through familiar genres and the construction of familial relations. Her circle shares images in a practice of imaging communities. Sometimes she just keeps the images for her own personal consumption, reminiscing and reflecting in rooms of nostalgia.

These are the micro-narratives of walking slowly – the poetics of delay – in an age dominated by rhetoric about immediacy. If mobile media practices such as location-aware gaming are any reflection, the future of mobile media is undoubtedly its ability to rehearse, re-enact and visit the haunting of previous geo-spatial, temporal and media politics and, at the same time, providing a space for new types of gendered storytelling, whether in the form of hyperfemininity *vis-à-vis* the ubiquity of social labour or via the actual micropractices of female users such as *keitai shôsetsu*, SMS and camera phone imagery. Amongst the full-time intimacy of the 'wireless leash' (Qiu 2007) that are indicative of the increasing exploitation of social labour and other hyperfeminine practices, we can see some practices of resistance that challenge the relationship between author and audience, industry and creative producer, as well as contesting stereotypes around gendered agency and media politics. These are the emerging stories of *Mobile Media in the Asia-Pacific*.

Throughout this book I have attempted to provide details of some of the emerging cartographies of personalisation that are challenging inherited notions of gendered, and specifically female, use of new technologies. These are the new imaging communities that are part of the unofficial discourse of the various imagined communities that constitute the Asia-Pacific which inevitably feed into official 'national' and transnational ideologies. This phenomenon not only speaks about women's changing socio-economic role in the region but, also, about the region's emerging visual and discursive economies of twenty-first century postmodernity. These imaging and imagined communities also reflect and contribute to the changing transnational dynamics in the region where the notion that there is only one centre for postmodernity has been completely disrupted.

In this last section, *Mobile media practices*, I have deviated into intersections between mobile media as new media in order to reflect more broadly on the 'art of being mobile'. As a relatively nascent area within mobile media – in comparison to much of the literature that has focused upon mobile media through sociological, anthropological or media and communication lenses – it is important to see some of the questions and challenges of mobile media through the lens of new media. New media can provide insight into current debates about mobile media authorship, 'produser' agency and emerging forms of creativity and expression that reflect the tenor and politics of the region. Mobile media as new media can also present possibilities for reimaging mobility and immobility, walking fast or slow, remembering and forgetting, immediacy and delay that typifies contemporary mobile media. This is but the beginning of the art of being mobile.

Notes

Introduction

1   Along with examples of global innovation and high penetration rates, the region also demonstrates the way in which the mobile phone can become a channel for other forms of mobilisation – most notably collective actions of religious and political agency. For example, in the Philippines, the role of the SMS has become so pivotal in everyday life that over 300 million text messages per day are sent, so that the Philippines has been described as texting capital of the world (Pertierra 2006). In a country with millions of Catholics, the mobile phone has given rise to the phenomenon where people text God (Ellwood-Clayton 2003), as well as becoming a much-cited symbol of people power in the demise of President Joseph Estrada (Pertierra 2006; Rafael 2003). This democratic potential of the mobile phone was also noted in the elections of President Roh in South Korea (Kim 2003), and the protests over the killing of young Korean females killed by US military personnel.

2   An important aspect to add to Butler's notion is the pivotal way in which gender is localised through a cultural context. With the rise of postcolonial feminists such as bell hooks, Trinh T. Min-ha, Rey Chow and Sarah Ahmed, the Western inflection underscoring much of the earlier work in gender studies is now challenged by new regional gender studies that witnessed debates about essentialism (McRobbie 1996).

3   In the region's shift from economic to ideological power, gender, as Khoo notes, figures prominently. Significantly the role of gender – both as a localised form of identity and as symbolic of tropes of femininity and masculinity – is embedded with the practice of mobile phone customisation. In sum, the region's multiple mobilities, signified by the practice and cultural index of the mobile phone, is also unmistakably a gender issue. Practices of co-presence, intimacy and identity are part of these gendered formations.

4   As an interdisciplinary framework, the domestic technologies approach (Silverstone & Haddon 1996; Haddon 1997; Miller 1987, 1988; Ling 2004) draws from anthropology (Douglas & Isherwood 1980), cultural studies (Hebdige 1988) and consumption studies (Miller 1987). A significant part of its lineage lies in anthropology and its commitment to analysing the processes of material cultures in everyday life. For example, in the earlier work of Miller (1987, 1988) into material cultures, we can find his perspicuous analysis emphasising the meaning of consumption practices in terms of appropriation and objectification.

5   In this study I was aware of the power relations implicit in the dependence upon the English language as a dominant mode for communication. Given this issue, I gave respondents the choice to communicate in their first language or English. Almost all respondents chose English, and this choice was no doubt influenced by

the respondents' cultural capital consisting of key factors such as tertiary education in which they were often trained as a matter of course in writing and speaking English.

6 For a comprehensive discussion of gender and feminism within Japan see Laura Miller (2006), Vera Mackie (1988), Sandra Buckley (Mackie & Buckley 1985) and Anne Allison (2000). In particular, Allison gives a wonderful contextualisation of gender and sexuality studies in the field of anthropology in *Permitted and Prohibited Desires: Mothers, Comics, and Censorship in Japan* (ibid.: 1–29).

4 Fast-forwarding to the present

1 These statistics, recorded in December 2007, can be found at: www.ja.wikipedia.org.
2 The first example of the mobile phone could be said to be Ericsson's phone in a car he built for his wife in 1919 (Agar 2003).
3 Zero is the honorific 'o'; 8 is 'hachi' or 'ha'; 4 is 'yon' or 'yo' and the final 0 makes it a yo long vowel, o-ha-yo-o.
4 See Chapter 7 for the failed i-mode adaptation in Australia.
5 The first BBS, 1 channel, was called amejo. It became unpopular after 1999 when it became a site for people displaying bad manners and using inappropriate language.
6 *Densha Otoko* is a popular example of the reconfiguration of Japanese male geeks (*otaku*) being viewed as 'good guys'. For a long time *otaku* had a negative connotation, particularly exemplified by the *hikikimori* syndrome in which predominantly young men locked themselves in their rooms for sometimes years. This phenomenon was seen as outcome of the demise of the *oyaji* as the national icon for Japan for many decades after World War II and before the 1997 financial crisis of the region. As the twenty-first century began, it was the otaku who became the new icon. This is highlighted by the fact that the *otaku* is only called 'train man' (*densha otoko*) and never given a personal name like the other characters such as Hermés.
7 Of the comic adaptations there are currently four versions: Hidenori Hara's *Densha Otoko – Net Hatsu, Kakueki Teisha no Love Story (English version – Train_Man: Densha Otoko)*; Mataru Watanabe's *Densha Otoko – Demo, Ore Tabidatsuyo*; Daisuke Dōke's *Densha Otoko – Ganbare Doku Otoko!*; and a *shōjo manga* by Machiko Ocha called *Densha Otoko – Bijo to Junjou Otaku Seinen no Net Hatsu Love Story* (English version – *Train_Man: A Shōjo Manga)*.
8 That is, the various modes of writing that express emotion and that are either deployed by females, *gyaru* or, due to their emotological nature, overtly feminised. See Katsuno & Yano (2002) for the discussion on these various forms of 'face marks'.
9 According to ja.wikipedia.org: 2,956 *keitai shōsetsu*; 3,074 *keitai* poems; 1,600 photos and 300 songs.

5 Engaging rings

1 On Internet researcher mail listing, AoIR (Association of Internet Researchers), there has been much discussion about the increasingly role SNS will play in the way that people glean information about others. Moreover, the fact that information about a user can still be traced on the Internet after they cancel their Facebook account, in an age where many go to the Internet as a source of information, suggests that the ephemeral nature of the digital is not necessarily applicable to the Internet.
2 The second longest was Japan with mixi, followed by Melbourne and Hong Kong.

7 Postal presence

1 I first began using this term to compare camera phone practices in Melbourne and South Korea in 'Snapshots' presented at *Cultural Space and the Public Sphere in Asia*, hosted by Asia's Futures Initiative, 15–16 March, Seoul, 2006. I then realised that the three Ds were particular to Australia's imagined community as played out by mobile media imaging communities. Minhee Son found that these three Ds could be reappropriated to discuss the role of ambivalence within camera phone practices in Cyworld mini-hompy in Seoul. For Son, however, the three Ds consisted of *delay, deference*, and *detachment* (Son 2007).

2 Like all of the sample groups, recruitment occurred in and around universities. This meant that the socio-economic capital of the group tended not to be too diverse. Future research in the area should extend beyond this limited demography. I intentionally kept the focus on university staff and students to keep this study consistent with my studies in the other locations. Limiting the respondents in this way was useful particularly in the non-English speaking locations, because tertiary-educated respondents felt more confident to speak in English and were better able to articulate their motivations and choices.

Bibliography

Abbas, A. (1997) *Hong Kong: Culture and the Politics of Disappearance*, Minneapolis: University of Minnesota.

Agar, J. (2003) *Constant Touch: a Global History of the Mobile Phone*, Cambridge: Icon Books.

Ahmed, S. (1999) 'Home and Away: Narrative of Migration and Estrangement', *International Journal of Cultures Studies*, 2: 329–247.

Allison, A. (2000) *Permitted and Prohibited Desires: Mothers, Comics, and Censorship in Japan*, Berkeley: University of California.

Allison, A. (2003) 'Portable monsters and commodity cuteness; *Pokémon* as Japan's new global power'. *Postcolonial Studies*, 6(3): 381–398.

Anderson, B. (1983) *Imagined Communities: Reflections on the Origin and Spread of Nationalism*, London: Verso.

Anderson, B. (1998) *The Spectre of Comparisons: Nationalism, Southeast Asia, and the World*, New York: Verso.

Appadurai, A. (1990) 'Disjuncture and difference in the global cultural economy' in M. Featherstone (ed.), *Global culture: nationalism, globalization and modernity*, Sage: London.

Appadurai, A. (1996) *Modernity at Large: Cultural Dimensions of Globalisation*, Minneapolis: University of Minnesota Press.

Appadurai, A. (2000) 'Grassroots globalisation and the research imagination', *Public Culture*, 12(1): 1–19.

Ariés, P. (1962) *Centuries of Childhood: a Social History of Family Life*, trans. R. Baldick, New York: Knopf.

Arnold, M. (2003) 'On the phenomenology of technology; the "Janus-faces" of mobile phones', *Information and Organization* 13: 231–256.

Arrighi, G. (1994) *The Long Twentieth Century: Money, Power, and the Origins of Our Times*, New York: Verso.

Arrighi, G. (1996) 'The rise of East Asia and the withering away of the interstate system', *Journal of World Systems Research*, 2(15), http://jwsr.ucr.edu/archive/vol2/v2_nf.php (assessed 5 July 2007).

Arrighi, G. (1998) 'Globalization and the rise of East Asia', *International Sociology*, 13(1): 59–77.

Arrighi, G. (2005) 'States, markets and capitalism, East and West', *Semináro Internacional Alernativas Globalização UNESCO*, Rio de Janeiro, 8–13 October, at http://bibliotecavirtual.clacso.org.ar/ar/libros/reggen/pp25.pdf (assessed June 30 2007).

Arrighi, G., Hamashita, T. and Selden, M. (eds) (2003) *The Resurgence of East Asia: 500, 150 and 50 Year Perspectives*, London: Routledge Asia's transformations series.

Attfield, J. (2000) *Wild Things*, London: Berg.

Australian Communications Authority (2003) http://www.acma.gov.au/WEB/LANDING/pc=TELECOMMUNICATIONS_MAIN (accessed 28 May 2008).

Bakardjieva, M. (2003) 'Virtual togetherness: an everyday-lifeperspective', *Media Culture Society*, 25, 291–313.

Bakardjieva, M. and Feenberg A. (2004) 'Virtual community: no "killer implication" ', *New Media & Society*, 6(1): 37–43.

Barraclough, G. (1967). *An Introduction to Contemporary History*, Harmondsworth: Penguin.

Bauman, Z. (1993) *Postmodern Ethics*, London: Polity Press.

Bauman, Z. (2000) *Liquid Modernity*, London: Polity Press.

Bauman, Z. (2003) *Liquid Love*, London: Polity Press.

Beaton J. & Wajcman J. (2004) 'The impact of the mobile telephone in Australia (a discussion paper)', Academy of the Social Sciences in Australia, Canberra.

Beatty, A. (2005) 'Emotions in the field: What are we talking about?', *Journal of the Royal Institute of Anthropology* 11: 17–37.

Beck, J. and Wade, M. (2003) *DoCoMo: Japan's Wireless Tsunami: How One Mobile Telecom Created a New Market and Became a Global Force*, New York: AMACOM.

Beck, U. and Beck-Gernsheim, E. (2002) *Individualization: Institutionalized Individualism and Its Social and Political Consequences*, London: Sage.

Befu, H. (2003) 'Globalization Theory from the Bottom Up: Japan's Contribution', *Japanese Studies*, 23(1): 1–22.

Bell, G. (2005) 'The age of the thumb: a cultural reading of mobile technologies from Asia', in P. Glotz and S. Bertschi (eds), *Thumb Culture: Social Trends and Mobile Phone Use*, Bielefeld: Transcript Verlag: 67–87.

Berg, S., Taylor, A.S. and Harper, R. (2005) 'Gift of the gab', in R. Harper, L. Palen and A. Taylor (eds), *The Inside Text: Social, Cultural and Design Perspectives on SMS*, Dordrecht: Springer, pp. 271–285.

Berlant, L. (1998) 'Intimacy: A special issue,' in Berlant, L. (ed.), *Intimacy*. Special issue of Critical Inquiry 24/2 (Winter): 281–288.

Berlant, L. (2000) (ed) *Intimacy*, Chicago: University of Chicago Press.

Berry, C., Martin, F. and Yue, A. (eds) (2003) *Mobile Cultures: New Media in Queer Asia*, Durham, NC: Duke University Press.

Bolter, J. and Grusin, R. (1999) *Remediation: Understanding New Media*, Cambridge, MA: MIT Press.

Bourdieu, P. (1984 [1979]) *Distinction: A Social Critique of the Judgment of Taste*, trans. R. Nice, Cambridge, MA: Harvard University Press.

Bourdieu, P. (1990) *Photography: A Middle-Brow Art*, London: Polity Press.

Bourriaud, N. (2002) *Relational Aesthetics*, trans. S. Pleasance and F. Woods. Dijon: Les Presses du Réel.

Boyd, J. (2005) 'The only gadget you'll ever need', *New Scientist*, 5 March: 28.

Brown, B., Green, N. and Harper, R. (eds) (2002) *Wireless World: Social, Cultural and Interactional Issues in Mobile Communications and Computing*, London: Springer-Verlag.

Bruns, A. and Jacobs, J. (eds) (2006) *Uses of Blogs*, New York: Peter Lang.

Butler, J. (1991) *Gender Trouble*, London: Routledge.

Cameron, D. (2005) 'Koreans cybertrip to a tailor-made world', *The Age*, 9 May.

Castells, M. (1996) *The Rise of the Network Society (The Information Age: Economy, Society and Culture, Volume 1)*. Malden, MA: Blackwell Publishers.

Castells, M. (2001) *The Internet Galaxy*, Oxford: Oxford University Press.

Castells, M., Fernandez-Ardevol, M., Qiu, J.L, and Sey, A. (2007) *Mobile Communication and Society: a Global Perspective*, Cambridge, MA: MIT Press.

Chan, D. (2006). 'Negotiating intra-Asian games networks: on cultural proximity, East Asian games design, and Chinese farmers', in *Fibreculture Journal*, 8, www.journalfibreculture.org (assessed 20 March 2007).

Chee, F. (2005) 'Understanding Korean experiences of online game hype, identity, and the menace of the "Wang-tta" ', presented at *DIGRA 2005 Conference: Changing Views – Worlds in Play*, Canada.

Chesher, C. (2004) 'Neither gaze nor glance, but glaze: relating to console game screens', *SCAN: Journal of Media Arts Culture*, 1(1), http://scan,net.au/scan/journal (assessed 10 February 2007).

Ching, L. (2000) 'Globalizing the regional, regionalizing the global: mass culture and Asianism in the age of late capital', *Public Culture* 12(1): 233–257.

Cho, H.J. (2000) ' "You are entrapped in an imaginary well": the formation of subjectivity within compressed development – A feminist critique of modernity and Korean culture', *Inter-Asia Cultural Studies* 1(1): 49–69.

Cho, H.J. (2004) 'Youth, Internet, and alternative public space', presented at *Urban Imaginaries: An Asia-Pacific Research Symposium*, Lingnan University, Hong Kong.

Cho, H.J. (2005) 'Reading the "Korean wave" as a global shift', *Korea Journal*, Winter 2005: 147–182.

Cho, H.J. (2007) 'Youth, the Internet and alternative public space', Yonsei University, September, Seoul, South Korea.

Choi, H.W. (2004) 'How do you make a "premium" cellphone?', *Wall Street Journal*, 15 April: B1.

Chow, R. (2000) 'Nostalgia of the New Wave: structure in Wong Kar-Wai's *Happy Together*', *Camera Obscurai*, 42: 31–47.

Chow, R. (2007) *Sentimental Fabulations, Contemporary Chinese Films*, New York: Columbia University Press.

Chua, B.H. (ed.) (2000) *Consumption in Asia*, London: Routledge.

Chua, B.H. (2006) 'East Asian pop culture: consumer communities and politics of the national', presented at *Cultural Space and the Public Sphere: An International Conference*, 15–16 March, Seoul, South Korea.

Chung, E.H.G. (2003), 'The Korean wireless experience – art or content?', in *Receiver* magazine, 9, www.receiver.vodafone.com/9/index.html.

Collins, J. (1989) *Uncommon Cultures: Popular Culture and Post-Modernism*, London: Routledge.

Colman, J. (1988) 'Social capital in the creation of human capital', *American Journal of Sociology*, 94: 95–120.

Cyworld mini-hompy factory, http://c2.cyworld.com/en/ (accessed 10 June 2006).

Davis, A. (2005), 'Mobilising Phone Art', in *Real Time*, May.

De Gournay, C. (2002) 'Pretence of intimacy in France', in J. Katz and M. Aakhus (eds), *Perpetual Contact*, Cambridge: Cambridge University Press, pp. 193–205.

Deleuze, G. and F. Guattari (1986) *A Thousand Plateaus*, Minnesota: University of Minnesota Press.

De Souza e Silva, A. (2004) 'Art by telephone: from static to mobile interfaces', *Leonardo Electronic Almanac*, 12.10, http://mitpress2.mit.edu/e-journals/LEA/TEXT/Vol_12/lea_v12_n10.txt (accessed 4 January 2006).

De Souza e Silva, A. (2006a) 'Interfaces of hybrid spaces', in A.P. Kavoori and N. Arceneaux (eds), *Cultural Dialectics and the Cell Phone*, New York: Peter Lang.

De Souza e Silva, A. (2006b) 'From cyber to hybrid: mobile technologies as interfaces of hybrid spaces', *Space & Culture*, 9 (3): 261–278.

Dirlik, A. (1992) 'The Asia-Pacific idea: reality and representation in the invention of a regional structure', *Journal of World History* 3:1, 55–79.

Dirlik, A. (ed.) (1993) *What is in a Rim? Critical Perspectives on the Pacific Region Idea*, Boulder: Westview Press.

Dirlik, A. (1995) 'Confucius in the Borderlands: global capitalism and the reinvention of Confucianism, *Boundary 2*, 22(3): 229–273.

Dirlik, A. (1999a) 'Culture against history? The politics of East Asian identity', *Development and Society*, 28(2): 167–190.

Dirlik, A. (1999b) 'Is there history after Eurocentricism?: Globalism, postcolonialism, and the disavowal of history', *Cultural Critique*, 42, Spring: 1–34.

Dirlik, A. (2000) 'Globalization as the End and the Beginning of History: The contradictory Implications of a New Paradigm', http://globalization.mcmaster.ca/wps/dirlik.PDF (accessed 20 June).

Dirlik, A. (2005) 'Asia Pacific studies in an age of global modernity', *Inter-Asia Cultural Studies*, 6(2): 158–170.

Dirlik, A. (2007) 'Global South: Predicament and Promise', *The Global South*, Winter 2007, 1(1): 12–23.

Douglas, M. and Isherwood, B. (1980) *The World of Goods: Towards an Anthropology of Consumption of Goods*, London: Routledge & Kegan Paul.

Du Gay, P., Hall, S., Janes, L., Mackay H. and Negus, K. (eds) (1997) *Doing Cultural Studies: the Story of the Walkman*, London: Sage.

Ehrenreich, B. and Hochschild, A. (eds) (2003) *Global Woman: Nannies, Maids and Sex Workers in the New Economy*, NY: Metropolitan Books.

Ellwood-Clayton, B. (2003) 'Virtual Strangers: young love and texting in the Filipino archipelago of cyberspace', in K. Nyírí (ed.), *Mobile Democracy: Essays on Society, Self and Politics*, Vienna: Passagen Verlag: pp. 35–45.

Erni, J. (2006) 'Enchantment and disenchantment', presented at *Cultural Space and the Public Sphere: An International Conference*, 15–16 March, Seoul, South Korea.

Fang, W. (2005) 'China's culture of the thumb', *Receiver magazine*, 13, www.receiver.vodafone.com

Featherstone, M. (1991) *Consumer Culture and Postmodernism*, London: Sage.

Fischer, C. (1994) *America Calling: A Social History of the Telephone to 1940*, Berkeley, California: University of California Press.

Flew, T. (2002) *New Media: An Introduction*, South Melbourne, Victoria: Oxford University Press.

Fortunati, L. (2002a) 'The mobile phone: towards new categories and social relations', *Information, Communication & Society*, 5: 513–528.

Fortunati, L. (2002b) 'Italy: stereotypes, true and false', in J.E. Katz and M. Aakhus (eds), *Perpetual Contact: Mobile Communications, Private Talk, Public Performance*, Cambridge: Cambridge University Press, pp. 42–62.

Fortunati, L. (2005a) 'Mobile phones and fashion in post-modernity', *Telekronikk* 3/4: 35–48.

Fortunati, L. (2005b) 'Mobile telephone and the presentation of the self', in R. Ling and P.E. Pederson (eds), *Mobile Communications: Re-negotiation of the Social Sphere*, London: Springer.

Fortunati, L. (2005c) 'Mobile phones and fashion', *Telektronikk* 3: 1–14.

Fortunati, L. (2007) 'Gender and the mobile phone', presented at *Mobile Media: An international conference* (organised by G. Goggin and L. Hjorth), University of Sydney, Sydney, Australia, July.

Fortunati, L. (2008) 'Gender and the mobile phone' in G. Goggin and L. Hjorth (eds), *Mobile Technologies*, London/New York: Routledge (forthcoming).

Fortunati, L. and Manganelli, A.M. (2002) 'Young people and the mobile telephone', *Revista de Estudios de Juventud*, 52: 59–78.

Fortunati, L., Katz, J.E. and Riccini, R. (eds) (2003) *Mediating the Human Body: Technology, Communication, and Fashion*, Mahwah, NJ: Lawrence Erlbaum.

Fujimoto, K. (2005) 'The third-stage paradigm: territory machine from the girls' pager revolution to mobile aesthetics', in M. Ito, D. Okabe and M. Matsuda (eds), *Personal, Portable, Pedestrian: Mobile Phones in Japanese Life*, Cambridge, MA: MIT Press, pp. 77–102.

Gai, B. (2007) 'The rising individualism in Beijing: a local study of camera phones', in G. Goggin and L. Hjorth (eds) *Mobile Media proceedings*, Sydney, University of Sydney.

Gaonkar, D. (1999) 'On Alternative Modernities', *Public Culture*, 11(1): 1–18.

Giddens, A. (1991) *Modernity and Self-identity: Self and Society in the Late Modern Age*, Cambridge: Polity Press.

Giddens, A. (1992) *Transformation of Intimacy: Sexuality, Love and Eroticism in Modern Societies*, Cambridge: Polity Press.

Glotz, P. and Bertschi, S. (eds) (2005) *Thumb Culture: Social Trends and Mobile Phone Use*, Bielefeld: Transcript Verlag.

Goffman, E. (1963) *Behaviour in Public Places: Note on the Social Organisation of Gatherings*, New York: Free Press.

Goffman, E. (1969) *The Presentation of Self in Everyday Life*, Harmondsworth: Penguin Books.

Goggin, G. (2006a) *Cell Phone Culture: Mobile Technology in Everyday Life*, London: Routledge.

Goggin, G. (2006b) 'Notes on the history of the mobile phone in Australia', *Southern Review*, 38(2): 4–22.

Gottlieb, N. and McLelland, M. (eds) (2003) *Japanese Cybercultures*, New York: Routledge.

Gregg, M. (2007) 'Work where you want: the labour politics of the mobile office', presented at *Mobile Media conference*, University of Sydney, Sydney, Australia. July.

Gye, L. (2007) 'Picture this', special issue (ed. G. Goggin) *Continuum*, 21(2).

Habuchi, I. (2005) 'Accelerated reflexivity', in M. Ito, D. Okabe and M. Matsuda (eds), *Personal, Portable, Pedestrian: Mobile Phones in Japanese Life*, Cambridge, MA: MIT Press: pp. 165–182.

Haddon, L. (ed.) (1997a) *Communications on the Move: the Experience of Mobile Telephony in the 1990s*, Brussels: COST 248.

Haddon, L. (1997b) *Empirical Research on the Domestic Phone: A Literature Review*, Brighton: University of Sussex.

Haddon, L. (2003) 'Domestication and mobile telephony', in J.E. Katz (ed.), *Machines That Become Us: the Social Context of Personal Communication Technology*, New Brunswick, NJ: Transaction Publishers, pp. 43–56.

Haddon, L. (2004) *Information and Communication Technologies in Everyday Life: a Concise Introduction and Research Guide*, Oxford and New York: Berg.

Haddon, L., Mante, E., Sapio, B., Kommonen, K-H., Fortunati, L., and A. Kant (eds) (2005) *Everyday Innovators: Researching the Role of Users in Shaping ICTs*, Dordrecht: Springer.

Hamashita, T (1994) 'The tribute trade system and modern Asia', in A.J.H. Latham and H. Kawakatsu, (eds), *Japanese Industrialization and the Asian Economy*, London and New York: Routledge, pp. 91–107.

Hamashita, T (1995) 'The intra-regional system in east Asia: 19th–20th centuries', paper prepared for the *Workshop Japan in Asia*, Cornell University, Ithaca, NY, March–April.

Hamill, L. and Lasen A. (2005) (eds.) *Mobile World: Past, Present and Future*, Berlin: Springer Verlag.

Hannam, K, Sheller, M. and J. Urry (2006) 'Editorial: mobilities, immobilities and moorings', *Mobilities*, 1(1): 1–22.

Haraway, D. (1985/1994) 'A manifesto for cyborgs: Science, technology and socialist feminism in the 1980s', in Seidman, S. (ed.), *The Postmodern Turn*, Cambridge: Cambridge University Press.

Haraway, D. (1991) Simians, Cyborgs, and women: the reinvention of nature, New York: Routledge.

Hardey, M. (2002) 'Life beyond the screen: embodiment and identity through the Internet', *The Sociological Review*, 570–585.

Harootunian, H. (1970) *Toward Restoration*, Berkeley, CA: University of California Press.

Harper, R., Palen, L. and Taylor, A. (eds) (2005) *The Inside Text: Social, Cultural and Design Perspectives on SMS*, Dordrecht: Springer.

Hebdige, D. (1988) *Hiding in the Light: on Images and Things*, London: Routledge.

Herzfeld, M. (1997) *Cultural Intimacy: Social Poetics in the Nation–State*, London: Routledge.

Hirsch, E. and Silverstone, R. (1992) (eds), *Consuming Technologies: Media and Information in Domestic Spaces*, London: Routledge.

Hjorth, L. (2003a) 'Kawaii@keitai', in N. Gottlieb and M. McLelland (eds), *Japanese Cybercultures*, New York: Routledge: 50–59.

Hjorth, L. (2003b) ' "Pop" and "Ma": the landscape of Japanese commodity characters and subjectivity', in F. Martin, A. Yue and C. Berry (eds), *Mobile Cultures: New Media in Queer Asia*, Durham, NC: Duke University Press, pp. 158–179.

Hjorth, L. (2005a) 'Odours of mobility: Japanese cute customization in the Asia-Pacific region', *Journal of Intercultural Studies*, 26: 39–55.

Hjorth, L. (2005b) 'Locating mobility: practices of co-presence and the persistence of the postal metaphor in SMS/MMS mobile phone customization in Melbourne', *Fibreculture Journal*, 6, http://journal.fibreculture.org/issue6/issue6_hjorth.html (assessed 10 December 2006).

Hjorth, L. (2005c) 'Postal presence: a study on mobile customisation and gender in

Melbourne', in P. Glotz and S. Bertschi (eds), *Thumb Culture: Social Trends and Mobile Phone Use*, Bielefeld: Transcript Verlag: 53–66.

Hjorth, L. (2006a) 'Fast-forwarding present: the rise of personalization and customization in mobile technologies in Japan', *Southern Review Journal*, 38(3): 23–42.

Hjorth, L. (2006b) 'Playing at being mobile: Gaming, cute culture and mobile devices in South Korea', in *Fibreculture Journal*, 8, http://journal.fibreculture.org/ (assessed 10 January 2007).

Hjorth, L. (2006c) 'Snapshots', *Cultural Space and the public sphere in Asia*, hosted by Asia's Futures Initiative, 15–16 March, Seoul.

Hjorth, L. (2007a) 'Snapshots of almost contact: case study on South Korea', special issue (ed. G. Goggin), *Continuum* 21(2): 227–238.

Hjorth, L. (2007b) 'Engagement rings: a cross-cultural analysis of camera phone genres, modes of sharing and digital storytelling', *The Future of Digital Media Culture: 7th International Digital Arts and Culture (DAC) Conference*, 15–18 September, Perth, Australia, http://www.beap.org/dac.

Hjorth, L. (2007d) 'Home and away: a case study of the use of Cyworld mini-hompy by Korean students studying in Australia', *Asian Studies Review*, The Internet in East Asia special issue (ed. A. McLaren), December, 31: 397–407.

Hjorth, L. and Kim, H. (2005) 'Being there and being here: gendered customising of mobile 3G practices through a case study in Seoul', *Convergence*, 11: 49–55.

Ho, K.C., Kluver, R. and K.C.C. Yang (2003) *Asia@com*, London: Routledge.

Hochschild, A.R. (1983). *The Managed Heart: Commercialization of Human Feeling*, Berkeley, CA: University of California Press.

Hochschild, A.R. (2000) 'Global care chains and emotional surplus value', in W. Hutton and A. Giddens (eds), *On The Edge: Living with Global Capitalism*, London: Jonathan Cape, pp. 130–146.

Hochschild, A.R. (2003) *The Commercialization of Intimate Life: Notes from Home and Work*, California: University of California Press.

Höflich, J. and Hartmann, M. (eds) (2006), *Mobile Communication in Everyday Life: An Ethnographic View*, Frank & Timme, Berlin.

Hong, J.-E. (2007) 'Mobile phone machine and its discourses analysis', Paper presentation for *Mobile Media seminar*, Yonsei University, Seoul, December.

Hopper, R. (1992) *Telephone Conversation*, Indiana: Indiana University Press.

Hughes, R. (1981). *The Shock of the New*, London: Thames and Hudson.

Huhh, J.-S. (2008) 'The culture and business of the PC bang', in special issue (ed. L. Hjorth) of *Games and Culture* on 'Gaming in the Asia-Pacific region', January, 3(1).

Huhtamo, E. (1997) 'From Kaleidoscomaniac to Cybernerd: Notes Toward an Archaeology of the Media'. *Leonardo*, 30(3).

International Labour Office (ILO) (2008) 'Global Employment Trends for Women', http://www.ilo.org/public/english/employment/strat/global.htm (accessed 2 March 2008).

Introna, L.D. and Ilharco, F.M. (2004) 'The ontological screening of contemporary life: a phenomenological analysis of screens', *European Journal of Information Systems*, 13: 221–234.

IPC Media (2006) http://www.ipcmedia.com/press/ (accessed 28 May 2008).

Ishii, K. (2004) 'Internet use via mobile phone in Japan', *Telecommunications Policy*, 28: 43–58.

Ito, J. (2004) 'Moblogging, blogmapping and moblogmapping related resources', http://joiwiki.ito.com/joiwiki/index.cgi?moblog (accessed 9 January 2006).

Ito, M. (2002) 'Mobiles and the appropriation of place', in *Receiver* magazine, 8, www.receiver.vodafone.com (10 December 2003) n. p.

Ito, M. (2005) 'Introduction: personal, portable, pedestrian', in M. Ito, D. Okabe and M. Matsuda (eds), *Personal, Portable, Pedestrian: Mobile Phones in Japanese Life*, Cambridge, MA: MIT Press, pp. 1–16.

Ito, M. and Okabe, D. (2003) 'Camera phones changing the definition of picture-worthy', *Japan Media Review*, http://www.ojr.org/japan/wireless/1062208524.php (accessed 10 June 2004).

Ito, M. and Okabe, D. (2005a) 'Intimate visual co-presence', presented at *Ubi-Comp 2005*, 11–14 September, Takanawa Prince Hotel, Tokyo, Japan, http://www.itofisher.com/mito/ (accessed 10 December 2005).

Ito, M. and Okabe, D. (2005b) 'Intimate connections: contextualizing Japanese youth and mobile messaging', in R. Harper, L. Palen and A. Taylor (eds), *The Inside Text: Social, Cultural and Design Perspectives on SMS*, Dordrecht: Springer.

Ito, M. and Okabe, D. (2005c) 'Mobile phones, Japanese youth, and the replacement of social contact', in R. Ling and P.E. Pederson (eds) *Mobile Communications: re-negotiation of the Social Sphere*, London: Springer.

Ito, M and Okabe, D. (2005d) 'Technosocial situations: emergent structuring of mobile e-mail use', in M. Ito, D. Okabe and M. Matsuda (eds) *Personal, Portable, Pedestrian: Mobile Phones in Japanese Life*, Cambridge, MA: MIT Press, pp. 257–276.

Ito, M., Okabe, D. and Matsuda, M. (eds) (2005) *Personal, Portable, Pedestrian: Mobile Phones in Japanese Life*, Cambridge, MA: MIT Press.

Iwabuchi, K. (2003) *Recentring Globalization: Popular Culture and Japanese Transnationalism*, Durham, NC: Duke University Press.

Jacobs, D. (2006) 'Cyworld lands on MySpace', *International Business Times*, 31 July, at http://ibtimes.com/articles/20060731/cyworld-myspace-sktelecom-newscorp.htm (accessed 10 August 2006).

Jamieson, L. (1998) *Intimacy: Personal Relationships in Modern Societies*, Cambridge: Polity Press.

Jamieson, L. (1999) 'Intimacy transformed: A critical look at the "pure relationship"' , *Sociology*, 33: 477–494.

Jenkins, H. (2005) 'Welcome to convergence culture', *Receiver*, 12, http://www.receiver.vodafone.com/12/articles/pdf/12_01.pdf (accessed 10 January 2006).

Jenkins, H. (2006) *Convergence Culture: Where Old and New Media Intersect*, New York: New York University Press.

Kanellos, M. (2006) 'Korean social-networking site hopes to nab U.S. fans', August 14. Retrieved August 21, 2006, from http://www.zdnetindia.com/news/communication/stories/151524.html.

Katsuno, H. and Yano, C.R. (2002) 'Face to face: on-line subjectivity in contemporary Japan', *Asian Studies Review*, 26(2): 205–231.

Katz, J.E. (1999) *Connections: Social and Cultural Studies of the Telephone in American Life*, New Brunswick, NJ: Transaction.

Katz, J.E. (ed.) (2003) *Machines That Become Us: the Social Context of Personal Communication Technology*, New Brunswick, NJ: Transaction Publishers.

Katz, J.E. and Aakhus, M. (eds) (2002) *Perpetual Contact: Mobile Communication, Private Talk, Public Performance*, Cambridge: Cambridge University Press.

Katz, J.E. and Sugiyama, S. (2005) 'Mobile phones as fashion statements: the co-creation of mobile communication's public meaning', in R. Ling and P.E.

Pederson (eds), *Mobile Communications: Re-negotiation of the Social Sphere*, London: Springer, pp. 63–81.

Kawakatsu, H. (1994) 'Historical background', in A.J.H. Latham and H. Kawa-katsu (eds) *Japanese Industrialization and the Asian Economy*, London: Routledge, pp. 4–8.

Khoo, O. (2007) *The Chinese Exotic*, Hong Kong: Hong Kong University Press.

Kim, S.D. (2002) 'Korea: personal meanings', in J.E. Katz and M. Aakhus (eds), *Perpetual Contact: Mobile Communications, Private Talk, Public Performance*, Cambridge: Cambridge University Press, pp. 63–79.

Kim, S.D. (2003) 'The shaping of new politics in the era of mobile and cyber com-munication', in Nyírí, K. (ed.), *Mobile Democracy*, Vienna: Passagen Verlag, pp. 317–326.

Kim, S.D. (ed.) (2004), proceedings of *Mobile Communication and Social Change conference*, Seoul, South Korea, October 18–19.

Kim, S.D. (ed.) (2005) *When Mobile Came: The Cultural and Social Impact of Mobile Communication*, Seoul: CommunicationBooks.

Kim, S.D. and Hjorth, L. (2005) 'Palm reading: possibilities for mobile TV aesthetics and reception', paper presented at Mobile Communication and Asian Modernities II conference, France Telecom and Beijing University, Beijing, October.

Kim, Y. (2008) (ed.) *Media Consumption and Everyday Life in Asia*, London: Routledge (forthcoming).

Kinsella, S. (1995) 'Cuties in Japan', in L. Skov and B. Moeran (eds), *Women, Media and Consumption in Japan*, Surrey, UK: Curzon Press, pp. 220–254.

Ko, Y.F. (2003) 'Consuming differences: "Hello Kitty" and the identity crisis in Taiwan', *Postcolonial Studies*, 6(2): 175–189.

Koch, P. et al. (2008) 'Beauty in the eye of the QQ beholder', in G. Goggin and M. McLelland (eds) *Internationalising Internet Studies*, London: Routledge (forthcoming).

Kogawa, T. (1984) 'Beyond electronic individualism', *Canadian Journal of Political and Social Theory/Revue Canadienne de Thetorie Politique et Sociale*, 8(3), Fall, http://anarchy.translocal.jp/non-japanese/electro.html (accessed 20 July 2007).

Kohiyama, K. (2005) 'A decade in the development of mobile communications in Japan (1993–2002)', in M. Ito, D. Okabe and M. Matsuda (eds), *Personal, Portable, Pedestrian: Mobile Phones in Japanese Life*, Cambridge, MA: MIT Press, pp. 61–74.

Kopomaa, T. (2000) *The City in Your Pocket: Birth of the Mobile Information Society*, Helsinki: Gaudemus.

Kopomaa, T. (2002) 'The reunited family of the media information society', *Receiver*, 6, www.receiver.vodafone.com (assessed 15 February 2007).

Kopytoff, I. (1986) 'The cultural biography of things: commoditization as process', in A. Appadurai (ed.), *The Social Life of Things: Commodities in Cultural Perspective*, Cambridge: Cambridge University Press: 64–91.

Koskela, H. (2004) 'Webcams, TV shows and mobile phones: empowering exhibition-ism'. *Surveillance & Society*, 1(2/3): 199–215.

Koskinen, I. (2007) 'Managing banality in mobile multimedia', in R. Pertierra (ed.), *The Social Construction and Usage of Communication Technologies: European and Asian Experiences*, Singapore: Singapore University Press, pp. 48–60.

Kristeva, J. (2002) *Intimate Revolt. The Power and Limits of Psychoanalysis*, II, trans. Jeanine Herman, New York: Columbia University Press.

Kundera, M. (1996) *Slowness*, trans. L. Asher, France: HarperCollins Publications.

Kusahara, M. (2001) 'The art of creating subjective reality: an analysis of Japanese digital pets', *Leonardo* 34 (4): 299–302.

Lantz, F. (2006) 'Big games and the porous border between the real and the mediated', *Receiver*, 16, http://www.receiver.vodafone.com/16/articles/index07.html (accessed 10 February 2007).

Latour, B. (1987) *Science In Action. How to Follow Scientists and Engineers through Society*, Cambridge MA: Harvard University Press.

Law, P., Fortunati, L. and Yang, S. (2006) *New Technologies in Global Societies*, New Jersey: World Scientific.

Lee, D.H. (2005) 'Women's making of camera phone culture', *Fibreculture Journal*, 6, http://journal.fibreculture.org (accessed 5 December 2005).

Lee, D.H. (2008) 'Re-imagining urban space: mobility, connectivity, and a sense of place', in G. Goggin and L. Hjorth (eds), *Mobile Technologies*, London/New York: Routledge (forthcoming).

Lee, D.H. and Sohn, S.H. (2004) 'Is there a gender difference in mobile phone usage?', in S.D. Kim (ed.), *Mobile Communication and Social Change conference proceedings*, South Korea, October: 243–259.

Lee, Y., Lee, I., Kim, J., and Kim, H. (2002) 'A cross-cultural study on the value structure of mobile internet usage: comparison between Korea and Japan', *Journal of Electronic Commerce Research*, 3(4): 227–239.

Lee, Y-J. (2000) 'Consumer culture and gender identity in South Korea', *Asian Journal of Women Studies*, 6(4): 11–38.

Leung, L. and Wei, R, (1999) 'Who are the mobile phone have-nots? Influences and consequences', *New Media and Society*, 1(2): 209–226.

Leung, L. and Wei, R. (2000) 'More than just talking on the move: Uses and gratifications of cellular phones', *Journalism and Mass Communication Quarterly*, 77 (2): 308–320.

Lin, A. (2004) 'New youth digital literacies and mobile connectivity: text-messaging among Hong Kong college students', presented at the *International Conference on Mobile Communication and Social Change*, 18–19 October, Korean Press Foundation, Seoul, South Korea.

Lin, A. (2005) 'Gendered, bilingual communication practices: mobile text-messaging among Hong Kong college students', *Fibreculture Journal*, 6, http://journal.fibreculture.org/issue6/issue6_lin.html (accessed 5 December 2005).

Lindgren, M., Jedbratt, J. and Svensson, E. (2002) *Beyond Mobile: People, Communications and Marketing in the Mobilised World*, Basingstoke: Palgrave.

Ling, L.H.M. (1999) 'Sex machine: global hypermasculinity and images of the Asian woman in modernity', *Positions*, 7(2): 227–306.

Ling, R. (2002) 'Adolescent girls and young adult men: two sub-cultures of the mobile telephone', *Revista de Estudios de Juventud*, 52: 33–46.

Ling, R. (2004) *The Mobile Connection*, San Francisco: Morgan Kaufmann Publishers.

Ling, R. and Pedersen, P. (eds) (2005) *Mobile Communications: Re-negotiation of the Social Sphere*, London: Springer-Verlag.

Ling, R. and Yttri, B. (2002) 'Hyper-coordination via mobile phones in Norway', in J.E. Katz and M. Aakhus (eds), *Perpetual Contact: Mobile Communication, Private Talk, Public Performance*, Cambridge: Cambridge University Press, pp. 139–69.

Liu, G. and Lau, J. (2006) 'Sexuality as public spectacle: the transformation of sex information and service in the age of the Internet', in P.L Law, L. Fortunati and S. Yang (eds) *New Technologies in Global Societies*, New Jersey: World Scientific.

Liu, M. and Zoninsein, M. (2007) 'These surfers do it their own way', *Newsweek*, 24 December, pp. 40–41.

Lo, S.-H. (1996) 'Hong Kong: Post-colonialism and political conflict', in R. Robison and D.S.G.. Goodman (eds), *The New Rich in Asia*, London: Routledge.

Luke, R. (2006) 'The phoneur: mobile commerce and the digital pedagogies of the wireless web', in P. Trifonas (ed.), *Communities of Difference: Culture, Language, Technology*, London: Palgrave, pp. 185–204.

Lupton, E (1994) 'Low and high: design in everyday life', in M. Bierut, W. Drenttel, S. Heller and DK Holland (eds) *Looking closer: Critical Writings on Graphic Design*, New York: Allworth Press.

Lury, C., (1996) *Consumer Culture*, New York: Polity Press and Rutgers Press.

Ma, E.K.-W. (2000) 'Re-nationalizing and me: My Hong Kong story after 1997', *Inter-Asia Cultural Studies*, 1(1): 173–179.

Ma, E.K.-W. (2005) 'Re-advertising Hong Kong: nostalgia industry and popular history', in J.N. Erni and S.K. Chua (eds), *Asian Media Studies*, London: Blackwell.

McGray, D. (2002) 'Japan's gross national cool', *Foreign Policy*, May/June: 44–54.

McKay, D. (2007) 'Sending dollars shows feeling – emotions and economies in Filipino migration', *Mobilities*, 2(2): 175–194.

MacKenzie, D. and Wajcman, J. (eds) (1999) *The Social Shaping of Technology*, 2nd edn, Buckingham, England, and Philadelphia, PA: Open University Press.

Mackie, V. and Buckley (1985) 'Women in the new Japanese state', in G. McCormack and Y. Sugimoto (eds), *Democracy in Japan*, Melbourne: Hale and Iremonger.

Mackie, V. (1988) 'Feminist politics in Japan', *New Left Review*, 167: 53–71.

McLelland, M. (2007) 'Socio-cultural aspects of mobile communication technologies in Asia and the Pacific: a discussion of the recent literature' in *Continuum*, 21(2): 267–77..

McLelland, M. and Goggin, G. (2008) (eds.) *Internationalising Internet Studies*, London: Routledge (forthcoming).

McLuhan, M. (1964) *Understanding Media*, New York: Mentor.

McRobbie, A. (1996) 'The E's and the anti-E's: questions for feminism and cultural studies', in M. Ferguson and P. Golding (eds), *Beyond Cultural Studies*, London: Sage.

McRobbie, A. (1999) *In the Culture Society*, London: Routledge.

McRobbie, A. (2000) *Feminism and Youth Culture*, Basington: Macmillan, 2nd edn.

McVeigh, B. (1996a) 'Cultivating "femininity" and "internationalism": rituals and routine at a Japanese women's junior college', *Ethos*, 24(2): 314–349.

McVeigh, B (1996b) 'Commodifying affection, authority and gender in the everyday objects of Japan', *Journal of Material Culture*, 1(3): 291–312.

McVeigh, B. (1997) 'Wearing ideology: how uniforms discipline minds and bodies in Japan', *Fashion Theory*, 1(2): 189–214.

McVeigh, B. (2000) 'How Hello Kitty commodifies the cute, cool and camp: "consumutopia" versus "control" in Japan', *Journal of Material Culture*, 5(2): 291–312.

McVeigh, B. (2003) 'Individualization, individuality, interiority, and the Internet', in N. Gottlieb and M. McLelland (eds), *Japanese Cybercultures*, London: Routledge, pp. 19–33.

McVeigh, B. (2004) *Nationalisms of Japan: Managing and Mystifying Identity*. Oxford, New York: Rowman and Littlefield.

Manovich, L. (2003) 'The paradoxes of digital photography', in Liz Wells (ed.), *The Photography Reader*, London: Routledge, pp. 240–249.

Mante-Meijer, E. and van de Loo, H. (1998) 'Blurring of the life spheres: Flexibility and teleworking', in A. Kant and E. Mante-Meijer (eds), *Blurring Boundaries: When Are Information and Communication Technologies Coming Home?*, Stockholm: Telia.

Margaroni, M. and Yiannopoulou, E. (2005) 'Intimate transfers: introduction', *European Journal of English Studies*, 9(3): 221–228.

Martin, M. (1991a) *Hello Central?: Gender, Culture, and Technology in the Formation of Telephone Systems*, Montreal: McGill-Queen's University Press.

Martin, M. (1991b) 'The culture of the telephone' in P.D. Hopkins (ed.), *Sex/ Machine: Readings in culture, gender and technology*, Indiana: Indiana University Press.

Massey, D. (1993), 'Questions of locality', *Geography*, 78: 142–149.

Massey, D. (1995) *Spatial Divisions of Labor: Social Structures and the Geography of Production*, London: Routledge.

Massey, D. (2005) *For Space*, London and Thousand Oaks, CA: Sage.

Matsuda, M. (2005) 'Discourses of *Keitai* in Japan', in M. Ito, Okabe, D. and M. Matsuda (eds), *Personal, Portable, Pedestrian: Mobile Phones in Japanese Life*, Cambridge, MA: MIT Press, pp. 19–40.

Matsuda, M. (2007) 'Mobile media and the transformation of family', presented at Mobile Media: An International Conference (organised by G. Goggin and L. Hjorth), University of Sydney, *Sydney, Australia, July*.

Mauss, M. (1954) *The Gift*, London: Kegan Paul.

Mäyrä, F. (2003). 'The City Shaman Dances with Virtual Wolves – Researching Pervasive Mobile Gaming' in *Receiver*, 12, www.receiver.vodafone.com.

Miller, D. (1987) *Material Culture and Mass Consumption*, London: Blackwell.

Miller, D. (ed.) (1988) *Material Cultures: Why Some Things Matter*, Chicago: University of Chicago Press.

Miller, D. and Horst, H. (2006) *Cell Phone*, Oxford and New York: Berg.

Miller, L. (2005) 'Bad girl photography,' in Bardsley and L. Miller (eds), *Bad Girls of Japan*, London: Palgrave Macmillan.

Miller, L. (2006) *Beauty Up*, Berkley: University of California.

Milne, E. (2004) 'Magic Bits of Paste-Board', *M/C Journal*, 7(1), http://journal.media-culture.org.au/ (Assessed 10 June 2004).

Ministry of Information Affairs and Communication (MIC) Japan. (2006, 2003, 2005) *White Papers*: http://www.johotsusintokei.soumu.go.jp/statistics/statistics05.html, http://www.johotsusintokei.soumu.go.jp/english/ (accessed 10 February 2007).

Ministry of Information and Communication (MIC) Republic of Korea (2006) *IT Statistics*, Korea: Ministry of Information and Communication.

Mitomo, H, Liqiao, W. and Otsuka, T. (2005) 'How the university students in East Asia utilize mobile phones: empirical findings from the Asian Metropolitan Mobile Survey with emphasis on the mobile phone choice in Beijing', presented at Mobile Communication and Asian Modernity II, France Telecom, 20–21 October, Beijing.

Mori, Y. (2005a) 'An analysis on dual network structure on a SNS', presented for the 2005 annual meeting of *The Japan Society of Information and Communication Research*, June 25, Graduate Institute of Information Security, Yokohama, Japan.

Mori, Y. (2005b) 'Toward the information distribution simulator based on a real human relationship structure from Social Networking Services', presented for *the Social Informatics Fair 2005*, Joint Annual Meeting of the Japan Association of Social Informatics & the Japan Society for Socio-Information Studies, Social network analysis session, 13 September, 2005, Kyoto University, Kyoto, Japan.

Mori, Y. (2005c) 'Marketing opportunities on Social Networking Services', presented for the *Next Generation Internet Services* symposium by Japan Research Institute for New Systems of Society, October 27, 2005, Meiji Kinenkan, Tokyo, Japan.

Morley, D. (2007) *Media, Modernity and Technology: The Geography of the New*, London: Routledge.

Morley, D. (2003) 'What's "home" got to do with it?' in *European Journal of Cultural Studies*, 6(4): 435–458.

Morley, D. and Robins, K. (1995) *Spaces of Identities: Global Media, Electronic Landscapes and Cultural Boundaries*, New York: Routledge.

Morris, M. (1988) 'Banality in cultural studies', *Discourse*, 10(2): 3–29.

Morse, M. (1998) *Virtualities: Television, Media Art, and Cyberculture*, Bloomington: Indiana University Press.

Mosco, V. (2004) *The Digital Sublime: Myth, Power, and Cyberspace*, Cambridge MA: MIT Press.

Moyal, A. (1992) 'The gendered use of the telephone', in S. Jackson and S. Moores (eds), *The Politics of Domestic Consumption*, Hemel Hempstead: Harvester Press.

Moyal, A. (1995) 'The gendered use of the telephone: an Australian case study', *Media, Culture and Society*, 14: 51–72.

Murakami, T. (2000) *Superflat*, exhibition catalogue, Parco gallery, Tokyo, Japan.

Murphie, A. (2007) 'Mobility, work and love', http://researchhub.cofa.unsw.edu.au/ccap/2007/07/10/mobility-work-and-love/ (assessed 30 August 2007).

Murphie, A., Hjorth, L., Fuller, G., and S. Buckley (2005) (eds) Mobility issue, *Fibreculture Journal*, 6, at www.journal.fibreculture.org/issue6/index.html.

Na, M. (2001) 'The cultural construction of the computer as a masculine technology: an analysis of computer advertisements in Korea', *Asian Journal of Women Studies*, 7(3): 93–114.

Nakamura, L. (2002) *Cybertypes: Race, Ethnicity, and Identity on the Internet*, New York: Routledge.

NIDA (National Internet Development Agency of Korea) (2006) *Netizen Internet Usage*, Press release: http://www.nida.go.kr/doc/issue–sum–report.pdf (accessed December 2006).

NTT DoCoMo (2003), at http://www.nttdocomo.com/pr/2003/.

Nyíri, K. (ed.) (2002) *Mobile Democracy: Essays on Society, Self and Politics*, Vienna: Passagen Verlag.

Nyíri, K. (ed.) (2003a) *Mobile Communication: Essays on Cognition and Community*, Vienna: Passagen Verlag.

Nyíri, K. (ed.) (2003b) *Mobile Learning: Essays on Philosophy, Psychology and Education*, Vienna: Passagen Verlag.

Nyíri, K. (ed.) (2005) *A Sense of Place: the Global and the Local in Mobile Communication*, Vienna: Passagen Verlag.

Okabe, D., Ito, M., Shimizu, A. and J. Chipchase (2008) 'Purikura as a social management tool', in G. Goggin and L. Hjorth (eds) *Mobile technologies*, London/New York: Routledge (forthcoming).

Okada, T. (2005) 'Youth culture and the shaping of Japanese mobile media: personalization and the *Keitai* Internet as multimedia' in M. Ito, D. Okabe and M. Matsuda (eds), *Personal, Portable, Pedestrian: Mobile Phones in Japanese Life*, Cambridge, MA: MIT Press, pp. 41–60.

Ong, A. (1999) *Flexible Citizenship: The Cultural Logic of Transnationality*, Durham and London: Duke University Press.

Ong, A. (2006) 'Mutations in citizenship', *Theory Culture Society*, 23: 499–531.

Ong, A. and Gates Peletz, M. (1995) *Shelf Mark: Bewitching Women, Pious Men: Gender and Body in Southeast Asia*, California: University of California Press.

Organisation for Economic Co-operation and Development (OECD)(2006) *OECD Broadband Statistics*, http://www.oecd.org/sti/ict/broadband (accessed December 2006).

Palmer, D. (2005) 'Mobile art', presented at the *Vital Signs conference*, ACMI, Melbourne, 8 September.

Papastergiadis, N. (2003) 'South-South-South', N. Papastergiadis (ed.) *Complex Entanglements: Art, Cultural Difference & Globalization*, London: Rivers Oram Press, pp. 156–177.

Papastergiadis, N. (2005) 'Mobility and the nation: skins, machines and complex systems', *Willy Brandt Series of Working Papers in International Migration and Ethnic Relations*, Malmö University, Sweden, www.mah.se.

Parikka, J. and J. Suominen (2006) 'Victorian Snakes? Towards A Cultural History of Mobile Games and the Experience of Movement', *Games Studies: the International Journal of Computer Game Research*, 6(1), December, http://gamestudies.org/0601 (retrieved April 2007).

Parreñas, R (2001) *Servants of Globalization: Women, Migration, and Domestic Work*, Stanford, CA: Stanford University Press.

Pertierra, R. (2002) *The Work of Culture*, Manila: De La Salle University Press.

Pertierra, R (2003) *Science, Technology and the Culture of Everyday Life in the Philippines*, Manila: Ateneo de Manila University.

Pertierra, R (2005a) 'Mobile phones, identity and discursive intimacy', *Human Technology*, 1: 23–44.

Pertierra, R (2005b) 'Without a room of your own? buy a cell phone', in A. Lin (ed.), proceedings of International Conference on Mobile Communication and Asian Modernities, 7–8 June, City University of Hong Kong, Kowloon.

Pertierra, R (2006) *Transforming Technologies: Altered Selves*, Philippines: De La Salle University Press.

Pertierra, R (ed.) (2007) *The Social Construction and Usage of Communication Technologies: European and Asian Experiences*, Singapore: Singapore University Press.

Plant, S. (1998) *Zeros and Ones*, London: Fourth Estate.

Plant, S. (2002), 'On the mobile', at http://www.motorola.com/mediacenter/news/detail/0,1958,534_308_23,00.html (accessed 10 December 2003).

Putnam, R. (2000) *Bowling Alone*, New York: Simon and Schuster.

Qiu, J. (2007) 'The wireless leash: mobile messaging service as a means of control', *International Journal of Communication*, 1: 74–91.

Qiu, J. (2008) 'Wireless working-class ICTs and the Chinese informational city', Special Issue of *Journal of Urban Technology* on 'Mobile Media and Urban Technology', (eds) Alice Crawford, Gerard Goggin & Larissa Hjorth, forthcoming.

Rafael, V. (2003) 'The cell phone and the crowd: messianic politics in the contemporary Philippines', *Popular Culture*, 15(3): 399–425.

Raiti, G. (2007) 'Mobile intimacy: theories on the economics of emotion with examples from Asia', *M/C Journal*, L. Hjorth and O. Khoo (eds), special issue on mobility in the Asia-Pacific, 10(1), at http://journal.media-culture.org.au/0703/02-raiti.php (assessed 10 March 2007).

Rakow, L.F. (1992) *Gender on the Line: Women, the Telephone, and Community Life*, Chicago: University of Illinois Press.

Rakow, L.F. and Navarro, V. (1993) 'Remote mothering and the parallel shift: women meet the cellular telephone', *Critical Studies in Mass Communication*, 10(2): 144–157.

Rao, M. (ed.) (2004) *News Media and New Media: The Asia-Pacific Internet Handbook*, Singapore: Times Academic Press.

Rao, M. and Mendoza L. (2004) *Asia Unplugged: The Wireless and Mobile Media Boom in Asia-Pacific*, New Delhi: Response Books.

Rheingold, H. (2002) *Smart Mobs: the Next Social Revolution*, Cambridge MA: Perseus.

Rheingold, H. (2003) 'Moblogs seen as a crystal ball for a new era in online journalism', *Online Journalism Review USC Annenberg*, 9 July, at http://www.ojr.org/ojr/technology/1057780670.php (accessed 22 July 2005).

Richardson, I. (2007) 'Pocket technoscapes: the bodily incorporation of mobile media', in *Continuum: Journal of Media & Cultural Studies*, 21(2): 205–216.

Robison, R. and Goodman, D.S.G. (eds) (1996) *The New Rich in Asia*, London: Routledge.

Roessler, P. and Höflich, J.R. (2005) 'More than a telephone: mobile phone and the usage of the short message service (SMS) by German adolescents', in S.D. Kim (ed.), *When Mobile Came*, Seoul: CommunicationBooks, pp. 104–134.

Said, E. (1978) Orientalism, New York: Random Books.

Sawhney, H. (2004) 'Mobile communication: new technologies and old archetypes', in A. Lin (ed.), proceedings of the *Mobile Communication and Asian Modernities I* in Hong Kong at City University of Hong Kong, June.

Scifo, B. (2005) 'The domestication of the camera phone and MMS communications: the experience of young Italians', in K. Nyîri (ed.), *A Sense of Place: the Global and the Local in Mobile Communication*, Vienna: Passagen Verlag, pp. 363–373.

Seltzer, M. (1992) Bodies and Machines, New York and London: Routledge.

Sen, K. and Stivens, M. (1998) (eds), *Gender and Power in Affluent Asia*, London: Routledge.

Shade, L.R. (2007) 'Feminizing the mobile: gender scripting of mobiles in North America', in special issue: Mobile Phone Cultures, G. Goggin (ed.), *Continuum: Journal of Media and Cultural Studies*, 21(2): 179–190.

Shamir, R. (2005) 'Without Borders? Notes on Globalization as a Mobility Regime', *Sociological Theory*, 23(2): 197–217.

Silverstone, R. and Hirsch, E (eds) (1992) *Consuming Technologies: Media and Information in Domestic Spaces*, London: Routledge.

Silverstone, R. and Haddon, L. (1996) 'Design and domestication of information and communication technologies: technical change and everyday life', in R. Silverstone and R. Mansell (eds), *Communication by Design: The Politics of Information and Communication Technologies*, Oxford, UK: Oxford University Press, pp. 44–74.

Skov, L. and Moeran, B. (1995) *Women, Media and Consumption in Japan*, Hawaii: Curzon and Hawaii University Press.

Slater, D. and Ritzer, G. (2001) 'Interview with Ulrich Beck', *Journal of Consumer Culture*, 1(2): 261–277.

Solis, R. (2007) 'Texting love: an exploration of text messaging as a medium for romance', *M/C Journal*, L. Hjorth and O. Khoo (eds), special issue on mobility in the Asia-Pacific, 10(1), http://journal.media-culture.org.au/ (assessed 10 March 2007).

Son, M.H. (2007) 'Cultures of personal ambivalence: a case study of college students' uses of the camera phone' and Cyworld's mini-hompy', *Seoul Symposium on Mobile Communication*, SK Telecom, October, Seoul, South Korea.

Sørensen, C. (2002) 'Digital nomads and mobile services', *Receiver*, 6, at www.receiver.vodafone.com/6 (accessed 10 December 2005).

Spigel, L. (1992) *Make Room for TV: Television and the Family Ideal in Postwar America*, Chicago: University of Chicago Press.

Stone, A.R. (1995) *The War of Desire and Technology*, Cambridge, MA: MIT Press.

Sullivan, J. (2008) 'Thumbs down to "the Nokia novel"' , *The Age*, A3 section, Saturday 9 February, 28.

Sutton-Smith, B. (1997) *The Ambiguity of Play*, London: Routledge.

Tapscott, D. (1995) *The Digital Economy: Promise and Peril in the Age of Networked Intelligence*, New York: McGraw-Hill Books.

Taylor, A. and Harper, R. (2002) 'Age-old practices in the "New World": A study of gift-giving between teenage mobile phone users', in *Changing Our World, Changing Ourselves* (proceedings of the *SIGCHI* Conference on Human Factors in Computing Systems, Minneapolis): 439–446.

Taylor, A. and Harper, P. (2003) 'The gift of gab? a design oriented sociology of young people's use of mobiles', *Journal of Computer Supported Cooperative Work*, 12: 267–296.

Telecoms InfoTechnology Forum (2004) *Hong Kong as Asia's Wireless Development Center*, Telecoms InfoTechnology Forum, Telecommunications Research Project, University of Hong Kong, http://www.trp.hku.hk/tif/papers/2004/mar/briefing_0 40325.pdf (accessed 25 May 2005).

Telstra, www.telstra.com (assessed 10 November 2005).

Toffler, A. (1980) *The Third Wave*, New York: William Morrow and Company.

Truong, T.D. (1999) 'The underbelly of the tiger: gender and demystification of the Asian miracle', *Review of International Political Economy*, 6:2: 133–165.

Tsuji, D. and Mikami, S. (2001) 'A preliminary student survey on the e-mail uses of mobile phones', presented at *JSICR*, Tokyo, June.

Turkle, S. (1995) *Life on the Screen: Identity in the Age of the Internet*, New York: Simon and Schuster.

Turner, B. (2007) 'The enclave society: towards a sociology of immobility', *European Journal of Social Theory*, 10(2): 287–303.

Urban Vibe (2005) Nabi Media Centre, http://eng.nabi.or.kr/project/view.asp?prj learn_idx=119 (accessed 28 May 2008).

Urry, J. (2000a) *Sociology Beyond Societies: Mobilities for the Twenty-first Century*, London: Routledge.

Urry, J. (2000b) 'Mobile sociology', *British Journal of Sociology* 51(1): 185–203.

Urry, J. (2002) 'Mobility and proximity', *Sociology*, 36(2): 255–274.

Urry, J. (2003) 'Social networks, travel and talk', *British Journal of Sociology* 54(2).

Van House, N., Davis, M., Ames, M., Finn, M. and Viswanathan, V. (2005) 'The uses of personal networked digital imaging: an empirical study of cameraphone photos and sharing', in *Extended Abstracts of the Conference on Human Factors in Computing Systems (CHI 2005)*, Portland, Oregon, ACM Press: 1,853–1,856.

Wajcman, J. (1991) *Feminism Confronts Technology*, Pennsylvania: Pennsylvania State University Press.

Wajcman, J. (2004) *Technofeminism*, Cambridge: Polity.

Wajcman, J. and Haddon, L. (eds) (2005) 'Technology, time and everyday life', Forum Discussion Paper No. 7, London: Oxford Internet Institute.

Wajcman, J., Bittman, M. and Brown, J. (2008) 'Intimate connections: the impact of the mobile phone on work–life boundaries', in G. Goggin and L. Hjorth (eds), *Mobile Technologies*, London/New York: Routledge (forthcoming).

Walher, B.K. (2006) 'Pervasive gaming: format, rules and space'. *Fibreculture Journal*, 8, www.journalfibreculture.org (assessed 20 December 2006).

Waters, P. (2004) 'Mobile competition: How many is too many?', ITU Telecom Asia 2004, *ITU Daily*, 10 September, at http://www.itudaily.com/new/print article.asp?articleid=4091001 (accessed 25 May 2005).

Wellman, B. (2002) 'Little boxes, glocalization, and networked individualism', in M. Tanabe, van den Besselaar, P. and T. Ishida (eds), *Digital Cities II: Computational and Sociological Approaches*, Berlin: Springer-Verlag, pp. 10–25.

West, D.M. (2006) *Global e-government, 2006*. Providence, Rhode Island: Center for Public Policy, Brown University.

White, M. (1993) *The Material Child: Coming of Age in Japan and America*, New York: Free Press.

Whittier Treat, J. (1996) 'Introduction', in J. Whittier Treat (ed.), *Contemporary Japanese Popular Culture*, Honolulu: University of Hawaii Press, pp. 1–16.

Wilhelm, A, Y. Takhteyev, R. Sarvas, N. Van House, and M. Davis. (2004) 'Photo annotation on a camera phone', presented at *CHI2004*, 24–29 April, Vienna, Austria: 1,403–1,406.

Williams, R. (1974) *Television: Technology and Cultural Form*, London: Fontana.

Williams, R. (1983) 'Mobile privatization', in reprinted version of P. du Gay, S. Hall, J. Lanes, H. Mackay and K. Negus (eds) (1997) *Doing Cultural Studies: the Story of the Sony Walkman*, London: Sage.

Wilson, A. (2004) *The Intimate Economies of Bangkok*, California: University of California Press.

Wilson, R. (2000) 'Imagining "Asia-Pacific": forgetting politics and colonialism in the magical waters of the Pacific. An Americanist critique', *Cultural Studies*, 14(3/4): 562–592.

Wilson, R. and Dirlik, A. (eds) (1995) *Asia/Pacific as Space of Cultural Production*, Durham: Duke University Press.

Wilson, S. (ed.) (2002) *Nation and Nationalism in Japan*, New York: Routledge-Curzon.

Yan, X. and Pitt, D. (1999) 'One country, two systems: contrasting approaches to telecommunications deregulation in Hong Kong and China', *Telecommunications Policy*, 23(3/4): 245–260.

Yang, S.J. and Park, Y.J. (2005) 'A study on the motivation and the influence of using personal community', *Journal of Consumer Studies*, 16(4): 129–150.

Yeh, M.-J. (2004) 'A preliminary study on SMS use of youth tribes', *Information Society Research*, 6: 235–282.

Yoo, S. (2008) 'Online community and community capacity', in G. Goggin and M. McLelland (eds), *Internationalising Internet Studies*, London: Routledge.

Yoon, K. (2003) 'Retraditionalizing the mobile: young people's sociality and mobile phone use in Seoul, South Korea', *European Journal of Cultural Studies*, 6: 327–43.

Yoon, K. (2006) 'The making of neo-Confucian cyberkids: representations of young mobile phone users in South Korea', *New Media and Society*, 8(5): 753–771.

Yoshimi, S. (1999) ' "Made in Japan": the cultural politics of "home electrification" in postwar Japan', *Media, Culture and Society*, 21: 149–171.

Yoshino, K. (1999) 'Rethinking theories of nationalism: Japan's nationalism in a marketplace perspective', in K. Yoshino (ed.), *Consuming Ethnicity and Multiculturalism: Asian Experiences*, Honolulu: University of Hawaii Press.

Yu, L. and Tng, T.H. (2003) 'Culture and design for mobile phones for China' in J.E. Katz (ed.), *Machines That Become Us: the Social Context of Personal Communication Technology*, New Brunswick, NJ: Transaction Publishers, pp. 197–200.

Yue, A. (2000) 'Asian–Australian cinema, Asian–Australian modernity' in H. Gilbert, T. Khoo, and J. Lo (eds), *Diaspora: Negotiating Asian Australia*, St. Lucia: University of Queensland Press.

Yue, A (2005) 'Migration-as-transition: Hong Kong Cinema and the ethics of love in Wong Kar-Wai's *2046* in B.P. Lorente (ed.), *Asian Migrations: Sojourning, Displacement, Homecoming And Other Travels*, Singapore: National University of Singapore Press.

Yue, A. and Hawkins, G. (2000) 'Going south', *New Formations*, 40, Spring: 49–63.

Yung, V. (2005) 'The construction of symbolic values of the mobile phone in the Hong Kong Chinese print media', in R. Ling and P. E. Pederson (eds), *Mobile Communications: Re-negotiation of the Social Sphere*, London: Springer, pp. 351–366.

Zelizer, V. (2005) *The Purchase of Intimacy*, New Jersey: Princeton University Press.

Zhao, Y. (2007) 'After mobile phones, What? Re-embedding the social in China's "digital revolution"' , *International Journal of Communication*, 1: 92–120.

Index

Aakhus, M. 29, 35
accessories 70–1, 89
advertising: Australia 194, 197–9;
    Hong Kong 154, 162; Korea
    148–9
Agar, J. 30, 247
Allison, A. 98, 99–100, 101
alternative modernities 63–4
Anderson, B. 60
Appadurai, A. 52, 64
Arnold, M. 191, 249
Arrighi, G. 64–5, 82
Ars Electronic 268
art 244
Art Center Nabi *see* Nabi
Asia-Pacific region 2–3, 19–20; imaging
    communities and rethinking the region
    234–7; locating the mobile in the
    region 4–6; modernity 63–6; regional
    mobilities 32–7
Asian financial crisis 119, 120, 121, 156,
    168
Asian miracle 67
Atari video games 10
Attfield, J. 49
Australia 11, 15, 24, 75, 157, 188–225,
    232, 234; case study of camera phone
    practices 212–23; case study of sample
    group 200–11; location in Asia-Pacific
    context 190–5; mobile media in
    Melbourne 195–200
avatars 131, 132

back-stage 165
Bakardjieva, M. 60, 61
banality 40, 41, 43, 243, 269
Bauman, Z. 56
Beaton, J. 24

Beck, J. 87, 95
Beck, U. 53
Beck-Gernsheim, E. 53
Bell, G. 6, 25, 49, 57, 58–9, 73
Berlant, L. 41, 43, 58
Berry, C. 34
Bertschi, S. 30
BlackBerry 157, 165, 166, 168, 169
Blast theory 244, 254
Bolter, J. 40–1, 246
Bourdieu, P. 54, 114, 221
Bourriaud, N. 244, 260
Boyd, J. 37
Britain 155–6, 165–6, 168
Brown, B. 29–30
Bruns, A. 3
bus man video 153–4, 269
Butler, J. 7

camera phones 39–40, 43–5; Australia
    208, 210–11, 212–23, 224–5; Hong
    Kong 171–83; images *see* images,
    camera phone; Japan 92–3, 112–14;
    Korea 127–8, 129, 133–48
Cameron, D. 129–30
capital mobility 49, 54–9
capitalism 64–5, 236
cartographies of personalisation 4, 13,
    15–16, 63, 75–6, 229, 236, 267
case studies: Australia 200–23;
    contextualising locations 10–16; Hong
    Kong 172–83; Japan 106–14; Korea
    133–48
Castells, M. 35, 38, 39, 49, 51, 53, 56, 62,
    226
casual work 68
Chee, F. 128, 258
Chesher, C. 46, 243

children 83, 228
China 5, 51; Hong Kong's modernity and 155–61; return of Hong Kong to 153, 156, 168
Ching, L. 2
Cho, H.J. 121, 122–3, 165, 266
Chow, R. 150, 152–3
Chua, B.H. 20, 71–3, 127
Chung, E.H.G. 255–6
cinema/film 72, 151, 235
circuits of culture model 9–10, 31
Code Division Multiple Access (CDMA) 196
Collins, J. 269
colourings 141, 176, 205
communities, online *see* online communities
communities of consumers 72–3, 127, 186
communities of co-presence 110, 192
community, sense of 146–7
compatibility between phones 180, 208–9
conferences 30, 35–6, 38
Confucius revivalism 51, 65, 81
consumption 14, 20, 53, 62–76, 235–6, 247; changing modes of 70–4; communities of consumers 72–3, 127, 186; cross-cultural 183–7; gendered modernities 67–70; imaging communities 74–6; and intimacy 56–7; Japan 89
convenience 107–8
convergence 15, 37–40; divergence and 242–5
co-presence 248–9; communities of 110, 192
cosmopolitanism 53
couples: Hong Kong 175, 185; Korea 124, 137–8, 141, 143, 148, 149–51, 231
creativity 108, 115–117, 232–4
crime 83, 228
cross-cultural consumption 183–7
cultural artefact 1–2, 24–5, 31
cultural capital 114, 163; global 25–6
culture, adaptation to a new 139–40
customisation 38–9; Australia 191, 202–7, 224–5; feminisation of 75; Hong Kong 155, 162–6, 167, 172, 173, 180–1, 184–5; Japan 88–93, 97–103, 107, 109–14, 114–19; Korea 123–4, 137–8, 141, 143–4, 146–8, 150–1
cute: cute culture in Korea 118, 132–3, 138, 139, 147; Hong Kong camera

phone users 179–81; *kawaii* culture *see kawaii* culture; techno-cute 83–4, 96
Cyworld mini-hompy 118, 120, 129–33, 144–7, 231–2

Daum 131; *YouthVoice Media Conference* 261
Debord, G. 244
decentring 189, 192, 193, 224
deference 189, 192, 193, 224
delay 42–3, 189, 192, 193, 224, 250, 265–71; delay/immediacy paradox 42–3, 248, 250, 257, 268–70
democracy 125
democratisation 251, 267
Densha Otoko ('Train Man') 101–2, 273
Digital Multimedia Broadcasting (DMB) (mobile TV) 125, 126, 126–7
Dirlik, A. 3, 51, 64, 65–6, 81
discontinuous innovation 200
distance 249
divergence 242–5
DoCoMo i-mode 36, 79, 86–7, 93–5, 157, 194, 195
documentary 135, 137
domestic technologies approach 9–10, 36–7, 42, 272; mobility 59–61; and new media 241–52
domesticating cartographies 15
Dotplay 255, 259–63, 264
'Dotplay Telecom' 261, 263
Du Gay, P. 9, 31
Du Gournay, C. 22

E-toy 262
elections 58, 123
electronic individualism 26, 82, 92
Emobile 86
emotional labour 27–8, 139
etomologies 84, 181, 233; emerging in mobile Internet usage 103–6; *see also kawaii* culture
enclave society 50, 61
Estrada, President 58
everyday images 141, 142, 143, 177–9
everyday poetics images 217
exchange students 139–41
exports 89–90

facades 214
face marks (*kaomoji*) 98, 103–6

family: images of 135; mothers and children 83, 228; and nationalism in Korea 122–3, 148–51
Fang, W. 159–60
fashion 70–1, 214–15, 228
fast-forwarding present 44, 145
Feenberg, A. 60
feminisation of labour 22–3
feminist movement 121
Filipino immigrant workers 22, 69, 169–70
film/cinema 72, 151, 235
Fischer, C. 28
Flew, T. 41
flows 52
food 136, 141–2, 214
Fortunati, L. 22, 27, 30, 227
four-tier structure 68–9
friends: Cyworld mini-hompy 131–2; images of 135, 215–16; phones chosen to match friends' phones 180, 208–9
front-stage 165
Fujimoto, K. 60, 89–91, 95, 100, 103–4
futurism 246, 251

gaming 202, 253–4; Korea 128, 253, 258; pervasive (location aware) 244–5, 254–5
gatekeeper role 139, 143, 231
Gates Peletz, M. 70
gender 26–7, 226–37; gendered modernities 67–70; gendered nationalism 81–5; gendered scripting 6, 197–9; nationalism, family and in Korea 148–51; new mobilities and immobilities 6–8; politics of gendered labour and media 229–34; re-imaging communities 234–7
gender performativity 82
gentleness 98–9
Geocaching 244
geographical mobility 48–54
geo-political imagery 161–6
Giddens, A. 56–7
gift-giving 97, 132–3, 144, 193
Gillard, P. 27
Girl Friday 200, 201, 232, 233
glaze 46, 243
global cultural capital 25–6
global order, new 66
Global System for Mobiles (GSM) 196
globalisation 22–3, 48–9, 52–5
Glotz, P. 30

Goggin, G. 10, 30–1, 32, 35, 36, 38, 169–70, 188, 196, 267–8
Goh, E. 130
Goodman, D.S.G. 3, 19, 20, 71
Gottlieb, N. 34
Green, N. 30
Grusin, R. 40–1, 246
Gye, L. 148–9

Habuchi, I. 97
hactivism, mobile 257, 259–63, 264
Haddon, L. 28, 30, 247
Hamill, L. 30
Hannam, K. 49
Happy Together 151
haptic, the 243–4, 248, 249–50, 251–2
Hardey, M. 57–8
Harootunian, H. 82
Harper, R. 28–9, 30, 193
Hartmann, M. 30
Hello Kitty 10, 83, 99, 117, 166, 185
Hemmer, R.L. 243, 254
Herzfeld, M. 60
Hilton, P. 262
Hirsch, E. 9, 29
Hjorth, L. 128
Ho, K.C. 34
Hochschild, A.R. 27, 55
Höflich, J. 30
Hong, J.E. 120, 121, 231, 258–9
Hong Kong 5, 11, 14–15, 151–87, 189, 232–3, 234; case study of users 172–83; geo-political imagery 161–6; imaging communities 153, 183–7; mobility 166–72; reterritorialisation 155–61; *wenqing zhuyi* 151, 158, 172–83, 183–7
Hong Kong and Shanghai Bank advertisement 162
Hopper, R. 28
Horst, H. 4, 30, 31
Howard, J. 194
Hughes, R. 250
Huhh, J.-S. 128
Huhtamo, E. 42, 246
Hutchison 157, 165, 195
hyperfemininity 62, 67–9, 226–7, 234; customisation 117–18
hypermasculinity 67–9, 234
hypermobility 229–34

i-mode 36, 79, 86–7, 93–5, 157, 194, 195
identity 1, 203–4

images, camera phone: Australia 214–22; Hong Kong 177–83; Japan 112–14; Korea 135–43
imagined community 60–1, 74
imaging communities 15, 16, 74–6, 143, 233–4, 271; Australia 223; Hong Kong 153, 183–7; rethinking the region 234–7
immediacy: delay/immediacy paradox 42–3, 248, 250, 257, 268–70; waiting for 48–54, 189, 265–6
individualism 45, 53, 54; electronic 26, 82, 92; Japan 26, 82, 91–2, 97; social capital, intimacy and 54–9
industrialisation 22, 67, 71
INP 15, 255, 257; *Dotplay* 255, 259–63, 264; *Urban vibe* 254, 256, 260
Internal Affairs 72
International Monetary Fund (IMF) 20, 119, 121
Internet 32, 34, 79; Australia 212–13; China 161; Hong Kong 174–5; i-mode 36, 79, 86–7, 93–5, 157, 194, 195; Korea 120, 128; mobile Internet usage in Japan 79, 86–7, 103–6; online communities *see* online communities; online gaming *see* gaming
Internet cafés 131
intimacy 41–2, 44–5; individualism, social capital and 54–9
Ish Media 200, 232
Ishii, K. 106
IT Strategic Headquarters (ITSHQ) 104
Ito, J. 44
Ito, M. 30, 32, 40, 51, 52, 79, 87, 191–2
Iwabuchi, K. 99, 100

Jacobs, D. 130–1
Jamieson, L. 41, 58
Japan 5, 11, 14, 32, 75, 79–119, 119, 234; case study of sample respondents 106–14; gaming 253; high-quality products 72; Hong Kong and 165; *kawaii* culture *see kawaii* culture; mobile Internet usage 79, 86–7, 103–6; nationalism 25–6, 81–5; portable novels 115, 116, 230–1; rise of female consumer 7, 87–8; rise of mobile technologies 86–95; techno-orientalism 25–6
Jenkins, H. 37–8, 241
JUNET 105

kaomoji (face marks) 98, 103–6

Kart Rider 147
Katsuno, H. 98
Katz, J. 28, 29, 30, 35, 39
kawaii culture 83–4, 93; changing nature of customisation 114–19; history of and migration into mobile phone cultures 95–103; in Hong Kong 163, 165
Kawakatsu, H. 82
KDDI 86
Keating, P. 194
keitai shôsetsu (portable novels) 115, 116, 230–1
Khoo, O. 7, 36, 51
Kill Bill films 72
Kim, H. 128
Kim, S.D. 30, 32–3, 123
Kim, Y. 70
Kinsella, S. 96–7
kitten writing 83, 96–7, 101
Kluver, R. 34
Koch, P. 175
Kodak camera 148–9
Kogawa, T. 26, 82
kôgyaru (young urban trendy female) 87–8, 97
Kohiyama, K. 93
Koizara 116
Kopomaa, T. 29, 41, 247–8
Kopytoff, I. 49
Korea, South 5, 11, 14, 32–3, 75, 119–51, 234; camera phone practices case study 133–48; contextualising the socio-technological space of Seoul 257–9; cute culture 118, 132–3, 138, 139, 147; Cyworld mini-hompy 118, 120, 129–33, 144–7, 231–2; gaming 128, 253, 258; mobile media as new media 255, 257–63; nationalism 122–3, 148–51; role of technology in economic growth 2, 21; technoscape 125–9
Korean War 119
Korean wave 21, 165
Koskela, H. 43
Koskinen, I. 40, 41, 43, 44, 243, 269
Kundera, M. 265, 266

labour 67–70; emotional 27–8, 139; feminisation of 22–3; politics of gendered labour and media 229–34; reproductive 27–8; social 27
landscape images 221, 222

language 11–12, 207; and SMS in Hong Kong 170–1
Lantz, F. 244
Lasen, A. 30
Latour, B. 30
Law, J. 35
Law, P. 30, 35
Lee, D.H. 45, 129, 232
LG 36, 169, 194, 199
Life of Mobile Data conference 30
lifestyle 71
Lin, A. 170–1
Lindgren, M. 29–30
Ling, L.H.M. 67, 124
Ling, R. 28, 30–1, 55–6, 247
liquid modernity 56
Little Smart 158, 159
Liu, G. 35
Liu, M. 161
Lo, S.-H. 166–8
locality 48–54
location aware gaming 244–5, 254–5
Luke, R. 256
luxury goods 90–1
Lyon, D. 30

Ma, E.K.-W. 150, 151, 162, 164, 185
McGray, D. 22, 25–6, 55, 69, 170, 227
McLelland, M. 8, 34, 35
McLuhan, M. 40, 224, 246
McVeigh, B. 23, 71, 81–2, 82–3, 89, 97–8
male-centred culture 122–3
maho-island 115, 116
Manovich, L. 43, 248
Mante-Meijer, E. 28
Margaroni, M. 44
Martin, M. 27
Massey, D. 52
materialism 23
maternalism 23
Matsuda, M. 83, 88, 108, 228
Mäyrä, F. 244
Melbourne *see* Australia
memento images 219–21
Mendoza, L. 34
Miller, D. 4, 29, 30, 31
Miller, L. 87
Milne, E. 41, 42
mini-hompy 118, 120, 129–33, 144–7, 231–2
mini-room 130, 131
Ministry of Information Affairs and Communication (MIC) 105–6
missy syndrome 121–2

mixi 101–3, 104–6, 109, 110–12, 132
MMOs (massively multiplayer online games) 253, 258
mobile communication 13–14, 19–47; convergence 37–40; current literature 24–6; domesticating new media 40–7; history of rise of mobile communication studies 26–32; regional mobilities 32–7
Mobile Communication and Asian Modernities conferences 36
Mobile Communication and Social Change conference 30, 35
mobile cultures 8–10
mobile hactivism 257, 259–63, 264
mobile media: convergence 37–40; multi-traditions of 245–50; as new media 15, 253–64; politics of 74–6
mobile privatisation 53–4, 254
mobile television (digital multimedia broadcasting) 125, 126, 126–7
mobilism 60
mobility 14, 48–61, 69–70, 254–5; capital 49, 54–9; cultures of 25; domestic 59–61; geographical 48–54; in Hong Kong 166–72; hypermobility 229–34; inertia of 255–7; new mobilities and immobilities 6–8; politics of gendered labour and media 229–34; regional mobilities 32–7; technics of 162, 229
mobility regime 50
modernity 48–9; in the Asia-Pacific 63–6; gendered modernities 67–70
Moeran, B. 8
mogi game 244, 254
moral economy 29
Mori, Y. 102–3
Morley, D. 53–4, 229, 247
Morris, M. 41
Morse, M. 246
mothers 23, 83, 228
Motorola 169, 194, 199
Moyal, A. 27
Murakami, T. 100
Murphie, A. 161–2, 229

Nabi 255, 256
nagara (multi-tasking) mobilism 60, 89, 96
Nam June Paik 266
nation-state 50–1, 60–1
national culture 33
nationalism 52; Japan 25–6, 81–5; Korea 122–3, 148–51

Navarro, V. 28
networked sociality 59
new media: domestic technologies
    approach and 241–52;
    domesticating 40–7; mobile media as
    15, 253–64
new rich 166–8
Newly Industrialised Countries (NICs)
    2
Nintendo DS 246
Nokia 148–9, 165, 169, 197–8
nostalgia 82; Hong Kong 151–3, 161–6;
    see also wenqing zhuyi
novels, portable (*keitai shôsetsu*) 115,
    116, 230–1
NTT 91; DoCoMo i-mode 36, 79, 86–7,
    93–5, 157, 194, 195
Nyíri, K. 29

odourless culture 99, 101
OFTA 172–3
Oh Soojung (*Bride stripped bare by the
    bachelors, even*) 142
Okabe, D. 40
Okada, T. 91–3
Ong, A. 50, 63–4, 70
online communities: Cyworld mini-
    hompy (Korea) 118, 120, 129–33,
    144–7, 231–2; Hong Kong 174; Japan
    101–3, 104–6, 109, 110–12, 132
Opium Wars 155–6
Optus 195
Orientalism 51, 66

PacManhattan 244, 254
pagers 7, 93
Palmer, D. 45, 260
Pan-Asian cinema 72
paparazzi images 112–13, 142, 179,
    181–3, 213, 215, 216–17
Papastergiadis, N. 50, 189, 190
paradoxes 42–3, 244, 248–50, 251–2;
    delay/immediacy 42–3, 248, 250, 257,
    268–70; reel/real 42–6, 248
Parikka, J. 42, 246
Parreñas, R. 55
partners *see* couples
PC bangs (PC rooms) 128, 132, 253,
    258
Pedersen, P. 30
performativity 146
personal consumption, images for 112,
    177, 218–19
personal essentials 214, 215–16

personal space images 216, 217
personalisation 91–2, 93; cartographies
    of 4, 13, 15–16, 63, 75–6, 229, 236, 267;
    see also customisation
Pertierra, R. 31, 32, 35, 39, 224
pervasive gaming 244–5, 254–5
Philippines 272; *see also* Filipino
    immigrant workers
phone-game hybrids 46
phoneur 256
photography 108, 177; *see also* camera
    phones
Pitt, D. 156
place: camera phones and performing
    place in Australia 223; mobility of
    48–54
Plant, S. 29, 86, 140, 168, 205, 216
Pokémon 83, 98, 99–101
portable novels (*keitai shôsetsu*) 115, 116,
    230–1
postal cartographies 195–200
postal presence 192, 193, 224, 225
postcard 45; images 44, 177–9, 216
postmodern tourism 56
postmodernity 62, 100
PostPet 84
pragmatism 217–18
primary essentials 214–15
privatising technologies 53–4
Proboscis *Urban Tapestries* 254,
    256
production 235–6; and consumption
    62–76
'produsers' 13, 21
Promise, The 160
'prosumers' 13, 39
publicness 43
pure relationships 57
purikura 88, 93
Putnam, R. 54, 55

Qiu, J. 157, 158–9, 160
QQ 174, 175

Raiti, G. 54, 57
Rakow, L. 28
Rao, M. 33, 34
reel/real paradox 42–6, 248
regional mobilities 32–7
relational aesthetics 244, 260
relational architecture 243, 254
remediation 40–1, 45, 93–4, 245–50
reproductive labour 27–8
research methodology 11–13

reterritorialisation 74–5, 158, 177, 187; Australia 222–3, 224; Hong Kong 155–61
Rheingold, H. 30
Riccini, R. 30
Richardson, I. 46, 249
ring tones 92, 203–4, 204–5
Robison, R. 3, 19, 20, 71
Roh, President 58, 123
routine images 135
Rudd, K. 194

Samsung 2, 33, 36, 123, 127, 169, 194, 198–9; Anycall 120, 121, 149, 231
Sawhney, H. 93–4
Scifo, B. 45
screen cultures 43
self-expression 39
self-Orientalism 51, 66
self-paparazzi 142
self-portraiture 136
Sen, K. 7, 22, 69, 70
Seoul *see* Korea, South
Shade, L.R. 197
Shamir, R. 50
sharing 75; Australia 211, 214, 218–19; Hong Kong 179, 184, 186; Japan 113–14; Korea 127–8, 140, 143
Sheller, M. 49
Shift_JIS (SJIS) 104, 105
shôjo (young Japanese female) 87–8, 97
silent mode 205
Silverstone, R. 9, 29
Situationist International (SI) 244
Skov, L. 8
slowness 265, 266–7; *see also* delay
smileys 98
SMS: Australia 191, 192–3, 203, 206–8, 209–10, 211, 224–5; China 159–61; Hong Kong 157, 159, 169–71, 173–4; Japan 93
social capital 54–9
social confirmation 108
social expected images 216–17
social labour 27
social networking sites (SNS): Hong Kong 174; mixi 101–3, 104–6, 109, 110–12, 132; 2ch 101–2, 104, 105
socio-technologies 257–9
soft power 26
SoftBank 86, 105
Sohn, S.H. 129
'solitary owl/calm dove' 140–1
Sony Ericsson 194, 198

Soo, H.S. 142
Sørensen, C. 52–3
space 52
special occasion images 135, 141, 142, 221, 222
stereotypes 202
Stivens, M. 7, 22, 69, 70
stories, portable (*keitai shôsetsu*) 115, 116, 230–1
students: Australia 212–23; Korea 133–48
Sugiyama, S. 39
Sullivan, J. 231
Suominen, J. 42, 246
'swift talker' 140

Taylor, A. 28–9, 193
tea 90
technics of mobility 162, 229
techno-cute 83–4, 96; *see also kawaii* culture
techno-nationalism 128; China 161
techno-orientalism 25–6
television 235, 268; mobile (DMB) 125, 126, 126–7
Telstra 194, 195, 196
Tencent QQ 174, 175
texting *see* SMS
the-phone-book Limited 256
third space 247–8, 258
third-way politics 56–7, 58
3 (Hutchison) 157
Three Day Growth 232
Toffler, A. 3, 39
Tokyo *see* Japan
tourism, postmodern 56
tourist images 44, 177–9, 216
toys 10
Train Man (*Densha Otoko*) 101–2, 273
transnationalism 61; transnational consumer communities 72–3
Truong, T.S. 22, 67, 68
Tsunoyama, S. 90
turbo-capitalism 121–2
Turner, B. 50, 61
2ch (Channel 2) 101–2, 104, 105

un-distance 249
uniforms 71, 89
unofficial sites 94
upward mobility 166–72
Urban Tapestries see Proboscis
Urban vibe 254, 256, 260

Urry, J. 48, 48–9, 59

values, contestation of 73
van de Loo, H. 28
van Oost, E. 197
Vodafone 165, 195

Wade, M. 87, 95
waiting for immediacy 48–54, 189,
    265–6
Wajcman, J. 22, 24, 27
wallpaper 171–2, 180
Waters, P. 150, 168, 172–3
wenqing zhuyi (warm sentiment-ism) 151,
    158, 172–83; and cross-cultural
    consumption 183–7
WILLCOM 86
Williams, R. 53
Wilson, A. 58
Wilson, R. 2, 3, 65
Wong Kar-wai 151

work-life balance 24

Yan, X. 156
Yang, K.C.C. 34
Yang, S. 30
Yano, C. 98
Yiannopoulou, E. 44
Yoon, K. 33, 51, 123, 125, 258
young female user 3–4, 6–7, 21–2, 73,
    226; Japan 7, 87–8, 97; Korea 125, 136,
    138–9, 140–1, 148
Young Hae Chang Heavy Industries 261,
    266
youth 228
youth cultures 28, 125
Yttri, B. 28
Yue, A. 158, 190

Zelizer, V. 55
Zhao, Y. 158
Zoninsein, M. 161